Introgression from Genetically Modified Plants into Wild Relatives

Contents

First-named Contributors

Abbott, A.G., Department of Genetics and Biochemistry, Clemson University, 100 Jordan Hall, Clemson, SC 29634, USA.

Bleeker, W., University of Osnabrück, Department of Systematic Botany, Barbarastraße 11, D-49076 Osnabrück, Germany.

Chèvre, A.-M., INRA-Station Amélioration des Plantes, PO Box 35327, Domaine de la Motte, F-35653 Le Rheu cedex, France.

Cooper, J.I., CEH Oxford, Mansfield Road, Oxford OX1 3SR, UK.

Cuguen, J., UMR CNRS 8016, Laboratoire de Génétique et Evolution des Populations Végétales, Université de Lille 1, bât. 59655 Villeneuve d'Ascq, France.

den Nijs, H.C.M., Institute for Biodiversity and Ecosystem Dynamics, University of Amsterdam, Kruislaan 318, 1098 SM Amsterdam, The Netherlands.

Dorokhov, D., Centre 'Bioengineering' of Russian Academy of Sciences, Pr. 60-letiya Oktyabrya, 7/1, Moscow, Russia.

Hails, R.S., NERC Centre for Ecology and Hydrology, Mansfield Road, Oxford OX1 3SR, UK.

Hauser, T.P., Plant Research Department, Risø National Laboratory, PO Box 49, DK-4000 Roskilde, Denmark.

Jacot, Y., Botanische Garten, Universitat Bern, Altenbergrain 21, CH-3013 Bern, Switzerland.

Jørgensen, R.B., Plant Research Department, PLE-309, Risø National Laboratory, DK-4000 Roskilde, Denmark.

Lavigne, C., Laboratoire Ecology, Systématique et Evolution, UMR 8079, CNRS/Université Paris-Sud, Centre d'Orsay, Bâtiment 360, F-91405 Orsay Cedex, France.

Lu, B.-R., Ministry of Education Key Laboratory for Biodiversity and Ecological Engineering, Institute of Biodiversity Science, Fudan University, Shanghai, 200433, China.

Norris, C., NIAB, Huntington Road, Cambridge CB3 0LE, UK.

Pallett, D.W., CEH Oxford, Mansfield Road, Oxford OX1 3SR, UK.

Papa, R., Dipartimento di Biotecnologie Agrarie ed Ambientali, Università Politecnica delle Marche, Ancona, Italy.

Pilson, D., School of Biological Sciences, 348 Manter Hall, University of Nebraska, Lincoln, NE 68588-0118, USA.

Slyvchenko, O., Institute for Sugar Beet UAAS, Klinichna, 25, 03110 Kiev, Ukraine.

Soukup, J., Department of Agroecology and Biometeorology, Faculty of Agronomy and Natural Resources, Czech University of Agriculture, Kamýchá 957, 165 21 Praha 6-Suchdol, Czech Republic.

Stewart, C.N., Department of Plant Sciences, University of Tennessee, Knoxville, TN 37996–4561, USA.

Strauss, S.H., Oregon State University, Corvallis, OR 97331, USA.

Sweet, J., NIAB, Huntingdon Road, Cambridge CB3 0LE, UK.

van de Wiel, C., Plant Research International, PO Box 16, 6700 AA Wageningen, The Netherlands.

Van Dijk, H., Laboratoire de Génétique et Evolution des Populations Végétales, UMR CNRS 8016, Université de Lille, bât. SN2, 59655 Villeneuve d'Ascq, France.

van Tienderen, P.H., Institute for Biodiversity and Ecosystem Dynamics, University of Amsterdam, Kruislaan 318, 1098 SM Amsterdam, The Netherlands.

Warwick, S.I., Agriculture and Agri-Food Canada (AAFC), Ottawa, Ontario, Canada K1A 0C6.

Wennström, A., Department of Ecology and Environmental Science, Umeå University, SE-901 87 Umeå, Sweden.

Wilkinson, M., School of Plant Sciences, The University of Reading, Reading RG6 6AS, UK.

Acknowledgements

We would like to express our thanks to the European Science Foundation for valuable financial support for this conference, and the Technische Universität Aachen, the Robert Koch Institut Berlin, as well as the University of Amsterdam for investment of the time necessary for the successful organization of the meeting. We also thank the many reviewers who helped us to compile this proceedings, which hopefully will contribute to a sound and solid science-based discussion on the introduction of GM crops.

Special thanks are due to the conference office, Lidy Groot Congress Events, for their professional, careful and successful preparation and logistical guidance of the event. We further realize that the meeting would not have run so smoothly if we had not had such a dedicated group of our faculty's biology students who were in charge of so many housekeeping matters and who performed these tasks excellently; we wish to thank them in particular.

We are also very much obliged to the University of Amsterdam, in particular the Institute for Biodiversity and Ecosystem Dynamics, for secretarial and logistical support during the preparatory phase of this proceedings volume. Also, Marieke de Boer, in charge of so much of the pre-publisher editing, did an excellent job.

Finally, we thank all those who contributed to this meeting and report with papers, posters, chairing sessions and contributions to the scientific debate.

1

Introduction and the AIGM Research Project

JEREMY SWEET[1], HANS C.M. DEN NIJS[2]
AND DETLEF BARTSCH[3]

[1]NIAB, Huntingdon Road, Cambridge CB3 0LE, UK, E-mail: jeremy. sweet@niab.com; [2]Institute for Biodiversity and Ecosystem Dynamics, University of Amsterdam, Kruislaan 318, 1098 SM Amsterdam, The Netherlands, E-mail: nijs@science.uva.nl; [3]Robert Koch Institute, Center for Gene Technology, Wollankstraße 15–17, D-13187 Berlin, Germany; E-mail: bartschd@rki.de

Abstract

As a sequel to a series of workshops organized by the European Science Foundation programme Assessment of the Impacts of Genetically Modified Plants (AIGM), this conference aimed at exploring the ecological and evolutionary effects of the introduction of transgenes into wild relatives of crops, by integrating and discussing current knowledge on crop–wild relative hybridization, introgression and measuring and predicting its consequences, and by identifying areas that needed further elaboration. Of particular importance was the need to discuss methods for determining the consequences of gene introgression in relation to the *Monitoring guidance notes* that accompany the requirement in the EC Directive 2001/18 for monitoring of the post-commercialization impacts of GM crops.

Introduction

The European Science Foundation research project 'Assessment of the Impacts of Genetically Modified Plants' (AIGM) focused on the potential environmental and agronomic impacts of the introduction of GM crops into Europe. AIGM examined risk assessments and determined the necessary elements required for ecologically responsible introduction of new GM crops. AIGM included a programme of workshops and informal working group meetings to examine a range of issues. These have included:

- The wider environmental implications of genetically modified plants (Cambridge, UK, January 2000, contact: jeremy.sweet@niab.com).

- Environmental implications of genetically modified plants with insect resistance genes (Bern, Switzerland, September 2000, contact: yolande. jacot@ips.unibe.ch).
- Environmental implications of genetically modified plants with fungal disease resistance (Roskilde, Denmark, November 2000, contact: hanne. oestergaard@risoe.dk).
- The environmental implications of gene flow from genetically modified sugarbeet to seabeet (Venice, Italy, May 2001, review paper published by Bartsch *et al.*, 2003).
- Interspecific gene flow from oilseed rape to weedy species (Rennes, France, 11–13 June 2001, summary published by Chèvre *et al.*, Chapter 18, this volume).
- Risk assessment methods for genetically modified plants – current trends and new developments (Ceske Budejovice, Czech Republic, September 2001, contact: ray@umbr.as.cz).
- Genetic interactions (Lisbon, Portugal, September 2001, contact: maria. pais@fc.ul.pt).
- Estimating and managing gene flow and dispersal in GM crops (Lille, France, May 2002, contact: Joel.Cuguen@univ-lille1.fr).
- Decision-making processes in plant biotechnology (Rome, Italy, September 2002, contact: mari@mlib.cnr.it).
- Ecological concepts and techniques to assess the impact of genetically modified plants in natural and crop ecosystems (Korsør, Denmark, November 2002, contact: Gabor.Lovei@agrsci.dk).
- New science for increasing biosafety of genetically modified plants (Braunschweig, Germany, March 2003, contact: J.Schiemann@bba.de).
- The impact of genetically modified plants on microbial communities in agricultural soils (Tromso, Norway, June 2003, contact: Jan. Husby@DIRNAT.NO).
- Risk assessment of transgenic long-lived plants; ecology and social aspects (Umeå, Sweden, February 2004, contact: anders.wennstrom@eg. umu.se).
- Modelling the effects of GM-introgression into wild relatives (Dorchester, Dorset, UK, 9–12 March 2004, contact: tienderen@science.uva.nl).

An important feature of these workshops has been studies on gene flow within crops and from crops into wild and weedy relatives in adjacent or distant ecosystems. Many studies have been performed to assess the frequency of gene flow from GM crop fields to wild relatives. Hybridization with crop relatives will occur at different rates depending on the relative genetic, ecological and physical distance between the populations. Knowledge of the frequency of crop–wild relative hybridization will indicate how risks should be managed. However, there is also a feeling among many researchers that one should assume that hybridization will occur with certain wild relatives and therefore research should concentrate on determining the likelihood of introgression and of a transgene becoming established in a wild plant population.

After initial interspecific hybridization, the critical issue is whether or not the hybrids are viable and fertile enough to establish and to backcross with the wild parental taxon, or the crop. Often little information is available, and it is important that research focuses on this lacuna in our knowledge. Further gene flow from an initial hybrid population back to non-GM crops may lead to the introduction of transgenes into this crop. This could result in problems with certification of non-GM seed and/or with placing products above thresholds for non-GM purity. These are essentially agronomic concerns, and special attention to this aspect had been given in the AIGM workshop 'Estimating and managing gene flow and dispersal in GM crops', in Lille, France, in May 2002.

Backcrossing with wild relative taxons may lead to introgression of crop genes into wild plant populations. The important questions are whether the gene will be taken up into a significant proportion of the population and what the effects are of this introgression into the populations. Transgene influx in wild relatives may affect their genetics and ecology, and thus influence their evolutionary and ecological behaviour and characteristics. At present, the consequences of this process for transgenes are less well known, and a number of features need to be determined.

Any effect will undoubtedly be strongly influenced by the particular crop trait(s) conferred by the transgene, and this necessarily means that all impacts (and assessments) are case specific. Recently, Ellstrand *et al.* (1999) presented a review with many baseline data for a series of important crops. In this review, hybridization and introgression events from conventional crop races into wild relatives were reported from the literature. Of a total of 13 world crops, 12 were found to hybridize at least somewhere in their cultivation range with one or another wild relative taxon. On the basis of these data, Ellstrand *et al.* (1999) argued that many wild relatives are not evolutionarily independent of the crops, and that the outcome of the genetic contacts could be twofold. First, a sort of hybrid vigour effect may lead to formation of new weeds, or increasing aggressivity in existing ones. The weed beet (*Beta*) is an example of this in Europe (Boudry *et al.*, 1993; Van Dijk and Desplanque, 1999; Mücher *et al.*, 2000). Secondly, hybridization and introgression may lead to genetic erosion of the recipient wild taxon. However, in the case of gene flow from cultivated beet to wild beet, gene flow and introgression did not reduce genetic diversity (Bartsch *et al.*, 1999; Desplanque *et al.*, 1999).

Some recent examples of stable incorporation of crop genes in the wild genome after backcrossing are *Helianthus annuus* (sunflower; Whitton *et al.*, 1997) and *Raphanus raphanistrum* (radish; Snow *et al.*, 2001), and it is likely to be the case in *Daucus carota* (wild carrot; Hauser and Bjørn, 2001) and *Medicago sativa/falcata* complex (lucerne and wild relatives; Rufener Al Mazyad and Ammann, 1999; Muller *et al.*, 2001). Whether or not this process will influence the performance of the wild plants and their evolutionary perspectives will largely depend on the effect of the introgressed genes on fitness components.

Features of this Conference

The conference was divided into four main sessions, and these sessions are reported as sections in these proceedings as follows:

1. Hybridization in crop–wild relative complexes: the baseline, Chapters 2–13

In two introductory chapters, van Tienderen and Bleeker discuss the process of hybridization as an evolutionary principle, and the role of hybridization in the evolution in the genus *Rorippa*. Other authors then discuss the extent to which introgression has already taken place in a range of wild species and crops. They describe studies of historical processes and current experiments by examining plant morphology and genome banding patterns in crop–wild relative complexes of *Daucus carota*, *Beta vulgaris*, *Triticum*, *Fragaria*, *Populus*, *Brassica*, *Phaseolus*, *Oryza*, *Glycine* and *Lactuca*.

2. Gene flow: introgression and adoption of genes, Chapters 14–20

After having established the incidence of the initial step of hybridization or interbreeding, a crucial next question is whether or not the crop genes establish in the wild-type genome, and are consecutively able to spread in its population complex. Further to this, to be able to assess any ecological impact of introgression, studies of performance and fitness analyses are necessary. In this section of the volume, case studies are presented of key crops, for which such data are becoming available: *Beta*, *Brassica* and *Helianthus*.

Fitness differences brought about by the influx of transgene traits may take several or more generations for incorporation into a significant proportion of the wild population. This series of chapters reports that fitness genes can increase the fecundity and reproduction of hybrid and backcross generations, making it more likely that the traits will become incorporated (*Helianthus*, Pilson *et al.*, Chapter 17, this volume; *Brassica*, Jørgensen *et al.*, Chapter 19, this volume). However, in order to assess their final effects on the viability and behaviour of a population (i.e. on the evolutionary scale), demographic and matrix modelling tools are needed that integrate as many data as possible from short-term empirical and experimental (field) studies. Suggestions for such modelling are presented in the last section (4) of this volume.

3. Impact and consequences of novel traits, Chapters 21–24

In this section, the objective was to assess the extent to which knowledge is available on new and novel traits that are being introduced, such as insect, virus and fungus resistance. The four chapters generally conclude that we often lack appropriate data from sound ecological studies to make solid evaluations. The authors generally considered that, although a fair basis of knowledge is available, one has to acknowledge that a better insight into environmental processes and effects is necessary. This is particularly the case for taxa that as yet have low levels of domestication, such as perennial crops and forest trees, where knowledge of ecological interactions of their wild relatives with pathogens is limited.

4. Monitoring: field studies, modelling and scientific standards for regulation, Chapters 25–27

Data on interactions between genotypes and their environment described in the sections above need to be supplemented by surveys and monitoring studies that examine and map the distribution of both crop and related wild plant species. Monitoring of GMO is conducted to achieve any of four specific objectives: (i) confirm compliance with regulatory requirements; (ii) collect information necessary for controlling and managing potentially adverse environmental situations or systems; (iii) assess environmental quality; and (iv) detect 'unexpected' and potentially damaging effects (Suter, 1993). As such, monitoring may be recommended to reduce uncertainty on the impact of large-scale releases remaining from environmental risk assessment (ERA). In addition, monitoring can be designed to confirm conclusions of ERA with additional data and modelling, or provide informational feedback on system status or condition. Monitoring is not a substitute for biosafety research or ERA. Rather, it is integrated with research and risk assessment to ensure that ecological systems and processes of value are being protected. In addition, some papers reported biogeographical models that were developed to estimate and predict the impact of a novel gene introgression into wild species. Special attention is given to an evaluation of different sorts of models, which can help in producing accurate predictions of various kinds. EU directive 2001/18 requires mandatory monitoring schemes to accompany the introduction of each GM crop. The chapters in this section establish some parts of a framework of what and how to monitor, especially in relation to gene flow and introgression, and how to analyse and report the monitoring data.

The Future

The AIGM Project is nearing completion and its programme of meetings has enabled a considerable body of European and other scientists to share their experiences, learn from each other and coordinate their research activities in this important area. In its final meeting in 2004, the AIGM Project will review many of the major research programmes currently being conducted in Europe. However, it is important that, after the AIGM Project, there continues to be a system for encouraging integrated research programmes on a European scale on the environmental impacts of GM crops. We look to the European Commission and member states to ensure that appropriate measures are included in both national and European framework research programmes.

References

Bartsch, D., Lehnen, M., Clegg, J., Pohl-Orf, M., Schuphan, I. and Ellstrand, N.C. (1999) Impact of gene flow from cultivated beet on genetic diversity of wild sea beet populations. *Molecular Ecology* 8, 1733–1741.

Bartsch, D., Cuguen, J., Biancardi, E. and Sweet, J. (2003) Environmental implications of gene flow from sugar beet to wild beet – current status and future research needs. *Environmental Biosafety Research* 2, 105–115.

Boudry, P., Mörchen, M., Saumitou-Laprade, P., Vernet, P. and Van Dijk, H. (1993) The origin and evolution of weed beets: consequences for the breeding and release of herbicide-resistant transgenic sugar-beets. *Theoretical and Applied Genetics* 87, 471–478.

Desplanque, B., Boudry, P., Broomberg, K., Saumitou-Laprade, P., Cuguen, J. and Van Dijk, H. (1999) Genetic diversity and gene flow between wild, cultivated and weedy forms of *Beta vulgaris* L. (*Chenopodiaceae*), assessed by RFLP and microsatellite markers. *Theoretical and Applied Genetics* 98, 1194–1201.

Ellstrand, N.C., Prentice, H.C. and Hancock, J.F. (1999) Gene flow and introgression from domesticated plants into their wild relatives. *Annual Review of Ecology and Systematics* 30, 539–563.

Hauser, T.P. and Bjørn, G.K. (2001) Hybrids between wild and cultivated carrots in Danish carrot fields. *Genetic Resources and Crop Evolution* 48, 499–506.

Mücher, T., Hesse, P., Pohl-Orf, M., Ellstrand, N.C. and Bartsch, D. (2000) Characterization of weed beets in Germany and Italy. *Journal of Sugar Beet Research* 37, 19–38.

Muller, M.H., Porsperi, J.M., Santoni, S. and Ronfort, J. (2001) How mitochondrial DNA diversity can help to understand the dynamics of wild–cultivated complexes. The case of *Medicago sativa* in Spain. *Molecular Ecology* 10, 2753–2763.

Rufener Al Mazyad, P. and Ammann, K. (1999) The *Medicago falcata/sativa* complex, crop–wild relative introgression in Switzerland. In: van Raamsdonk, L.W.D. and den Nijs, J.C.M. (eds) *Plant Evolution in Man-made Habitats*. Proceedings of the VIIth International Symposium IOPB Amsterdam, Hugo de Vries Laboratory, Amsterdam, pp. 271–286.

Snow, A.A., Uthus, K.L. and Culley, T.M. (2001) Fitness of hybrids between cultivated radish and weedy *Raphanus raphanistrum*: implications for rapid evolution in weeds. *Ecological Applications* 11, 934–943.

Suter, G.W. (1993) Environmental surveillance. In: Suter, G.W. (ed.) *Ecological Risk Assessment*. Lewis Publishers, Chelsea, Michigan, pp. 377–383.

Van Dijk, H. and Desplanque, B. (1999) European *Beta*: crops and their wild and weedy relatives. In: van Raamsdonk, L.W.D. and den Nijs, J.C.M. (eds) *Plant Evolution in Man-made Habitats*. Proceedings of the VIIth International Symposium IOPB Amsterdam, Hugo de Vries Laboratory, Amsterdam, pp. 257–270.

Whitton, J., Wolf, D.E., Arias, D.M., Snow, A.A. and Rieseberg, L.H. (1997) The persistence of cultivar alleles in wild populations of sunflowers five generations after hybridization. *Theoretical and Applied Genetics* 95, 33–40.

2

Hybridization in Nature: Lessons for the Introgression of Transgenes into Wild Relatives

PETER H. VAN TIENDEREN

Department of Experimental Plant Systematics, Institute for Biodiversity and Ecosystem Dynamics (IBED), University of Amsterdam, Kruislaan 318, 10985M Amsterdam, The Netherlands, E-mail: tienderen@science.uva.nl

Abstract

Many factors determine the chance of transfer of genes from crops to wild relatives: the distance and divergence between them, the extent of gene flow, and the fitness of hybrids. The spectrum of genes that can be transferred to a target species by genetic modification is much wider than that of traditional breeding, and the effects of gene exchange between GM crops and wild relatives may therefore be quite different. Important questions need to be answered: could the transgene escape and, if so, will it stay confined to the initial hybrids, for instance due to hybrid inviability or infertility? Or will it introgress into the wild relative, perhaps changing the genetic diversity of the wild species? What is the phenotypic effect of the transgene, and is it something entirely new to the recipient plant? Does it increase the fitness of F_1 hybrids, backcrosses and later generations? Will the transgene stay intact and functional or will changes take place?

The focus of this chapter is to see what we know of hybridization in nature, in order to identify what factors determine the exchange of genes between and within plant species and the success of hybrids. The goal is to identify the main processes involved, and see what this tells us about the study of the potential risks of escape of transgenes from GM crops to wild relatives. Although hybrids are present in many taxa, natural hybridization as a speciation mode is subject to discussion, except when it is coupled to polyploidization. It is also evident that a lot of baseline data on putative recipient wild relatives is still missing, hampering the evaluation of the risks of introgression of genes from GM crops to wild relatives.

Introduction

In the painting 'Les grâces naturelles', The Natural Graces, by René Magritte, the Belgian painter depicts an organism that is half plant, half dove (see cover). It is quite different from the picture of Frankenstein's monster, the cult horror figure that has become the metaphor for the dangers of genetically manipulated foods. Magritte's organism, while much nicer to look at, is, just like Frankenstein's monster, a figment of the imagination. However, at the same time, the image nicely illustrates the meaning of hybridization: the characteristics of different organisms are blended to yield something quite new, with characteristics that have never been combined before, and perhaps even never been thought of except in an artist's mind. Public concern about the introduction of GM food has focused mainly on aspects of food safety and consequences for agronomic practice rather than hybridization of transgenic crops with wild relatives. The consequences of introgression of genes from GM crops to wild relatives has been primarily the topic of a scientific debate, although the public awareness of the potential risks of transgenes escaping to wild populations is increasing, and was picked up by organizations such as Greenpeace. The main question is whether this constitutes a real problem or, more precisely, under what circumstances the potential effects of escape of transgenes are so serious that the risks outweigh the benefits of growing GM crops. Will Magritte's graces be able to take off and spread and, if so, will they do any harm?

The goal of this chapter is to discuss the role of hybridization and gene exchange in nature, in the hope that it may contribute to a better understanding of the potential effects of the introduction of GM crops on wild relatives. First, a brief overview of the evidence for hybridization in nature is given. The process of gene exchange is discussed, explaining the fundamental differences between crosses within and between species. In the subsequent sections, the potential effects of hybridization on fitness and the distribution of species are introduced, as well as the stability of genes in hybrids. Finally, the conclusions on what we can learn from hybridization in nature are drawn.

How Common is Hybridization in Plants?

In their much-cited review, Ellstrand *et al.* (1996) showed that hybridization is a common process, albeit not as ubiquitous as previously stated. For instance, the flora of the British Isles lists 2950 'true' species and 642 species hybrids. Moreover, hybrids are present in many taxa; 56 out of 164 British families contain hybrids (34%), and 143 out of 918 genera (16%). These percentages for families and genera are similar for flora elsewhere, whereas the number of listed hybrids is somewhat lower than that for the British flora (Ellstrand *et al.*, 1996). Thus there is no reason to doubt that hybridization in plants occurs, and that reproductive isolation between species is often not complete. Nevertheless, hybridization is absent in the majority of plant genera, so the obvious question is whether it is relevant for crop species. The answer is yes, and hybrids are found among many crops and their wild relatives (Raybould and Gray, 1993; Ellstrand *et al.*, 1999), as other chapters of this volume will also show.

Ellstrand *et al.* (1996) also investigated the traits that were associated with the occurrence of hybrids. Hybrids were more frequent among clonal and perennial species, in outcrossers and also in asexual species. Hybridization appears to be promoted by a shortage of compatible, conspecific mating partners, for instance by lack of variation in self-incompatibility genes in obligate outcrossers that form extensive stands by clonal growth. Asexual reproduction and clonal growth can also play a role in stabilizing hybrids by adopting a reproductive mode that does not involve meiosis (Grant, 1981).

In the genus *Rorippa* (yellow cress), for instance, hybrids are recorded between related species (Jonsell, 1968). All three native species in The Netherlands (*Rorippa sylvestris*, *R. palustris* and *R. amphibia*) hybridize, and also with the invasive species *R. austriaca*. Indeed, except for *R. palustris*, the species are clonal perennials and self-incompatible, in accordance with the trend observed by Ellstrand *et al.* (1996), and hybrid formation is linked to disturbances (Bleeker and Hurka, 2001; Bleeker, Chapter 3, this volume). These examples suggest that we can learn from the biology of natural hybrids, by identifying taxa prone to hybridization, such as the Poaceae, and groups in which hybrids are seldom found, such as the Solanaceae (Ellstrand *et al.*, 1996). However, although some of these trends are significant, hybridization is not so predictable that it can be used to assess the chances of an exchange between crops and wild relatives on the basis of their natural frequency and distribution, mode of reproduction, or other characteristics. Moreover, the existing hybrid record is also a reflection of the botanist's interest in studying and collecting strange looking plants. The abundance of hybrids in the British flora could also be the result in part of centuries of naturalist's studies, with often only a few plants found sometime, somewhere (Stace, 1975).

The observed frequency of hybrids is only one aspect of hybridization in natural systems. Introgression of genes from one species to another may not be reflected by high frequencies of detectable hybrids in the field. Thus, there appear to be three main themes to consider when attempting to assess the success of hybridization in nature: (i) what is the potential for sexual exchange between and within species?; (ii) what are the fitnesses of the various types of hybrids (F$_1$s, backcrosses)?; and (iii) what is the fate of the introgressed genes in the different genetic backgrounds during the process of hybrid formation? These three questions must be addressed for a thorough analysis of the role of hybridization in nature, and what it might say about the effects of the introduction of GM crops.

Sexual Exchange Within and Between Species

In general, gene exchange between taxa is expected to be easier if the taxa are more closely related. Thus, we need to know how much genome divergence separates taxa, and how this affects the initial formation of hybrids and their viability and fertility. Three aspects will be considered: the relatedness of the taxa and its consequences; the frequency of hybrids and backcrosses in the field; and the impact of differences in ploidy levels between taxa. However, first some definitions have to be given to avoid later confusion. If two plants

(-groups) are conspecific, the terms *interbreeding* and *gene flow* are used to describe the process of gene exchange. If, on the other hand, they belong to different species, the terms *hybridization* and *introgression* would be more appropriate. However, it is obvious that a strict separation of these two extremes is not possible, unless one chooses to adopt a very dogmatic species concept.

Relatedness and exchange

If the parental lines that come into contact, due to either a natural cause or human action, are closely related/belong to one species, there will be no specific mechanism that prevents gene flow among them, although some reproductive isolation due to spatial or phenological separation may exist. Moreover, the genomes will be co-linear and, at least if ploidy differences are absent, problems in sexual reproduction (notably meiosis) are not to be expected. A good example is *Beta vulgaris*, a species in which wild forms (subspecies *maritima*) and cultivated forms (subspecies *vulgaris*, e.g. sugar-beet) can freely interbreed (Bartsch and Pohl Orf, 1996). Consequently, after a number of generations, any focal gene is expected to be present in many different genetic backgrounds. Once the gene is assimilated in the recipient species, its spread is determined mainly by its average effect on fitness; it is possible that some initial outbreeding depression occurs, if new combinations of genes have a reduced fitness compared with the co-adapted gene complexes of the original parental lines, as will be discussed later. It should be noted that extrapolations from gene exchange in nature, or from gene exchange between non-GM crops and wild relatives, may not be representative of exchange between a GM crop and a recipient population. Bergelson *et al.* (1998) found some evidence that transgenic lines of *Arabidopsis thaliana* were more likely to sire seeds of neighbouring plants than expected from the normal (low) incidence of outcrossing.

If, in contrast, the parental lines that mate are less closely related, all sorts of additional complications may arise. First, the combined genomes of the two may not give rise to viable offspring, for instance due to sterility factors or incompatible gene combinations, as found in *Helianthus* sunflowers (Rieseberg *et al.*, 1999; Rieseberg, 2000). Also, the genomes may not be co-linear, leading to problems with the pairing of homologous chromosomes in meiosis, and subsequent infertility of hybrids due to distorted segregation. Both may lead to strong initial selection against some hybrid forms. Such forms of selection can be considered 'intrinsic' as they are not related to specific environmental conditions in the habitats of the plants.

Distribution and frequency of hybrids

Hybridization among related taxa can have different outcomes (e.g. Allendorf *et al.*, 2001).

1.   Incidental hybrids, or localized hybrid zones: hybrids occur in a specific area of contact between genetically distinct populations. Due to the geographic confinement, effects on the genetic integrity of the parental populations are limited or absent.

2.   Hybrid swarm: a population of interbreeding hybrid individuals (F_1s, backcrosses with parental types) that is less localized and may affect the genetic integrity of parental populations by genetic introgression. The degree of admixture is the proportion of alleles that originate from each of the parental taxa. Introgression is not always symmetrical, and factors such as spatial distribution of parents and hybrids, (in-)compatibilities, pollen fertility and habitat requirements may all affect the direction of introgression.

3.   Hybrid taxa: independently evolving population(s) with traits distinct from the parental taxa from which it was derived. Reproductive isolation between the new taxon and the parental taxa may be reinforced by differences in phenology, habitat use or other pre-zygotic isolation mechanisms.

These three situations are three points out of a continuum of possibilities, and intermediate situations may also exist. The most relevant category for GM crops could be the middle category: in the first case, hybrid formation is a rare event and contained; the third case appears to be a very specific situation applicable to few species. However, in cases of hybrid swarms, the amount of gene exchange can be extensive, with increased probabilities that transgenes will also travel from crop to wild relative.

Polyploidy

Ploidy differences may play an important role in the strength of reproductive isolation. Polyploidization within a species can be seen as an instant production of new, reproductively isolated groups (Ramsey and Schemske, 2002), albeit that it is common practice to group plants with different ploidy levels into the same species if they are morphologically similar. For instance, *R. amphibia* contains diploid ($2n = 16$) and tetraploid ($2n = 32$) forms. However, the amount of reproductive isolation between the two levels may be comparable with that among different species, for instance between tetraploid *R. amphibia* and *R. sylvestris* (the *R. × anceps* hybrid). Hybrid sterility is expected in crosses among different ploidy levels, associated with problems of pairing of homologous chromosomes in meiosis in F_1 hybrids between ploidy levels, and this can be an effective isolation mechanism. Moreover, the ploidy level that is rare in a population always has a disadvantage, since it receives mostly pollen of the other type, and usually cannot produce fertile offspring. This so-called 'Minority Cytotype Exclusion' principle means that geographic and/or phenological differences are common, and sympatry of different ploidy levels is exceptional (Levin, 1975).

Many crops are polyploids (e.g. canola, cotton, maize, potatoes, wheat) and may have several diploid wild relatives. The best-studied system of effective ploidy barriers among species and crops is presumably that of *Medicago falcata* (sickle medic) and *Medicago sativa* (lucerne) in Switzerland. The native *M. falcata* occurs in two ploidy levels that are geographically isolated (Savova *et al.*, 1996; Rufener Al Mazyad and Ammann, 1999) and rarely form hybrids. The diploid form is only found in the Unterengadin region in eastern Switzerland; the tetraploids in the other regions. In this case, geographic isolation contributes to the limited gene exchange between the two ploidy levels of *M. falcata*. On the other hand, lucerne (*M. sativa*) is a tetraploid relative that is grown all over Switzerland and also has established feral populations. Evidence for hybrid formation and genetic erosion of wild-type *M. falcata* was found only in regions where *M. sativa* and the tetraploid cytotype of *M. falcata* co-occur.

Fitness and Habitat Use of Hybrids

The second aspect that determines the success of hybridization concerns the fitness of hybrid offspring. In this section, I focus on extrinsic factors that contribute to fitness, while the intrinsic factors due to sterility and gene incompatibilities are discussed in the next section. The fitness of first-generation hybrids, specific backcrosses and later generations may all differ, and result from interactions between intrinsic factors due to sterility and incompatible gene combinations, in addition to the extrinsic factors that have to do with the functioning of the hybrid plants in the wild. This complicates any analysis of the fitness of hybrids, as it may not be the average fitness that matters, but the fitness of some rare, elite plants that do not suffer from intrinsic deleterious effects (Arnold, 1997). The impact of hybrid fitness can be subdivided into three aspects, the effects on the population dynamics; the effects on the ecological distribution of the plants; and, finally, the long-term evolutionary consequences.

Population dynamics and selection

The spread of hybrids and the genes contained in hybrids depends on the lifetime fitness of hybrids in the field, which implies that fitness has to be defined in a life history framework. For genes assimilated into many genetic backgrounds (i.e. when there are no reproductive, intrinsic barriers, and the species are conspecific or very closely related), the *fitness effect of the gene* is what will matter most for its spread and persistence in a recipient population (Rieseberg and Burke, 2001). In order to elucidate the fitness effect of the transgene apart from the background effect of any other gene, controls with transgenes need to be included in scientific studies.

Lifetime fitness is made up of different components such as germination rate, juvenile and adult survival, and pollen and seed production. Matrix projection models are often used as a framework to model life cycles (Caswell, 2000). In such a model, a population is divided into different age or stage groups, and a transition matrix is constructed that describes the changes in the vector of age/stage groups in one cycle, usually 1 year. The transition matrix contains the vital rates such as age-/stage-specific survival and fecundity (Box 2.1), and together they determine the expected population growth rate λ. The transition matrix is then characterized by the sensitivities and elasticities of the vital rates. Elasticities have the nice feature that the effects of different components are expressed on a standardized scale, and that the sum of all elasticities equals 1 (de Kroon *et al.*, 2000). For instance, an elasticity of 0.2 for survival indicates that a 1% increase in survival results in a 0.2% increase in population growth rate.

Box 2.1.   Finding the sensitive phases in the life cycle of a plant.

Consider the following hypothetical example of the life cycle of a perennial plant with three stages, seeds (S), juvenile (non-reproductive) plants (J) and reproductive adults (A). The arrows in Fig. 2.1 correspond to transitions between the stages from one year to the next, and the values need to be estimated in the field.

   This could for instance correspond to the following matrix projection:

$$\begin{pmatrix} S \\ J \\ A \end{pmatrix}_{t+1} = \begin{pmatrix} 0.2 & 0 & 100 \\ 0.01 & 0 & 1.5 \\ 0 & 0.3 & 0.3 \end{pmatrix} \begin{pmatrix} S \\ J \\ A \end{pmatrix}_{t}$$

This matrix is derived from the following transitions: a fraction (0.2) of the seeds in the seed bank is still in there in the following year (the rest germinates or dies). Of all seeds produced per adult per year, 100 end up in the seed bank the next year (this combines two processes: seed yield *per se* and the survival of seeds in the soil) and on average 1.5 seeds succeed to become a juvenile in the next year. The latter value is the seed yield per adult times the fraction of seeds that both germinate and are able to establish themselves. A fraction (0.01, i.e. 1%) of the seeds in the seed bank become juveniles in the next year and, finally, the annual survival rates of juveniles and adults are both 30%. The matrix entries are called the vital rates of the life cycle. Analysis of this simple matrix, for instance using POPTOOLS (http://www.cse.csiro.au/poptools), indicates that this population has an expected population growth rate (λ) of 1.056, so it is almost stable. Elasticities and sensitivities are used to describe the potential impact of a change in a vital rate on the population growth rate. A sensitivity is the increase in λ per unit increase in a vital rate a (mathematically the partial derivative of λ with respect to the vital rate a, i.e. $\partial\lambda/\partial a$), whereas an elasticity is the proportional increase in λ with a proportional increase in vital rate (mathematically the derivative of λ with respect to the vital rate a both expressed on a log scale, i.e. $\partial \ln \lambda/\partial \ln a$).

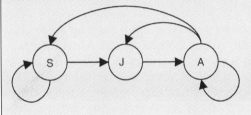

Fig. 2.1.   Hypothetical life cycle of a perennial plant with a seed bank.

continued

Box 2.1. *Continued.*

The elasticities for all transitions are as in Table 2.1.

Table 2.1.   Elasticities for all transitions according to the vital rates of the matrix above.

From	To	Value	Elasticity
Seed	Seed	0.2	0.035
Adult	Seed	100	0.149
Seed	Juvenile	0.01	0.149
Adult	Juvenile	1.5	0.192
Juvenile	Adult	0.3	0.341
Adult	Adult	0.3	0.135
			1.000

This means that the transition from juveniles to adults has the biggest impact on the population growth rate (elasticity 0.341). Also, the direct production by adults of seeds that immediately become juveniles in the following year has a bigger impact (0.192) than the production of seeds that first enter the seed bank (0.035), and may or may not become juveniles in a later year. Consequently, one would predict that hybrid plants with a certain proportional increase in the transition from juvenile to adult have a much higher lifetime fitness than hybrids with a similar proportional increase in survival of seeds in the soil seed bank. Likewise, transgenic traits that affect juvenile to adult survival (e.g. herbivore resistance) are expected to have a higher chance of introgressing into this recipient population than traits affecting seed survival in the soil.

There are two, related applications in the analysis of hybrid fitness. First, we can estimate vital rates for the population of hybrids, and compare the outcome (in terms of population growth rate) with similar estimates for one or both parents. Thus, different matrices are constructed and analysed. Alternatively, we could be interested in the chances of the 'successful' introduction of hybrid material in a target population. This could be an endemic species threatened by the introduction of relatives from elsewhere (Rhymer and Simberloff, 1996; Huxel, 1999), but also a wild relative of a crop that picks up (trans-)genes from the crop. In such cases, only the matrix of the receiving population is created, and from this we can already see what the most sensitive phases are in the life cycle of the recipient. We can then focus our attention on the question of how these vital rates are affected by the hybrid, or the genes carried by the hybrid. Such analyses are helpful for three reasons. First, they identify those transitions that have a large impact on population growth. If adult survival has a much higher elasticity than fertility, for instance, a proportional increase in survival has a much higher impact on population growth rate than an increase in fertility. Consequently, hybrids with a higher survival are then expected to increase more rapidly than hybrids with a comparable increase in fecundity. Secondly, it allows us to evaluate and integrate fitness components over the entire life cycle. Should a hybrid differ in several vital rates (e.g. have a higher survival but a lower fecundity), the model allows us to calculate the net effect of the two components. The effects of a correlation between the two traits can also be taken into account (van Tienderen, 1995). Thirdly, sensitivities and elasticities are also closely related to the intensity of selection in a quantitative genetic framework (Caswell, 2000; van Tienderen, 2000). This means that transitions that have a high elasticity are also traits that are under strong directional selection. Thus, if a donor species has characteristics with an effect on such a vital rate, it can be expected that the genes involved will increase in frequency in the recipient population.

The usage of such models is much more complicated if F_1s, backcrosses and other generations have to be dealt with separately, for instance due to clear differences in their performance, or specifically if the initial phase of an escape of a gene is relevant (as could be the case for the spread of transgenes). There is no easy solution for this, and more complicated models may need to be created, for instance by linking separate matrices for the different classes, together with the transition probabilities. In the context of the spread of transgenes into wild relatives, there appear to be no examples of the application of these methods, and matrix modelling is only used to assess the risk of the invasiveness of the crop itself (see Bullock, 1999).

A direct test of the fitness of transgenes brought into wild relatives under field conditions would of course be very informative if performed on an adequate scale with sufficient statistical power, and superior to any indirect assessment of the potential effects using models. Recently, three studies have been published, and the results were mixed: an increase in fitness due to the *Bt* transgene was found in *Helianthus annuus* (Snow *et al.*, 2003), and no difference in two other cases; a gene that increases white mould resistance (OxOx, oxalate oxidase), also in *H. annuus* (Burke and Rieseberg, 2003), and a gene providing resistance to a viral disease (cpBNYVV, rhizomania coat protein gene) in *Beta* (Bartsch *et al.*, 2003).

A gene that is subject to selection also affects genetic variation in the rest of the genome. Positive selection, if individuals containing the gene are favoured, will alter the genome by a *selective sweep* and flanking genes are dragged along in the slipstream of the gene, increasing in frequency. If the selection is negative, the genome will be subject to *background selection* and flanking genes will also be affected. In both cases, selection reduces the diversity around the focal gene. They can, in principle, be distinguished based on the frequency spectrum of DNA sequence variants around the putative gene under selection (Charlesworth *et al.*, 1995; Cummings and Clegg, 1998). The proportion of the genome subject to this reduction (called the 'genomic window') will vary according to the level of recombination around the genes. For outcrossing individuals, characterized by high levels of heterozygosity and effective recombination, the region of the genome that remains linked around the transgene can be very small, small meaning between 500 and 10,000 bp in maize (Wang *et al.*, 1999, 2001; Remington *et al.*, 2001; see Gepts and Papa, 2003). In predominantly selfing organisms, the linked region (said to be in 'linkage disequilibrium' with the transgene) is expected to be larger. Nordborg *et al.* (2002) showed that in *A. thaliana* global samples of natural populations, disequilibrium decayed within approximately 1 cM or 250 kb.

In early generation hybrids between crops and wild relatives, the parental chromosomes are still largely intact, and the genomic window is wide. With subsequent further introgression and recombination of the parental genomes, the genomic window slowly contracts. This suggests that the number of flanking genes sweeping through the recipient population can be quite high for transgenes that increase fitness.

For vegetatively propagated organisms, the entire genome is in linkage disequilibrium, regardless of whether the marker loci are located on the same chromosome as the transgene or on different chromosomes (Gepts and Papa, 2003).

Effects of hybridization and introgression on species distributions

A second effect of hybridization concerns the distribution of the recipient species. Given the claims made on the importance of hybridization in creating evolutionary novelties, surprisingly little is known about the fitness of between-species hybrids in the field and the habitat specificity of their distribution (Arnold *et al.*, 2001). If creation of novel traits is indeed important, one would expect that hybrids occupy different niches from those occupied by the parental lines. Reciprocal transplant experiments in which the performances of hybrids across a range of natural conditions are tested are scarce. Emms and Arnold (1997) studied Louisiana irises (*Iris fulva, I. hexagona* and their hybrids), and found some site-specific effects. In most cases, however, hybrid studies have been performed under controlled conditions rather than across a range of field conditions.

The most detailed study of the impact of crossing *within* species was done with *Chamaecrista fasciculata* (Fabaceae) (Fenster and Galloway, 2000); they found hybrid vigour in the first-generation hybrids, and reduced fitness of F_3s relative to the F_1s, which could be due to the breakdown of co-adapted gene complexes in later generations, or reduced heterozygosity. The net result for the comparison between parental lines and later generations is a balance between hybrid vigour and outbreeding effects, and only in cases where parents grew more than 1000 km apart did outbreeding depression dominate. This example shows that even in conspecific crosses, there are potential differences in the fitness of first- and later generation hybrids, as the genes of the parental lines are reshuffled and create new combinations, some good, some bad.

In the context of potential effects of GM plants on wild relatives, it is often stated that the main point is to assess the effects of the trait involved, and not so much the mechanisms by which it is introduced. Transgenes may give a plant a novel phenotype, but, as a recent report of the US National Research Council (NRC, 2002) concludes, 'the transgenic process presents no new categories of risks compared to conventional methods of crop improvement, but specific traits introduced by both approaches can pose unique risks'. Obviously, the spectrum of traits that becomes available through genetic modification greatly exceeds that of traditional breeding programmes that are constrained by the diversity in the crop and its nearest relatives. Especially novel traits associated with stress tolerance (salt, cold and drought) could affect habitat use. In such cases, a species could extend its range at the cost of a decline in other species. For example, crops adapted to increased soil salt content may alter the establishment and spread of transgenic feral populations or wild relatives to halophyte stands (Bartsch, 2000): tolerance against salinity is one of the determining factors controlling the survival in salt marsh and salt desert environments. Competition is hypothesized to play a key role in determining both the upper and lower limits of species distribution along a salinity gradient (Ungar, 1998). The relationship between the level of salt tolerance of species and their ability to compete with glycophytes in less saline habitats seems to be reciprocal. Halophytes – sugarbeet is a good example – are not competitive in non-saline habitats. Their competitiveness increases in saline habitats such as seashores.

For herbivore/pathogen resistance that is introduced into wild relatives, a lot will depend on the effects on the natural enemies of the plant, and if these enemies have an impact on the dynamics of plant populations, or, vice versa, if the natural enemies depend critically on the plant as hosts. What is needed to assess the potential impact of a trait is an analysis of what factors set the margins of the current distribution of a species, which will be a challenging task for ecologists. Other traits that may have a great impact on the performance and distribution of a recipient population are traits that change the breeding system, such as male sterility, asexual reproduction by apomixis or clonality, and increased weediness (Goodman and Newell, 1985).

Effects of hybridization on long-term evolution

The relevance of hybridization as a creative process is contentious. On the one hand, hybridization brings together genomes that have been separated by many generations of independent evolution, carrying independent adaptations, and thus their combination could be a rich source for evolutionary novelties. It would give rise to a reticulate (network) process of evolution, rather than the tree-like process by the accumulation of independent adaptations in separate lineages. Thus, hybridization is seen as a unique creative process, unlike that of a steady accumulation of beneficial mutations over evolutionary time, and a process with great ramifications for evolution in the groups involved (Arnold, 1997). In well-studied systems such as Louisiana irises, and sunflower species, the history of the species and their hybrids is unravelled in great detail using modern molecular techniques. The *Helianthus* work, for instance, tells a fascinating story of sterility genes, strong selection and intensive chromosomal rearrangements (Rieseberg *et al.*, 1999; Rieseberg, 2000).

Schemske (2000) is one of the scientists who takes a stand against the view of hybridization being an important creative process. He does not refute the claim that hybridization occurs in nature, nor that reproductive isolation is often not complete if tested in the laboratory. However, there are many different mechanisms that prevent between-species hybridization. Separation of habitats in space or time is a mechanism that effectively reduces gene exchange between species, and also means that selection for better isolation is not operative. Hybridization, Schemske argues, is confined mostly to cases where there is some human disturbance that brings species into contact and opens up the possibilities of hybrid formation. This argument, that can be traced back to Anderson's work (1948), indeed appears to apply to many hybrids, including the well-studied systems such as the irises and sunflowers. If hybrids are mostly found in disturbed situations where reproductive isolation is rendered ineffective, it is argued that the process of hybridization has been of limited importance for evolution and speciation (Schemske, 2000). Indeed, for the *Rorippa* mentioned above, there are indications that the hybrids mainly occur after disturbances, for instance in riparian habitats (Bleeker and Hurka, 2001). Rieseberg (1997), in his review on hybrid origins of plant species, lists eight species for which there is unequivocal evidence that homoploid hybridization occurred, three *Helianthus* species, the Louisiana irises, two *Peaonia* species (-groups), *Pinus densata* and *Stephanomeria diagensis*; in all eight cases, the parental lines were separated by eco-geographic (pre-mating) isolation, and, in five out of eight, post-mating isolation due to genic or chromosomal sterility factors was found (Rieseberg, 1997).

One important exception must be made, however: hybridization coupled to polyploidization. Combined genomes stabilized by duplication are considered to be a common and important process in plant speciation (Levin, 1983; Ramsey and Schemske, 1998, 2002; Soltis and Soltis, 2000), and the *Brassica* crops and wild relatives are an excellent example of the extent of the process (Fig. 2.2; also see Chèvre *et al.*, Chapter 18, this volume). The potential role of neopolyploidization has not been studied extensively in a GMO context, except for systems such as the *Brassica* where the allopolyploids are already present. Presumably, the process of a GM crop becoming involved in a *new* polyploidization event is considered too unlikely, as it involves two improbable events: the formation of a hybrid with the GM crop, and the duplication of the genomes by one of the possible mechanisms that can lead to polyploids (Ramsey and Schemske, 2002).

Stability of Genes in Hybrids

One final issue that may affect the chance of successful hybridization concerns the genetic stability of hybrids and, for the GM effects in particular, what will happen with a transgene after escape to a wild relative. Two aspects need to be considered: (i) genetic changes in DNA sequences, for instance the loss of a gene from the genetic material or fixation of a gene; and

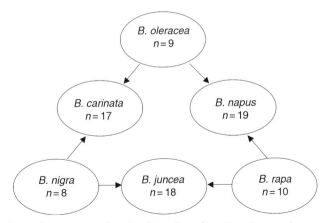

Fig. 2.2.   The well-known U-triangle of interbreeding *Brassica* species, named after its creator (U, 1935). Three diploid species with haploid chromosome numbers of 8, 9 and 10 and their allopolyploid hybrids with haploid chromosome numbers 17, 18 and 19.

(ii) changes in gene expression. The *Helianthus* work clearly showed that hybridization among taxa can be accompanied by extensive genetic changes and reshuffling, so there is a chance that during this process the transgene will be lost (Rieseberg, 2000). Also, in polyploid hybrids, there is a lot of evidence of reshuffling of genomes after the initial production, in both autopolyploids and allopolyploids (Soltis and Soltis, 2000; Ramsey and Schemske, 2002). This suggests that the initial phase is characterized by intense selection, with only few of the hybrids surviving. An additional effect is that a large impact of the number of formed hybrids is likely. Up-scaling effects after a widespread introduction of a GM crop could tip the balance from a selective bottleneck too severe to overcome, towards a 'successful' introgression in which some particular hybrids with a specific genetic constitution happened to be formed and increase in frequency. Such rare events are difficult to predict, given the limited knowledge we have on crop–wild relative combinations. A second consequence of bottlenecks and strong selection in the initial phase is that the genome location where the transgene has been introduced may be of importance. The *Helianthus* work also showed that successful hybrids always contain particular genomic regions from the respective parental species (Rieseberg, 2000). Thus, a transgene that happens to be inserted in such a region would hitchhike along and be transmitted preferentially to later generations. Alternatively, if the transgene is inserted close to a gene causing hybrid infertility, the chances of further spread are lower and contingent on a recombination event between the two genes. Obviously, the transgene's position relative to genes that affect intrinsic hybrid fitness is usually difficult to assess, and the genetic background may also affect the outcome (Burke and Rieseberg, 2003).

Whether introduced genes will go to fixation and hence change the evolutionary trajectory once and for all is more difficult to judge. For endemic species, introgression of genetic material from introduced species into the native population is considered to be one of the main threats to the biodiversity on islands (Levin *et al.*, 1996). Once established, the process is irreversible, and the endemic lost forever. Whether transgenes introduce a risk that is analogous to the situation of endemics will depend on all the different factors introduced in the previous sections, but, in particular, traits that cause changes in their habitat requirements (e.g. stress-related genes, weediness) need to be taken seriously.

More is known about gene silencing and epigenetic effects after the introduction of transgenes (for example, see Matzke *et al.*, 2000). Silencing is considered to have evolved as a protective defence against the expression of alien DNA, such as transposable elements, viruses and bacterial DNA. Much has been learned about the process in recent years, after it became a nuisance to the genetic engineering of stably expressed transgenes (Kooter *et al.*, 1999). Two different processes appear to be responsible for lowered expression of transgenes. The first acts at the DNA level and is thought to be caused by *de novo* DNA methylation and/or chromatin reconfiguration preventing transcription (TGS; transcriptional gene silencing). The second mechanism involves cytoplasmic degradation of RNA products produced by alien DNA (PTGS; post-transcriptional gene silencing). The process has several consequences for the chances of the spread of a gene into wild populations. First, genetic engineers will try to produce GM lines with a stable expression of the transgene. It is not known what this means for a transgene introduced into a wild relative, but it seems plausible – at least in the initial phases of hybridization – that the gene will also be stably expressed then, as the location and nature of the inserted DNA is not a target of TGS or PTGS. What will happen later may also depend on reshuffling of DNA during the process of introgression. Thus, the phenotypic effects of the transgene may be transient, reducing the probability of changes in the host population and of any unwanted effects on other species. However, at the same time, extra care needs to be taken in the estimation of the fitness effects of a transgene, since their expression can be variable.

Conclusions

A number of conclusions can be drawn from the knowledge of hybridization in nature, and how it affects any study of the potential effects of the introduction of GM crops. Little is known about the stability and expression of transgenes outside the crop in which they were introduced. We have to assume that a transgene is also expressed in wild relatives, which may have positive or negative effects on the recipient species, as well as its competitors, herbivores, pathogens, etc. Thus, the likelihood of hybridization and its consequences need to be established. Hybridization occurs in many plant taxa, and the nature of the process depends strongly on the relatedness of the involved taxa. For close relatives (conspecifics; e.g. *B. vulgaris*), reproductive barriers are minimal, some outbreeding depression may occur and overrule the effects of hybrid vigour, and the main aspects to consider are the phenotype and fitness of the hybrids. This requires ecological information on the sensitive parts in the life cycle of the recipient species, its habitat requirements and the factors that limit its current range. The focus of research should be to assess which traits and which genes may have a large impact on the species demography, distribution and genetic composition. Stress tolerance, insect and pathogen resistance and reproductive mode (e.g. apomixis) are certainly traits with potential fitness effects in recipient species.

For crosses among diverged species (e.g. *Brassica* species), the situation can be quite different. Interspecific hybrids occur in many taxa and, although significant differences among taxa exist in the frequency of hybridization, most crops are interfertile with wild relatives at least somewhere in their range. However, while many hybrid offspring can be produced, only a very small fraction becomes stabilized, and initial selection can be intense, leading to bottlenecks and genetic changes. The scale of GM introduction and hence hybrid formation may thus matter for the probability of obtaining successful hybrids. It would be advisable to have a much better picture of the extent of the genetic barriers that separate a crop from its wild relative, which involves more than data on phenology, spatial distribution and estimates for the outcrossing rate. This will determine whether only the effects on the initial hybrid generation need to be considered, or also the potential effects of further introgression into the wild population. The latter can be accompanied by intense selection and (cyto-)genetic changes, and could also affect the chances of the stacking of different transgenes.

Transfer of a transgene into a new genetic background may also affect genes flanking the gene, that, for instance, are dragged along from crop to wild relative in the case of a selective sweep. The position of the transgene determines which genes these will be, and generalizations on the potential effects are difficult. The reverse situation may also occur, and transgenes may be assimilated into the recipient species due to positive selection on nearby genes; for instance, genes preventing hybrid sterility. Again, the insert position of the transgene will be of importance. Of course, the phenotypic and fitness effects will also play an important role for interspecific crosses, similar to that for crosses among closer relatives.

It appears that a lot of baseline data for the putative recipients of transgenes from GM crops are still missing, and it would benefit all stakeholders to invest in better data, with information on the following:

- Life cycles of species, and in particular the most sensitive phases as inferred from demographic models;
- The ecological and spatial distributions of species;
- Intrinsic factors affecting gene exchange (ploidy levels, viability and sterility of hybrids);
- Current levels of exchange between non-GM crops and wild relatives, for instance making use of marker-assisted studies of introgression; and
- Breeding systems, selfing rates and amount of vegetative propagation.

Some of the required data can be obtained from existing sources (e.g. the 'Botanical Files' of de Vries *et al.*, 1992; see also Louwers *et al.*, 2002), but additional work is needed for more thorough analyses. Without this, the possible short-term effects are difficult, and long-term evolutionary success even impossible to predict. However, even with more data available, the great challenge for the coming years is to provide a thorough GM risk assessment given the trial and error character of the process of hybridization and introgression, and to get a better understanding of when it is safe to unleash Magritte's graces.

Acknowledgements

I would like to thank Jan Kooter and Alan Musgrave for helpful suggestions and discussion, and the editors for their support and patience.

References

Allendorf, F.W., Leary, R.F., Spruell, P. and Wenburg, J.K. (2001) The problems with hybrids: setting conservation guidelines. *Trends in Evolution and Ecology* 16, 613–622.

Anderson, E. (1948) Hybridisation of the habitat. *Evolution* 2, 1–9.

Arnold, M.L. (1997) *Natural Hybridisation and Evolution*. Princeton University Press, Princeton, New Jersey.

Arnold, M.L., Kentner, E.K., Johnston, J.A., Cornman, R.S. and Bouck, A.C. (2001) Natural hybridisation and fitness. *Taxon* 50, 1–12.

Bartsch, D. (2000) Ecological considerations for genetically modified halophytes. In: Lieth, H. and Moschenko, M. (eds) *Cashcrop Halophytes, Potential, Pilot Projects, Basic and Applied Research on Halophytes and Saline Irrigation*. Conference Proceedings, University Botanical Garden, Osnabrück, p. 20.

Bartsch, D. and Pohl-Orf, M. (1996) Ecological aspects of transgenic sugar beet, transfer and expression of herbicide resistance in hybrids with wild beets. *Euphytica* 91, 55–58.

Bartsch, D., Hoffmann, A., Lehnen, M. and Wehres, U. (2003) Ecological consequences of gene flow from cultivars to wild relatives, Rhizomania resistance genes in the genus *Beta*. In: Lelley, T., Balazs, E. and Tepfer, M. (eds) *Ecological Impact of GMO Dissemination in Agro-ecosystems*. Facultas Verlags- und Buchhandelsgesellschaft AG, Wien, Austria, pp. 115–130.

Bergelson, J., Purrington, C.B. and Winchman, G. (1998) Promiscuity in transgenic plants. *Nature* 395, 25.

Bleeker, W. and Hurka, H. (2001) Introgressive hybridization in *Rorippa* (Brassicaceae): gene flow and its consequences in natural and anthropogenic habitats. *Molecular Ecology* 10, 2013–2022.

Bullock, J.M. (1999) Using population matrix models to target GMO risk assessment. *Aspects of Applied Biology* 53, 205–212.

Burke, J.M. and Rieseberg, L.H. (2003) Fitness effects of transgenic disease resistance in sunflowers. *Science* 300, 1250.

Caswell, H. (2000) *Matrix Population Models: Construction, Analysis, and Interpretation*. Sinauer Associates, Sunderland, Massachusetts.

Charlesworth, D., Charlesworth, B. and Morgan, M.T. (1995) The pattern of neutral molecular variation under the background selection model. *Genetics* 141, 1619–1632.

Cummings, M.P. and Clegg, M.T. (1998) Nucleotide sequence diversity at the alcohol dehydrogenase 1 locus in wild barley (*Hordeum vulgare* ssp. *spontaneum*): an evaluation of the background selection hypothesis. *Proceedings of the National Academy of Sciences USA* 95, 5637–5642.

de Kroon, H., van Groenendael, J. and Ehrlén, J. (2000) Elasticities: a review of methods and model limitations. *Ecology* 81, 617–618.

de Vries, F.T., van der Meijden, R. and Brandenburg, W.A. (1992) Botanical Files. A study of the real chances for spontaneous gene flow from cultivated plants to the wild flora of The Netherlands. *Gorteria Supplement* 1, 1–100.

Ellstrand, N.C., Whitkus, R. and Rieseberg, L.H. (1996) Distribution of spontaneous plant hybrids. *Proceedings of the National Academy of Sciences USA* 93, 5090–5093.

Ellstrand, N.C., Prentice, H.C. and Hancock, J.F. (1999) Gene flow and introgression from domesticated plants into their wild relatives. *Annual Review of Ecology and Systematics* 30, 539–563.

Emms, S.K. and Arnold, M.L. (1997) The effect of habitat on parental and hybrid fitness: transplant experiments with Louisiana irises. *Evolution* 51, 1112–1119.

Fenster, C.B. and Galloway, L.F. (2000) Inbreeding and outbreeding depression in natural populations of *Chamaecrista fasciculata* (Fabaceae). *Conservation Biology* 14, 1406–1412.

Gepts, P. and Papa, R. (2003) Possible effects of (trans)gene flow from crops on the genetic diversity from landraces and wild relatives. *Environmental Biosafety Research* 2, 89–103.

Goodman, R.M. and Newell, N. (1985) Genetic engineering of plants for herbicide resistance: status and prospects. In: Halvorson, H.O., Pramer, D. and Rogul, M. (eds) *Engineered Organisms in the Environment: Scientific Issues*. American Society for Microbiology, Washington, DC, pp. 47–53.

Grant, V. (1981) *Plant Speciation*. Columbia University Press, New York.

Huxel, G.R. (1999) Rapid displacement of native species by invasive species: effects of hybridization. *Biological Conservation* 89, 143–152.

Jonsell, B. (1968) Studies in the North-West European species of *Rorippa* s. str. *Symbolae Botanicae Upsaliensis* XIX.

Kooter, J.M., Matzke, M.A. and Meyer, P. (1999) Listening to the silent genes: transgene silencing, gene regulation and pathogen control. *Trends in Plant Sciences* 4, 340–347.

Levin, D.A. (1975) Minority cytotype exclusion in local plant populations. *Taxon* 24, 35–43.

Levin, D.A. (1983) Polyploidy and novelty in flowering plants. *American Naturalist* 122, 1–25.

Levin, D.A., Francisco-Ortega, J. and Jansen, R.K. (1996) Hybridization and the extinction of rare plant species. *Conservation Biology* 10, 10–16.

Louwers, N., Brandenburg, W., Gilissen, L., Kleter, G. and Wagenaar, J. (2002) The Biosafety Files, a new link in biosafety information. *Biotechnology and Development Monitor* 49, 13–14.

Matzke, M.A., Mette, M.F. and Matzke, A.J.M. (2000) Transgene silencing by the host genome defense: implications for the evolution of epigenetic control mechanisms in plants and vertebrates. *Plant Molecular Biology* 43, 401–415.

Nordborg, M., Borevitz, J.O., Bergelson, J., Berry, C.C., Chory, J., Hagenblad, J., Kreitman, M., Maloof, J.N., Noyes, T., Oefner, P.J., Stahl, E.A. and Weigel, D. (2002) The extent of linkage disequilibrium in *Arabidopsis thaliana*. *Nature Genetics* 30, 190–193.

NRC (National Research Council) (2002) *Environmental Effects of Transgenic Plants: the Scope and Adequacy of Regulation.* National Academy Press, Washington, DC.

Ramsey, J. and Schemske, D.W. (1998) Pathways, mechanisms, and rates of polyploid formation in flowering plants. *Annual Review of Ecology and Systematics* 29, 467–501.

Ramsey, J. and Schemske, D.W. (2002) Neopolyploidy in flowering plants. *Annual Review of Ecology and Systematics* 33, 589–639.

Raybould, A.F. and Gray, A.J. (1993) Genetically modified crops and hybridisation with wild relatives: a UK perspective. *Journal of Applied Ecology* 30, 199–219.

Remington, D., Thornsberry, J., Matsuoka, Y., Wilson, L.M., Whitt, S., Doebley, J.F., Kresovich, S., Goodman, M.M. and Buckler, E. (2001) Structure of linkage disequilibrium and phenotypic associations in the maize genome. *Proceedings of the National Academy of Science USA* 98, 11479–11484.

Rhymer, J.M. and Simberloff, D. (1996) Extinction by hybridization and introgression. *Annual Review of Ecology and Systematics* 27, 83–109.

Rieseberg, L.H. (1997) Hybrid origins of plant species. *Annual Review of Ecology and Systematics* 28, 359–389.

Rieseberg, L.H. (2000) Crossing relationships among ancient and experimental sunflower hybrid lineages. *Evolution* 54, 859–865.

Rieseberg, L.H. and Burke, J.M. (2001) The biological reality of species, gene flow, selection and collective evolution. *Taxon* 50, 47–67.

Rieseberg, L.H., Whitton, J. and Gardner, K. (1999) Hybrid zones and the genetic architecture of a barrier to gene flow between two sunflower species. *Genetics* 152, 713–727.

Rufener Al Mazyad, P. and Ammann, K. (1999) The *Medicago falcata/sativa* complex, crop–wild relative introgression in Switzerland. In: van Raamsdonk, L.W.D. and den Nijs, J.C.M. (eds) *Plant Evolution in Man-made Habitats.* Proceedings of the VIIth International Symposium IOPB Amsterdam, Hugo de Vries Laboratory, Amsterdam, pp. 271–286.

Savova, D., Rufener Al-Mazyad, P. and Felber, F. (1996) Cytogeography of *Medicago falcata* L and *M. sativa* L in Switzerland. *Botanica Helvetica* 106, 197–207.

Schemske, D.W. (2000) Understanding the origin of species (book review). *Evolution* 54, 1069–1073.

Snow, A.A., Pilson, D., Rieseberg, L.H., Paulsen, M.J., Pleskac, N., Reagon, M.R., Wolf, D.E. and Selbo, S.M. (2003) A Bt transgene reduces herbivory and enhances fecundity in wild sunflowers. *Ecological Applications* 13, 279–286.

Soltis, P.S. and Soltis, D.E. (2000) The role of genetic and genomic attributes in the success of polyploids. *Proceedings of the National Academy of Sciences USA* 97, 7051–7057.

Stace, C.A. (1975) *Hybridization and the Flora of the British Isles.* Academic Press, London.

U, N. (1935) Genome analysis in *Brassica* with special reference to the experimental formation of *Brassica napus* and peculiar mode of fertilization. *Japanese Journal of Botany* 7, 389–452.

Ungar, I.A. (1998) Are biotic factors significant in influencing the distribution of halophytes in saline habitats? *Botanical Review* 64, 176–199.

van Tienderen, P.H. (1995) Life-cycle trade-offs in matrix population-models. *Ecology* 76, 2482–2489.

van Tienderen, P.H. (2000) Elasticities and the link between demographic and evolutionary dynamics. *Ecology* 81, 666–679.

Wang, R.L., Stec, A., Hey, J., Lukens, L. and Dobley, J. (1999) The limits of selection during maize domestication. *Nature* 398, 236–239.

Wang, R.L., Stec, A., Hey, J., Lukens, L. and Dobley, J. (2001) Correction: the limits of selection during maize domestication. *Nature* 410, 718.

3

Introgressive Hybridization Between Invasive and Native Plant Species – a Case Study in the Genus *Rorippa* (Brassicaceae)

W. BLEEKER

University of Osnabrück, Department of Systematic Botany, Barbarastraße 11, D-49076 Osnabrück, Germany, E-mail: bleeker@biologie. uni-osnabrueck.de

Abstract

Evolutionary processes associated with the introduction of new species or new varieties have only recently received wider attention. Molecular markers (chloroplast DNA: *trn*L intron; amplified fragment length polymorphism (AFLP)) provided substantial evidence for hybridization and bidirectional introgression between the invasive *Rorippa austriaca* and the native *R. sylvestris* in Germany. Three hybrid zones between the invasive and the native species were located in the Ruhr Valley (Mülheim) and at the river Main near Würzburg (Randersacker and Winterhausen). Different patterns of introgression, ploidy levels and hybrid fitness were detected in the different hybrid zones. It is concluded that hybridization between the invasive *R. austriaca* and the native *R. sylvestris*: (i) leads to different outcomes under different environmental conditions; (ii) results in hybrids which are not generally less fit than their parent species; (iii) may alter the invasive potential of *R. austriaca*; and (iv) may lead to the evolution of a new invasive species (*R. × armoracioides*).

Biological Invasions and Evolution

Analogous to genetically modified plants, invasive plant species are new elements of our flora interacting with native species. The spread of species beyond their natural ranges is not a new phenomenon and has always played a key role in the dynamics of biodiversity. However, the present rate of species exchange is unprecedented and has become one of the most intensively studied fields in ecology. Invasions of new territories follow accidental or deliberate introductions by man or are caused by climatic changes (Dukes and Mooney, 1999; Walther *et al.*, 2002; Parmesan and Yohe, 2003; Root *et al.*, 2003). The impact of biological invaders is manifold. They may lead to changes of ecosystem function, threaten the indigenous flora and fauna, add to its species enrichment, cause epidemic diseases and become pest organisms. Their ecological, sociological and economic impacts can be tremendous. Invasive species are topics of numerous documentation and monitoring programmes. However, evolutionary aspects of biological invasions have only recently received wider attention (Abbott, 1992; Mooney and Cleland, 2001; Hänfling and Kollmann, 2002; Lee, 2002; Hurka *et al.*, 2003). An invasive species must respond quickly and efficiently to new environmental conditions. Changes in the selective regime acting on the invader include both increased selection for adapted genotypes and relaxed selection for defence, because of the absence of co-evolved natural enemies (Blossey and Nötzold, 1995; Hänfling and Kollmann, 2002; Mitchell and Power, 2003). The genetic architecture of the invading species and its ability to respond to natural selection play a major role in the ultimate establishment of certain invasive species (Lee, 2002). Hybridization, either interspecific hybridization or hybridization between previously isolated populations of the same species (after multiple introductions), may be a source of genetic variation leading to increased invasiveness (Abbott, 1992; Ellstrand and Schierenbeck, 2000; Milne and Abbott, 2000). Interspecific hybridization is often regarded as an important speciation mechanism (Rieseberg, 1997) and may provide the raw material for adaptive evolution in rapidly changing environments (Anderson, 1949; Arnold, 1997).

The Genus *Rorippa*

The genus *Rorippa* Scop. comprises about 50–80 species and is represented by indigenous species on all continents except Antarctica (Al-Shehbaz, 1984). *Rorippa* is closely related to *Barbarea* R. Br., *Cardamine* L., *Nasturtium* R. Br. and *Armoracia* P. Gaertner, as revealed by DNA sequence analysis (Les, 1994; Franzke *et al.*, 1998; Koch *et al.*, 2001). Chloroplast DNA variation and biogeography of *Rorippa* was studied by Bleeker *et al.* (2002). Three *Rorippa* species are native in central European lowland areas: *R. amphibia* (L.) Besser, which comprises diploid and tetraploid cytotypes; the tetraploid *R. palustris* (L.) Besser; and *R. sylvestris* (L.) Besser, which includes tetraploids, pentaploids and hexaploids (Jonsell, 1968).

   The invasion of *R. austriaca* (Crantz) Besser into Central Europe provides excellent opportunities to analyse the impact of hybridization on the evolution of invasiveness. *R. austriaca* is a Sarmatic element which proceeds rather far into eastern Central Europe (Jonsell, 1973). The western boundary of its natural distribution area touches eastern Austria and Moravia (eastern Czech Republic; Fig. 3.1). For several decades, *R. austriaca* has been invasive in Germany. It has also been introduced to North America and is viewed as an exotic pest plant in several states of the USA. *R. austriaca* is a diploid ($2n = 16$), perennial species (Jonsell, 1968). Hybridization is a common phenomenon in *Rorippa*. The frequency and the evolutionary consequences of introgressive hybridization between the native *Rorippa* species turned out

Fig. 3.1. Western part of the natural distribution area of *Rorippa austriaca* and location of hybrid zones between *R. austriaca* and the native *R. sylvestris* in Germany. Modified from Base map Schweizer Weltatlas 2002, ©EDK.

to be environment dependent (Bleeker and Hurka, 2001). Here, I present results of a case study on hybridization between the invasive *R. austriaca* and one of the native species (*R. sylvestris*). Several hybrids between *R. austriaca* and the native *Rorippa* species have been reported in the literature (Jonsell, 1968; Javurkova-Kratochvilova and Tomsovic, 1972; Rich, 1991; Stace, 1997). *R. × armoracioides* (Tausch) Fuss (2n = 24, 32), the putative morphologically intermediate hybrid between *R. austriaca* and *R. sylvestris*, was recently found at several sites in northern Germany and the adjacent Netherlands (Bleeker, 2002). It forms large populations without its putative parent species *R. austriaca* and *R. sylvestris* growing nearby. The establishment of *R. × armoracioides* at sites where the putative parent species are absent or in competition with the latter led to the working hypothesis that hybridization may play an important role in the evolution of invasiveness in *Rorippa*.

Formation of Contact Zones Between the Invasive *R. austriaca* and the Native *R. sylvestris*

Biological invasions are leading to the breakdown of geographical isolation barriers. The formation of contact zones between invasive and native plant species depends on the habitat preferences of the species. The *R. austriaca* invasion in Germany has been studied in detail by Keil (1999) and Woitke (2001). The native *Rorippa* species *R. amphibia*, *R. palustris* and *R. sylvestris* are naturally occurring in various habitats of river systems, but are also found in man-made habitats. *R. austriaca* is invasive along riversides, rural areas and, locally, also vineyards, which causes problems in agriculture. Along riversides, it is invading river embankments and seasonally flooded grassland, habitats also typical for the native *R. sylvestris*. Contact zones between *R. austriaca* and *R. sylvestris* were located at the river Ruhr near Mülheim and at the river Main (two contact zones) near Würzburg (Fig. 3.1; Bleeker, 2003). *R. austriaca* and *R. sylvestris* can easily be distinguished based on fruit and leaf characters (Fig. 3.2). *R. austriaca* is characterized by globose fruits and entire leaves with auricles at the base, while *R. sylvestris* shows pinnate leaves without auricles at the base and linear fruits. Morphologically intermediate individuals (*R. × armoracioides*) were recognized in two of the three contact zones (the river Ruhr, Mülheim and the river Main, Randersacker).

Analyses of Interspecific Gene Flow Using Molecular Markers

R. austriaca and *R. sylvestris* form contact zones, but do they hybridize? Pre-mating isolation barriers (temporal isolation, ethological isolation) and/or post-pollination barriers (incompatibility, hybrid inviability) may prevent the production of hybrid progeny. Morphological markers are of limited value for detecting interspecific gene flow, since hybrids are not generally morphologically intermediate (Rieseberg and Ellstrand, 1993). In *Rorippa*, hybrids between *R. amphibia* and *R. palustris* are difficult to distinguish from hybrids between *R. amphibia* and *R. sylvestris*, which led

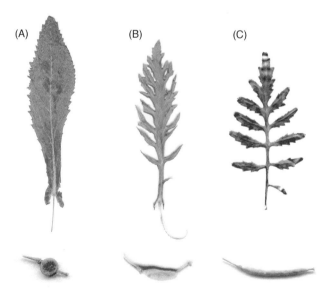

(A)   (B)   (C)

Fig. 3.2.   Leaf and fruit morphology of *Rorippa austriaca* (A), *R. sylvestris* (B) and their morphologically intermediate hybrid *R. × armoracioides* (C).

to a misinterpretation of hybridization patterns by taxonomists (Bleeker, 2001; Bleeker and Hurka, 2001).

Molecular markers provide powerful tools for analysing patterns of interspecific gene flow. Introgressive hybridization between the invasive *R. austriaca* and the native *R. sylvestris* has been studied using biparentally inherited nuclear AFLP markers and uniparentally inherited chloroplast DNA (*trn*L intron).

AFLP analysis

In the three contact zones between the invasive *R. austriaca* and the native *R. sylvestris* (the river Ruhr, Mülheim; and the river Main, Randersacker and Winterhausen), introgressive hybridization was clearly indicated (Bleeker, 2003). Hybrids were detected by additivity of diagnostic parental markers of the local populations. Markers were viewed as diagnostic when they occurred in all analysed individuals of one parent species but not in the other one. Table 3.1 exemplifies additive banding patterns in the hybrid zone at the river Ruhr (Mülheim). In the Mülheim contact zone, we were able to detect five diagnostic *R. austriaca* markers and 12 diagnostic *R. sylvestris* markers. The three hybrids showed all 17 markers (Table 3.1).

Table 3.1. Distribution of selected diagnostic *R. austriaca* and *R. sylvestris* AFLP markers in three hybrids collected in an invasive *R. austriaca* population near Mülheim (Ruhr).

Primer combination/ fragment length	R. austriaca				R. sylvestris
	Invasive	Hybrid 1	Hybrid 2	Hybrid 3	Native
*Mse*I-CTA/*Eco*RI-AAC					
52 bp	−	+	+	+	+
54 bp	−	+	+	+	+
64 bp	+	+	+	+	−
88 bp	−	+	+	+	+
98 bp	−	+	+	+	+
100 bp	+	+	+	+	−
117 bp	−	+	+	+	+
137 bp	−	+	+	+	+
194 bp	−	+	+	+	+
237 bp	−	+	+	+	+
*Mse*I-CTA/*Eco*RI-AAG					
65 bp	−	+	+	+	+
80 bp	+	+	+	+	−
108 bp	+	+	+	+	−
145 bp	−	+	+	+	+
154 bp	−	+	+	+	+
157 bp	+	+	+	+	−
268 bp	−	+	+	+	+

Eighteen AFLP markers indicated introgression of *R. austriaca* characters into *R. sylvestris* or of *R. sylvestris* characters into *R. austriaca* (Bleeker, 2003). Five markers (*Eco*RI-AAC/*Mse*I-CTA, 67, 173 and 177 bp; *Eco*RI-AAG/*Mse*I-CTA, 133 and 193 bp) were shared by *R. austriaca* and *R. sylvestris* in one or two of the hybrid zones but were absent in all other *R. austriaca* populations. The occurrence of 13 markers (*Eco*RI-AAC/*Mse*I-CTA, 55, 99, 112, 161, 165 and 285 bp; *Eco*RI-AAG/*Mse*I-CTA, 80, 93, 108, 130, 143, 149 and 261 bp) in *R. sylvestris* is most probably due to introgression of *R. austriaca* characters into *R. sylvestris* (Bleeker, 2003).

Figure 3.3 shows the results of a principal coordinates (PCO) analysis based on pairwise Euclidean squared distances between the 54 different phenotypes observed in the three hybrid zones. The first two axes accounted for 49.2% of the variation. Axis 1 (36%) separated *R. austriaca* from *R. sylvestris*. Seven different hybrid AFLP phenotypes were placed between *R. austriaca* and *R. sylvestris* due to the additivity of parental-specific markers (Fig. 3.3). Five of these hybrids (three from Mülheim and two from Randersacker) were morphologically intermediate (*R. × armoracioides*); three hybrids from Winterhausen (two of them showed the same AFLP phenotype!) were morphologically close to *R. sylvestris*, indicating introgression of *R. austriaca* markers into *R. sylvestris*. Axis 2 (13.2%) separated the hybrid zone Mülheim at the river Ruhr from the hybrid zones Winterhausen and Randersacker at the river Main. The hybrids are grouped according to their geographical origin, providing substantial evidence for independent multiple hybridizations (Fig. 3.3).

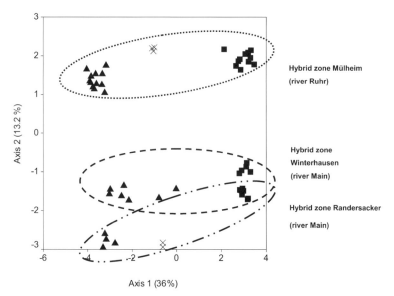

Fig. 3.3. Principal coordinates (PCO) analysis of 54 AFLP phenotypes detected in three hybrid zones between the invasive *R. austriaca* and the native *R. sylvestris* in Germany: triangles, *R. sylvestris*; squares, *R. austriaca*; crosses, *R. × armoracioides*. Modified from Bleeker (2003).

The hybrid status of six isolated *R. × armoracioides* populations, which were found without the parent species nearby, was confirmed in the AFLP analysis. All 45 individuals collected in these isolated populations showed a combination of *R. austriaca* and *R. sylvestris* characters (Bleeker, 2003).

Chloroplast DNA analysis

Chloroplast DNA analysis provided substantial evidence for introgression of the invasive *R. austriaca* chloroplast into the native *R. sylvestris*. The *trn*L intron of the chloroplast genome is one of the standard molecular markers in plant molecular sytematics (Gielly and Taberlet, 1994; Franzke *et al.*, 1998; Mummenhoff *et al.*, 2001). In *R. austriaca*, the *trn*L intron is characterized by a species-specific deletion. Nineteen native *R. austriaca* populations (46 individuals) from Russia, Kazakhstan, Bulgaria, the Czech Republic, Hungary and eastern Austria, and nine invasive *R. austriaca* populations (73 individuals) from Germany have been analysed so far (Bleeker *et al.*, 2002; Bleeker, 2003; Hurka *et al.*, 2003). All 119 individuals showed a species-specific *trn*L intron amplification product length variant of 420 bp. All other Eurasian *Rorippa* species are characterized by a *trn*L intron amplification product length variant of 590 bp (Bleeker *et al.*, 2002). Table 3.2 shows the distribution of the length variants in the three hybrid zones between *R. austriaca* and *R. sylvestris*. The length variant specific for *R. austriaca* (420 bp) was found in four individuals of *R. sylvestris* collected in the hybrid zone in Winterhausen near Würzburg at the river Main, providing evidence for introgression of the *R. austriaca* chloroplast into *R. sylvestris*. The other four *R. sylvestris* individuals sampled in the Winterhausen contact zone

Table 3.2. Distribution of two *trn*L intron amplification product length variants in *R. austriaca*, in three hybrid zones between *R. austriaca* and *R. sylvestris*, and in six isolated *R. × armoracioides* populations.

	n	420 bp	590 bp
R. austriaca (19 native and six invasive populations)	87	87	–
Hybrid zone Mülheim			
R. austriaca	12	12	–
R. sylvestris	18	–	18
R. × armoracioides	4	4	–
Hybrid zone Randersacker			
R. austriaca	4	4	–
R. sylvestris	4	–	4
R. × armoracioides	3	–	3
Hybrid zone Winterhausen			
R. austriaca	16	16	–
R. sylvestris	8	4	4
R. × armoracioides (six isolated populations)	38	27	11

and all *R. sylvestris* individuals collected in the hybrid zones in Randersacker (river Main) and Mülheim (river Ruhr) showed the 590 bp length variant (Table 3.2). The *R. sylvestris* individuals showing the *R. austriaca* chloroplast were also characterized by several *R. austriaca* markers in the AFLP analysis and were placed intermediate in the PCO analysis (Fig. 3.3). The morphologically intermediate hybrid *R. × armoracioides* showed the chloroplast either of *R. austriaca* (Mülheim hybrid zone) or of *R. sylvestris* (Randersacker hybrid zone). The same was true for the isolated *R. × armoracioides* populations, providing evidence that both parent species, *R. austriaca* or *R. sylvestris*, might serve as the maternal parent in hybridization events, since chloroplast DNA is maternally inherited in the Brassicaceae (Harris and Ingram, 1991).

A selection of fragments was purified and sequenced to test the homology of the 490 bp length variant within *R. austriaca* and among *R. austriaca* and *R. sylvestris* / *R. × armoracioides*. All sequences of the 420 bp length variant could be aligned non-ambiguously to sequences of other *Rorippa* species and showed a deletion of 169 bp at the same position. The homology of the deletion was confirmed in a phylogenetic analysis (Bleeker, 2003).

Evolutionary Consequences of Introgressive Hybridization Between Invasive and Native *Rorippa* Species

The results of our case study show that the invasive *R. austriaca* and the native *R. sylvestris* are likely to hybridize wherever they form contact zones. Introgressive hybridization is leading to bidirectional introgression and to the evolution of a new invasive hybrid taxon (*R. × armoracioides*).

Within the three analysed hybrid zones, different ploidy levels and different hybrid fitness were detected (Bleeker *et al.*, unpublished data). In the Mülheim hybrid zone (river Ruhr), hybridization between diploid *R. austriaca* and hexaploid *R. sylvestris* led to the formation of pentaploid hybrids, thus one of the parental species contributed unreduced gametes. The pentaploid hybrids showed high seed set and were not generally less fit than the parental species. In the hybrid zone Randersacker at the river Main, we detected diploid *R. austriaca*, tetraploid *R. sylvestris* and triploid hybrids. The triploid hybrids showed low pollen viability, and no seed set was observed in 2002 (Bleeker *et al.*, unpublished data). It can be concluded that the evolutionary outcome of hybridization between *R. austriaca* and *R. sylvestris* depends on the environment. The ploidy level of the hybridizing native species *R. sylvestris*, which is known to be environment dependent (Jonsell, 1968), plays a key role in determining the genetic constitution of the hybrids. Looking at the observed patterns in the Mülheim and the Randersacker hybrid zone, one would expect hybridization to run towards introgression in the Randersacker hybrid zone and towards speciation in the Mülheim hybrid zone.

Introgression and its consequences

Molecular data provide substantial evidence for unidirectional introgression of the invasive *R. austriaca* chloroplast into *R. sylvestris* (Winterhausen), and for bidirectional introgression of AFLP markers (Mülheim and Randersacker). Although the observed fitness in the triploid hybrids in Randersacker is extremely low, molecular data prove that introgression is occurring. In *R. austriaca*, genetic variation within the hybrid zones was higher than outside it (Bleeker, 2003). Interspecific gene flow between *R. austriaca* and *R. sylvestris* is one source of variation in the invasive *R. austriaca* and may contribute to the generation of novel genotypes increasing the invasion success of the species. However, intraspecific hybridization between previously isolated populations of the same species after multiple introductions and accumulation of genetic diversity within populations after the species has invaded an area are also likely mechanisms leading to additional variation within *R. austriaca*.

R. sylvestris is not a rare species and is certainly not endangered by hybridization with *R. austriaca*. However, numerous examples have been cited where rarer species are considered threatened by hybridization with a more common, related taxon (Rhymer and Simberloff, 1996). Levin *et al.* (1996) describe a number of cases of extinction by hybridization on islands. A loss of species integrity can follow from progressive introgression (Rieseberg and Wendel, 1993) as well as from genetic swamping of rare native species by abundant alien pollen (Ellstrand, 1992).

Hybridization and speciation

While hybridization with invaders can be a threat to species integrity, it can also be a source of new variation and the origin of new species. Our data provide evidence for the evolution of a new invasive taxon via hybridization between an invasive diploid (*R. austriaca*) and a native polyploid (*R. sylvestris*) species. The pentaploid hybrids in the Mülheim hybrid zone were morphologically intermediate and showed a high fitness, especially when compared with *R. austriaca*. The existence of isolated *R. × armoracioides* populations proved that the hybrid taxon is able to leave the hybrid zones and to found new populations. The isolated *R. × armoracioides* populations showed different ploidy levels (4x and 5x) and various degrees of pollen viability, seed set and seed germination (Bleeker *et al.*, unpublished data). *R. × armoracioides* is characterized by a combination of parental characters. It grows at rural sites and on abandoned agricultural fields in the area of rivers and is currently spreading in northern Germany, The Netherlands and in certain parts of eastern Germany (Bleeker, 2002). Some of the analysed populations of *R. × armoracioides* showed high levels of within-population variation (Bleeker, 2003). The variation at the molecular level coincides with morphological variation in the isolated hybrid populations and may play an important role in the invasion success of the hybrid. The question arises of whether the high AFLP phenotypic diversity is likely to be due to multiple origins or segregation following sexual reproduction. The molecular data showed that *R. austriaca* and *R. sylvestris* are likely to hybridize wherever they form contact zones. Multiple origins will thus be one explanation for high variation within the hybrid. The differences in population variation observed among the analysed populations are probably due to variation in sexual reproduction. Only some populations of *R. × armoracioides* reproduce sexually (Bleeker *et al.*, unpublished results).

R. × armoracioides is a striking example of ongoing speciation in our modern environment. Other examples of ongoing hybrid speciation in man-made habitats in Central Europe are from the genera *Nasturtium* (Bleeker *et al.*, 1999), *Viola* (Neuffer *et al.*, 1999), *Cardamine* (Urbanska *et al.*, 1997) and from hybridization between native *Rorippa* species (Bleeker and Hurka, 2001). A textbook example of the origin of an invasive new species by hybridization between an introduced and a native species is from the cordgrass *Spartina*. In the late 19th century, the highly aggressive *S. anglica* C.E. Hubbard originated in England via hybridization between native *S. maritima* (Curtis) Fern and introduced *S. alternifolia* Lois. (Gray *et al.*, 1991; Ferris *et al.*, 1997). Another example of the same phenomenon concerns the origin of the allopolyploid *Senecio cambrensis* Rosser in the British Isles following hybridization between the introduced *S. squalidus* L. and the native *S. vulgaris* L. Multiple origins and an extinction have been documented for *S. cambrensis* (Abbott and Forbes, 2002). Abbott (1992) and Ellstrand and Schierenbeck (2000) listed several examples for the evolution of invasiveness following hybridization.

Outlook

Invasive species will continue to be a major focus in evolutionary biology and ecology. The recently observed global climatic change is likely to promote the further spread of species beyond their natural range, including thermophilous plants that spread from gardens into surrounding countryside (Dukes and Mooney, 1999; Menzel and Fabian, 1999; Walther *et al.*, 2002). The question of whether hybridization between invasive and native plant species, or hybridization between newly introduced genetically modified plant species and their native relatives, will lead to the evolution of invasiveness/weediness is of great public interest. The results of the *Rorippa* case study show that each hybrid zone is unique. The evolutionary consequences of hybridization can only be studied case by case, which means hybrid zone by hybrid zone. Common garden experiments may help us in estimating the fitness of hybrids under different environmental conditions. However, extensive field studies are indispensable to detect the real-time consequences of hybridization and introgression over a longer time period.

Acknowledgements

The presented study is part of the project 'Evolutionary consequences of biological invasions', which is financed by the German Federal Ministry of Education and Research (BIOLOG 01C0012; project leader H. Hurka).

References

Abbott, R.J. (1992) Plant invasions, interspecific hybridization and the evolution of new plant taxa. *Trends in Ecology and Evolution* 7, 401–405.

Abbott, R.J. and Forbes, D.G. (2002) Extinction of the Edinburgh lineage of the allopolyploid neospecies, *Senecio cambrensis* Rosser (Asteraceae). *Heredity* 88, 267–269.

Al-Shehbaz, I.A. (1984) The tribes of Cruciferae (Brassicaceae) in the Southeastern United States. *Journal of the Arnold Arboretum* 65, 343–373.

Anderson, E. (1949) *Introgressive Hybridization*. John Wiley & Sons, New York.

Arnold, M.L. (1997) *Natural Hybridization and Evolution*. Oxford University Press, New York.

Bleeker, W. (2001) Verbreitung und Häufigkeit von *Rorippa amphibia × Rorippa palustris* in Nordwestdeutschland. *Floristische Rundbriefe* 35, 11–17.

Bleeker, W. (2002) Entstehung und Etablierung hybridogener Sippen: Drei Fallbeispiele aus Nordwestdeutschland. *Neobiota* 1, 205–216.

Bleeker, W. (2003) Hybridization and the *Rorippa austriaca* (Brassicaceae) invasion in Germany. *Molecular Ecology*, 12, 1831–1841.

Bleeker, W. and Hurka, H. (2001) Introgressive hybridization in *Rorippa* (Brassicaceae): gene flow and its consequences in natural and anthropogenic habitats. *Molecular Ecology* 10, 2013–2022.

Bleeker, W., Huthmann, H. and Hurka, H. (1999) Evolution of hybrid taxa in *Nasturtium* R. Br. (Brassicaceae). *Folia Geobotanica* 34, 421–433.

Bleeker, W., Weber-Sparenberg, C. and Hurka, H. (2002) Chloroplast DNA variation and biogeography in the genus *Rorippa* Scop. (Brassicaceae). *Plant Biology* 4, 104–111.

Blossey, B. and Nötzold, R. (1995) Evolution of increased competitive ability in invasive nonindigenous plants: a hypothesis. *Journal of Ecology* 83, 887–889.

Dukes, J.S. and Mooney, H.A. (1999) Does global change increase the success of biological invaders? *Trends in Ecology and Evolution* 14, 135–139.

Ellstrand, N.C. (1992) Gene flow by pollen: implications for plant conservation genetics. *Oikos* 63, 77–86.

Ellstrand, N.C. and Schierenbeck, K.A. (2000) Hybridization as a stimulus for the evolution of invasiveness in plants. *Proceedings of the National Academy of Sciences USA* 97, 7043–7050.

Ferris, C., King, R.A. and Gray, A.J. (1997) Molecular evidence for the maternal parentage in the hybrid origin of *Spartina anglica* C.E. Hubbard. *Molecular Ecology* 6, 185–187.

Franzke, A., Pollmann, K., Bleeker, W., Kohrt, R. and Hurka, H. (1998) Molecular systematics of *Cardamine* and allied genera (Brassicaceae). ITS and noncoding chloroplast DNA. *Folia Geobotanica* 33, 225–240.

Gielly, L. and Taberlet, P. (1994) The use of chloroplast DNA to resolve plant phylogenies: noncoding versus *rbc*L sequences. *Molecular Biology and Evolution* 11, 769–777.

Gray, D.F., Marshall, D.F. and Raybold, A.F. (1991) A century of evolution in *Spartina anglica*. *Advanced Ecological Research* 21, 1–64.

Hänfling, B. and Kollmann, J. (2002) An evolutionary perspective of biological invasions. *Trends in Ecology and Evolution* 17, 545–546.

Harris, S.A. and Ingram, R. (1991) Chloroplast DNA and biosystematics: the effect of intraspecific diversity and plastid transmission. *Taxon* 40, 393–412.

Hurka, H., Bleeker, W. and Neuffer, B. (2003) Evolutionary processes associated with biological invasions in the Brassicaceae. *Biological Invasions* 5, 281–291.

Javurkova-Kratochvilova, V. and Tomsovic, P. (1972) Chromosome study of the genus *Rorippa* Scop. in Czechoslovakia. *Preslia* 44, 140–156.

Jonsell, B. (1968) Studies in the north-west European species of *Rorippa* s. str. *Symbolae Botanicae Upsaliensis* 19, 1–221.

Jonsell, B. (1973) Taxonomy and distribution of *Rorippa* (Cruciferae) in the southern USSR. *Svensk Botanisk Tidskrift* 67, 281–302.

Keil, P. (1999) Ökologie der gewässerbegleitenden Agriophyten *Angelica archangelica ssp. litoralis*, *Bidens frondosa* und *Rorippa austriaca* im Ruhrgebiet. *Dissertationes Botanicae*, Vol. 321. J. Cramer, Berlin.

Koch, M., Haubold, B. and Mitchell-Olds, T. (2001) Molecular systematics of the Brassicaceae: evidence from coding plastidic *mat*K and nuclear *chs* sequences. *American Journal of Botany* 88, 534–544.

Lee, C.E. (2002) Evolutionary genetics of invasive species. *Trends in Ecology and Evolution* 17, 386–391.

Les, D.H. (1994) Molecular systematics and taxonomy of lake cress (*Neobeckia aquatica*; Brassicaceae), an imperiled aquatic mustard. *Aquatic Botany* 49, 149–165.

Levin, D.A., Francisko-Ortega, J. and Janssen, R.K. (1996) Hybridization and the extinction of rare plant species. *Conservation Biology* 10, 10–16.

Menzel, A. and Fabian, P. (1999) Growing season extended in Europe. *Nature* 397, 659.

Milne, R.I. and Abbott, R.J. (2000) Origin and evolution of naturalized material of *Rhododendron ponticum* L. in the British Isles. *Molecular Ecology* 9, 541–556.

Mitchell, C.E. and Power, A.G. (2003) Release of invasive plants from fungal and viral pathogens. *Nature* 421, 625–627.

Mooney, H.A. and Cleland, E.E. (2001) The evolutionary impact of invasive species. *Proceedings of the National Academy of Sciences USA* 10, 5446–5451.

Mummenhoff, K., Brüggemann, H. and Bowman, J.L. (2001) Chloroplast DNA phylogeny and biogeography of *Lepidium* (Brassicaceae). *American Journal of Botany* 88, 2051–2063.

Neuffer, B., Auge, A., Mesch, H., Amarell, U. and Brandl, R. (1999) Spread of violets in polluted pine forests: morphological and molecular evidence for the ecological importance of interspecific hybridization. *Molecular Ecology* 8, 365–377.

Parmesan, C. and Yohe, G. (2003) A globally coherent fingerprint of the climate change impacts across natural systems. *Nature* 421, 37–42.

Rhymer, J.M. and Simberloff, D. (1996) Extinction by hybridization and introgression. *Annual Review of Ecology and Systematics* 27, 83–109.

Rich, T.C.G. (1991) *Crucifers of Great Britain and Ireland*. Botanical Society of the British Isles, London.

Rieseberg, L.H. (1997) Hybrid origin of plant species. *Annual Review of Ecology and Systematics* 27, 359–389.

Rieseberg, L.H. and Ellstrand, N.C. (1993) What can molecular and morphological markers tell us about plant hybridization? *Critical Reviews in Plant Sciences* 12, 213–241.

Rieseberg, L.H. and Wendel, J.F. (1993) Introgression and its consequences in plants. In: Harrison, R.G. (ed.) *Hybrid Zones and the Evolutionary Process.* Oxford University Press, Oxford, pp. 70–109.

Root, T.L., Price, J.T., Hall, K.R., Schneider, S.H., Rosenzweig, C. and Pounds, J.A. (2003) Fingerprints of global warming on wild animals and plants. *Nature* 421, 57–60.

Stace, C. (1997) *New Flora of the British Isles,* 2nd edn. Cambridge University Press, Cambridge.

Urbanska, K.M., Hurka, H., Landolt, E., Neuffer, B. and Mummenhoff, K. (1997) Hybridization and evolution in *Cardamine* (Brassicaceae) at Urnerboden, Central Switzerland: biosystematic and molecular evidence. *Plant Systematics and Evolution* 204, 233–256.

Walther, G.-R., Post, E., Convey, P., Menzel, A., Parmesan, C., Beebee, T.J.C., Fromentin, J.-M., Hoegh-Guldberg, O. and Bairlein, F. (2002) Ecological responses to recent climate change. *Nature* 416, 389–395.

Woitke, M. (2001) Artenkombination, Etablierungsstadium und anthropogenes Störungsregime als Einflussfaktoren auf die Bestandsentwicklung der invasiven Brassicaceae *Bunias orientale* L. und *Rorippa austriaca* (Crantz) Besser in experimenteller Vegetation. PhD thesis, University of Würzburg, Germany.

4

Hybrids Between Cultivated and Wild Carrots: a Life History

Thure P. Hauser[1], Gitte K. Bjørn[2], Line Magnussen[1] and Sang In Shim[3]

[1]Plant Research Department, Risø National Laboratory, PO Box 49, DK-4000 Roskilde, Denmark; [2]Department of Horticulture, Danish Institute of Agricultural Sciences, Box 102, DK-5792 Årslev, Denmark; [3]Division of Plant Resources and Environment, Gyeongsang National University, Chinju, 660–701, Korea; E-mail thure.hauser@risoe.dk

Abstract

Cultivated and wild carrots can hybridize spontaneously, and transgenes may therefore potentially spread from genetically modified carrots into wild populations. This can happen by several different routes. Flowering cultivated carrots may send out pollen to neighbouring populations of wild carrots either from root production fields, where some plants flower despite being bred to be biennial, or from seed production fields, where a large number of flowering plants are concentrated. In addition, wild carrots growing in the vicinity of seed production fields may pollinate the seed plants and produce hybrid seeds. These may then be imported with the sowing seeds to other regions of the world. In the root production fields in Denmark, the hybrids flower in their first year and, if not weeded, they may pollinate each other and wild carrots in neighbouring populations.

F_1 hybrids (wild ♀ × cultivar) survive and reproduce as frequently, or almost as frequently, as wild carrots. This is surprising, because carrot hybrids are less tolerant to frost, and because other cultivar traits could be maladaptive in nature. Whether cultivar genes are in fact incorporated into the gene pool of wild carrots has not been documented, but recent studies indicate that some carrot individuals growing close to Danish fields are indeed hybrids.

Our results show that the hybridization between cultivated and wild carrots is very likely, and that it probably occurs via several different hybridization routes. Consideration of seed movement by trade, and of the management of seed production areas is thus extremely important for risk assessment of future genetically modified carrots.

Introduction

Do cultivated and wild carrots exchange genes through hybridization and introgression? Do their hybrids establish, survive and reproduce in nature? These are the questions that we will address in this chapter. In the context of this book, the reason to ask such questions is that we want to know whether genetically modified (GM) carrots may spread the inserted genes (transgenes) to wild carrots, growing as weeds in fields or in natural habitats. Several GM carrots have been field tested, with transgenes that make them more pest resistant, herbicide tolerant, or produce different sugar constituents (Information Systems for Biotechnology, 2003). However, no commercial cultivars are on the market yet.

Another reason to ask questions about hybridization in carrots is that it tells us something about the evolution and domestication of our crops and weeds. Crops that are known to hybridize with wild relatives include the major crops worldwide, and some of the most serious weeds of the world hybridize with crops (Ellstrand *et al.*, 1999; N.C. Ellstrand, Amsterdam, 2003, personal communication). In some of the species, exchange of genes between cultivars and wild plants is thought to have influenced the evolution of both the crop (at least before modern controlled plant breeding) and the weedy plants (Harlan, 1965). Several examples of such complexes of crops, weeds and wild relatives have been described and classified (Small, 1984; Van Raamsdonk and Van Der Maesen, 1996).

Cultivated and wild carrots (*Daucus carota* ssp. *sativa* and ssp. *carota*, respectively) are believed to form such a complex (Wijnheijmer *et al.*, 1989; Van Raamsdonk and Van Der Maesen, 1996). Cultivated carrots are grown worldwide as an important high-value vegetable (Stein and Nothnagel, 1995). Wild carrots come from Europe and western Asia (Naturhistoriska Riksmuseet, 2003), and have been introduced into North and South America, New Zealand and Australia (Dale, 1974; T. Hauser, unpublished data). Thus, there are ample opportunities for the exchange of genes in many regions of the world. The two subspecies hybridize without any serious crossing barrier, which is well known among plant breeders because pollen from wild carrots can contaminate cultivar seed lots if they grow close to seed production fields (Heywood, 1983; Small, 1984; D'Antuono, 1985). One hypothesis for the origin of western, orange-rooted carrots involves spontaneous hybridization between wild carrot subspecies and earlier yellow- and white-rooted cultivars (Heywood, 1983).

However, even if hybridization is possible, this does not necessarily lead to transfer of (trans)genes from cultivated to wild carrot. The hybrids of first and later generations must be able to survive and reproduce outside cultivation, and send their genes into new generations of hybrids until the genes are incorporated into true wild plants.

Hybridization and Introgression Routes

Cultivated and wild carrots can potentially hybridize and exchange genes by several different routes. These involve root or seed production fields in various parts of the world, where cultivar plants function as either fathers or mothers. Our focus here is mostly on the hybridization routes and conditions relevant for Denmark.

Pollen from root fields may move into neighbouring wild populations

Although cultivated carrots have been bred to be biennial, some plants do flower in root carrot fields, especially after a cold spell in the spring (Atherton *et al.*, 1990). In a survey of carrot fields in a major production area in Denmark, we found flowering individuals (bolters) in all fields (Fig. 4.1). Some of the flowering individuals had orange roots (Figs 4.2 and 4.3), indicative of cultivar plants (Hauser and Bjørn, 2001).

Fig. 4.1.   Flowering bolters within Danish carrot fields. Left: a field with just a few bolters. Right: a heavily infested field in the process of being weeded.

Fig. 4.2.   Morphology of bolters, showing a variety of root colours and shapes.

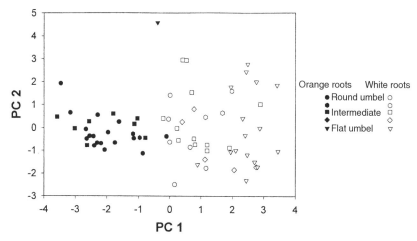

Fig. 4.3. Principal component analysis of traits from bolting individuals in carrot fields. Root colour and umbel shape, which had the highest correlations to the first axis, are indicated by symbols. The investigation included, in addition, the position in/out of rows, root shape, thickness, branching, surface and consistency, presence of small side roots, hairiness of flowering stem, flower colour, presence of central red flowerets and anthers, and pollen viability. Taken from Hauser and Bjørn (2001).

From the flowering cultivar plants, bees and other insects may transport pollen to surrounding wild carrots. Wild carrots are commonly found around Danish carrot fields (T. Hauser, personal observations), and because they do not affect root production in the fields there is no incentive for the farmers to remove them. Carrot flowers are highly attractive to bees, which may fly several kilometres to visit flowering carrots (Gary *et al.*, 1972, 1977). Wild carrots also strongly attract hover flies in natural populations (Memmott, 1999). The F_1 hybrid seeds, originating from pollen flowing out from fields, will thus develop on wild plants and disperse from there directly into the wild habitats.

Pollen from seed production fields may move into neighbouring wild populations

In the seed production areas, huge numbers of cultivated carrots flower at the same time, and insects may disperse their pollen to wild plants in the surroundings. In contrast to the root production areas, wild carrots are removed by intensive weeding to avoid inflow of wild pollen. Not all plants are removed, however, as shown by Wijnheijmer *et al.* (1989), who found wild plants close to Dutch seed fields. Our own studies clearly show that despite weeding, some wild plants still grow within pollination distance from seed fields (see below; Hauser and Bjørn, 2001).

The F_1 hybrid seeds that are produced by the pollen from seed fields will develop and be dispersed directly into the habitats of the wild parents. Subsequently, the hybrids may reproduce with local wild carrots to produce more advanced hybrids. An example of this is suggested in the study by Wijnheijmer *et al.* (1989), who found wild carrots growing in the verges of a seed field that had leaf shapes intermediate between cultivar and wild plants. However, their results should be viewed with caution, as the leaf shapes of pure wild plants were measured on plants from dune populations that may be genetically and ecologically different from pure wild plants closer to the cultivation area.

Pollen from wild carrots may move into seed production fields

In the seed production areas, pollen from wild plants growing in the vicinity may pollinate the seed-producing plants (Small, 1984). D'Antuono (1985) found that seeds harvested close to field margins develop higher percentages of white roots, indicating hybrid origin, than seeds harvested further inside the fields. This distribution suggests that white-rooted offspring mainly originate from pollen coming from outside the fields. Contamination of commercial seed lots with hybrid seeds is also suggested by tests of cultivars and seed lots. Different seed lots (of the same cultivar) and different cultivars, tested at several locations in Denmark, often produce significantly different frequencies of white-rooted bolters (Hauser and Bjørn, 2001). Such differences can only arise if there are differences in the amount of wild pollen reaching the seed plants in the seed production areas. Despite a low degree of contamination, the cultivars may still be certified for cultivation by the authorities; the threshold for impurity is 1% in Denmark at present.

From the seed production areas, the hybrids are imported to the cultivation regions. These may be close by, but also very distant. For example, in Denmark, seeds are not produced any more due to unstable yields. Instead, seeds are imported from seed production areas in southern Europe, California, New Zealand, Australia and other regions with a warmer climate. The same is the case for other cultivation areas in colder, more northern regions.

When cultivars are sown, any hybrid seeds present in the seed lots are sown as well. Most of the hybrid plants flower in their first summer, at least in Denmark, as shown by the cultivar tests where all white-rooted plants bolt (Hauser and Bjørn, 2001). In the southern seed production regions, wild carrots are annual, and their hybrid offspring probably inherit this trait. Similarly, in North America, wild carrots in the southern part of their range more frequently flower in their first year than do wild carrots from the northern parts (Lacey, 1988), suggesting that the timing of reproduction is an adaptation to latitudinal conditions (Lacey, 1988). The beneficial conditions within Danish carrot fields, compared with the tougher conditions in natural grasslands, allow the hybrids to grow large enough to flower; size has been shown to constrain flowering in wild carrots (Gross, 1981; Lacey, 1982). In contrast to our observations, Wijnheijmer *et al.* (1989) described cultivated ♀ × wild carrots as being biennial. However, their hybrids were produced by wild Dutch carrots that probably were more biennial than the southern wild carrots that produced the hybrids we studied in Danish fields. Even in The

Netherlands, however, seed production has moved to more southern regions, and it is likely that annuality of hybrids imported with seeds is becoming more common.

Weedy hybrid populations

Bolting individuals within carrot fields, of either pure cultivar or wild–hybrid origin, are able to pollinate each other and produce seeds (Hauser and Bjørn, 2001). If the seeds are allowed to ripen, they can survive in the seed bank until carrots are again cultivated in the same field. This also creates a winter survival niche for the hybrids. They probably inherit sensitivity to frost from their wild southern fathers, together with the seed dormancy. That seeds do survive in the seed bank is evident from the patches of bolters that can be seen in some fields, which shows that seed-shattering bolters grew there some time before. In a few worst cases, the numbers of bolters may be enormous and cover parts of a field with a white flowering blanket (Fig. 4.1). The bolters combine morphological traits from wild and cultivated carrots, grow both within and between the rows, and produce fully fertile pollen (Figs 4.2 and 4.3; Hauser and Bjørn, 2001). From these weedy stands, pollen and seeds may spread to wild plants in the surroundings.

Farmers weed bolters, both to avoid the 'weed' problem, and because bolters can interfere with harvest machinery. However, due to decreased farm labour, the weeding intensity has decreased. A survey of carrot fields in one of the major carrot-producing regions in Denmark showed that there were a few bolters in all the fields (Hauser and Bjørn, 2001).

Weed management practices obviously have a major influence on the probability of gene transfer from crop to wild populations. In some other cultivation regions in Denmark, carrots are only infrequently cultivated in the same fields, in contrast to our study area, and we suspect that there will be many fewer weedy carrots, as a seed bank will not build up. The potential for gene transfer to wild populations will therefore be less (even though bolting cultivar plants and imported bolting hybrids will still be there, if not weeded away).

Establishment and Survival of Hybrids in Natural Habitats

Even if hybrids are generated spontaneously, they may not be able to establish, survive and reproduce under conditions in wild habitats. If hybrids do not survive or reproduce, gene introgression will not take place.

Hybrids in Danish natural grasslands

To study the performance of hybrids under natural conditions, we sowed seeds of wild carrots and hybrids, produced by controlled pollinations (wild Danish ♀ × cultivar), into three grasslands of different ages that all contained wild carrots already (Hauser *et al.*, unpublished results). Emergence, survival and reproduction were followed for 3 years. Hybrids survived as well and flowered as frequently or more frequently than the wild carrots, but developed somewhat fewer and smaller flower umbels. In none of the plots did the hybrids flower in their first summer. This is consistent with our discussion on the annuality/bienniality of carrot hybrids above: in our field experiment, hybrids were produced by wild Danish mothers, and thus inherited a high degree of bienniality/perenniality (in contrast to the hybrids imported with sowing seeds).

F_1 hybrids between cultivated and wild carrots (wild ♀ × cultivar) thus seem to be able to survive and reproduce in natural grasslands in Denmark, and they may at least sometimes be as fit or nearly as fit as wild plants. This conclusion is perhaps a bit surprising, given that such F_1 hybrids are less tolerant to frost than wild plants (Hauser, 2002). By chance, our field experiment covered three relatively mild winters, but such sequences of mild winters occur regularly and seem to allow hybrids to survive and reproduce at least sometimes.

How do backcrosses and hybrids of more advanced generations survive and reproduce in natural habitats? That has never been studied, and we can thus only speculate. The relatively high fitness of the F_1 hybrids in our study (Hauser *et al.*, unpublished results) is probably caused by heterosis, but heterosis degrades in subsequent hybrid generations. Second and later generations of hybrids may be affected negatively from the break-up of favourable, parental gene combinations by recombination (an effect termed outbreeding depression; Lynch, 1991). We guess, however, that outbreeding depression is not very strong, judging from the weedy hybrid populations within carrot fields that are doing very well indeed. Many of these are at least second-generation hybrids.

Hybrids from other hybridization routes

The results on fitness of hybrids in natural Danish grasslands were obtained from wild ♀ × cultivar plants, and may not hold for hybrids from other hybridization routes. Cultivar ♀ × wild southern hybrids inherit adaptations to southern conditions from their wild fathers and may thus be less fit under Danish conditions. However, the F_1 hybrids (wild Danish ♀ × cultivar ♂) that we used in our field study were also offspring from an 'adapted' (wild ♀) and a 'maladapted' (cultivar ♂) parent, and they survived well. In the generations subsequent to F_1, offspring are expected to become gradually more locally adapted since an increasing proportion of their genome will come from the wild Danish plants.

Seeds dispersed from hybrid bolters in fields to neighbouring wild populations are probably not likely to give rise to further hybrid generations. Their only chance is to grow large enough to flower within their first summer, and to survive winters as seeds. This is not very likely, however, due to the harsh conditions and strong competition within natural grasslands.

Survival in other geographical regions

To what extent carrot hybrids are able to establish in natural habitats of other parts of the world is not known. In regions further to the north, frost probably prohibits their establishment. Even wild carrots do not grow further to the north in Europe than southern Norway, central Sweden and southern Finland (Naturhistoriska Riksmuseet, 2003). In areas further to the south, other environmental stresses such as heat and drought may limit survival of carrot hybrids. However, if the hybrids are annual, they may survive drought as seed, but no studies have addressed this.

Gene Establishment

If hybrids between cultivated and wild carrots do in fact survive and reproduce in natural habitats, does this lead to introgression of cultivar genes, and among them transgenes, into the wild carrot gene pool? Small (1984) argued that there has been no influence of gene flow from cultivated plants on wild carrots in North America, but he based this conclusion only on the lack of morphological cultivar traits in wild populations. In contrast, Wijnheijmer *et al.* (1989) suggested that hybridization had led to intermediate leaf morphologies of wild plants growing close to seed production fields.

In an ongoing study, we have genotyped by amplified fragment length polymorphism (AFLP) four cultivars commonly grown in Denmark, white-rooted bolters from five fields with varying densities of bolters, wild carrots from three wild populations adjacent to Danish carrot fields, and four wild populations away from cultivation areas (Magnussen and Hauser, unpublished results). Our results indicate that bolters contain markers from both cultivated and wild carrots and, more importantly, that a few of the wild plants adjacent to carrot fields are somewhat similar to white-rooted bolters (based on discriminant analysis and assignment tests where each individual is matched to the most similar population). This should be expected if the bolters do hybridize with plants in the nearby, wild populations. In contrast, none of the wild plants growing far away from fields do assign to the bolters. Even if a few plants adjacent to fields are indeed first- or later generation hybrids, hybridization has not been sufficiently frequent to influence the average genetic composition of their populations. Wild populations adjacent to carrot fields were not on average more similar to cultivars or bolters than wild populations distant from fields. Our results thus suggest that a few F_1 or probably later generation hybrids can be found in wild carrot populations close to cultivation.

Parallels to Other Crops

The different hybridization and introgression routes that exist for carrots can also be found in beets and seabeets, as described in other chapters of this book (Cuguen *et al.*, Chapter 15, this volume; Van Dijk, Chapter 5, this volume). These complicated hybridization routes have not, to our knowledge, been described for other crops. However, they may be there if looked for. The unifying characteristic of carrot and beet life histories is that both are cultivated for their vegetative parts, and therefore have been selected to be biennial (reproduction taps energy from vegetative parts and makes them smaller and of lower quality). Both have wild relatives in southern seed production areas that produce hybrid seeds, and the hybrids inherit a high degree of annuality that allow them to flower already in their first summer and to survive winters as seeds in more northern cultivation areas. Additionally, both crops also have wild relatives in the northern cultivation areas (for more information on beets, see also Desplanque *et al.*, 1999, 2002). These characteristics can probably be found in other crops, such as other vegetables.

Conclusion

Our studies, together with others, show that cultivated and wild carrots do hybridize and that the hybrids are sometimes able to establish and persist in natural habitats, at least in Denmark. However, as in many other crop × wild complexes, we do not have any hard proof that introgression of genes from the crop into the wild relative has taken place, and even less that introgressed genes have influenced how wild carrots behave and interact with other organisms. However, wild and cultivated carrots have a history of interaction as long as domestication: modern orange carrots may have originated partly by hybridization with wild carrots (Heywood, 1983), pollen from wild carrots has probably always contaminated seed production, and pollen has flown regularly from bolters into wild populations. With this information, it seems highly likely that cultivar genes and, in the future, transgenes, do flow into wild populations.

The question is whether the transfer of transgenes to wild carrots will cause novel and harmful environmental effects. We simply do not know. Wild carrots may become weedier, not only in carrot fields but also in other fields. Interestingly, wild carrots are considered noxious weeds in some parts of North America (Mitich, 1996; Stachler and Kells, 1997; National Plant Board, 2002); even if carrots are not serious weeds in, for example, Europe, they may become so with changed conditions. Transgenes conferring herbicide, stress or pest tolerance could perhaps cause such a change, even though different management methods are probably more important. Whether the possibility that wild carrots may become weedier is enough to ban the use of GM carrots is questionable and up to the value judgements that society has to make.

The hybridization routes between cultivated and wild carrots via the seed-producing areas have important implications when we consider the future management of GM carrots. Movement of seeds by trade will, in many cases, be as important or more important for the spread of transgenes to wild plants than the gene flow directly from flowering plants in fields to neighbouring wild populations. The consequence of this is that effects cannot be evaluated and regulated at a national level only, but have to incorporate the whole life history of both the crop and the recipient wild plants. If hybridization between GM and wild carrots does create problems (e.g. with weed evolution), this can most probably not be managed by regulating national farmers alone.

Acknowledgements

This study was supported by 'Centre for Effects and Risks of Biotechnology in Agriculture' (supported by the Danish Environmental Research Programme) and 'Centre for Bioethics and Risk Assessment' (funded by the Danish Research Agency).

References

D'Antuono, L.F. (1985) Studio sull'inquinamento genetico causato da polline de tipi spontanei in carota da seme. *Rivista di Agronomia* 19, 297–304.

Atherton, J.G., Craigon, J. and Basher, E.A. (1990) Flowering and bolting in carrot. I. Juvenility, cardinal temperatures and thermal times for vernalization. *Journal of Horticultural Science* 65, 423–429.

Dale, H.M. (1974) The biology of Canadian weeds. 5. *Daucus carota. Canadian Journal of Plant Science* 54, 673–685.

Desplanque, B., Boudry, P., Broomberg, K., Saumitou-Laprade, P., Cuguen, J. and Van Dijk, H. (1999) Genetic diversity and gene flow between wild, cultivated and weedy forms of *Beta vulgaris* L. (Chenopodiaceae), assessed by RFLP and microsattelite markers. *Theoretical and Applied Genetics* 98, 1194–1201.

Desplanque, B., Hautekèete, N. and Van Dijk, H. (2002) Transgenic weed beets: possible, probable, avoidable? *Journal of Applied Ecology* 39, 561–571.

Ellstrand, N.C., Prentice, H.C. and Hancock, J.F. (1999) Gene flow and introgression from domesticated plants into their wild relatives. *Annual Review of Ecology and Systematics* 30, 539–563.

Gary, N.E., Witherell, P.C. and Marston, J. (1972) Foraging range and distribution of honey bees used for carrot and onion pollination. *Environmental Entomology* 1, 71–78.

Gary, N.E., Witherell, P.C., Lorenzen, K. and Marston, J.M. (1977) The interfield distribution of honey bees foraging on carrots, onions, and safflower. *Environmental Entomology* 6, 637–640.

Gross, K.L. (1981) Predictions of fate from rosette size in four 'biennial' plant species: *Verbascum thapsus, Oenothera biennis, Daucus carota*, and *Tragopogon dubius*. *Oecologia* 48, 209–213.

Harlan, J.R. (1965) The possible role of weed races in the evolution of cultivated plants. *Euphytica* 14, 173–176.

Hauser, T.P. (2002) Frost sensitivity of hybrids between wild and cultivated carrots. *Conservation Genetics* 3, 75–78.

Hauser, T.P. and Bjørn, G.K. (2001) Hybrids between wild and cultivated carrots in Danish carrot fields. *Genetic Resources and Crop Evolution* 48, 499–506.

Heywood, V.H. (1983) Relationships and evolution in the *Daucus carota* complex. *Israel Journal of Botany* 32, 51–65.

Information Systems for Biotechnology (2003) International Field Test Sources. Blacksburg, Virginia. Available at: http://www.isb.vt.edu/CFDOCS/globalfieldtests.cfm

Lacey, E.P. (1982) Timing of seed dispersal in *Daucus carota*. *Oikos* 39, 83–91.

Lacey, E.P. (1988) Latitudinal variation in reproductive timing of a short-lived monocarp, *Daucus carota* (Apiaceae). *Ecology* 69, 220–232.

Lynch, M. (1991) The genetic interpretation of inbreeding depression and outbreeding depression. *Evolution* 45, 622–629.

Memmott, J. (1999) The structure of a plant-pollinator food web. *Ecology Letters* 2, 276–280.

Mitich, L.W. (1996) Wild carrot (*Daucus carota* L.). *Weed Technology* 10, 455–457.

National Plant Board (2002) State Regulated Noxious Weeds. USDA, Washington, DC. Available at: http://www.aphis.usda.gov/npb/statenw.html

Naturhistoriska Riksmuseet (2003) Morot, *Daucus carota*, Distribution Maps. Stockholm, Sweden. Available at: http://linnaeus.nrm.se/flora/di/apia/daucu

Small, E. (1984) Hybridization in the domesticated–weed–wild complex. In: Grant, W.F. (ed.) *Plant Biosystematics*. Academic Press, Toronto, pp. 195–210.

Stachler, J.M. and Kells, J.J. (1997) Wild carrot (*Daucus carota*) control in no-tillage cropping systems. *Weed Technology* 11, 444–452.

Stein, M. and Nothnagel, T. (1995) Some remarks on carrot breeding (*Daucus carota sativus* Hoffm.). *Plant Breeding* 114, 1–11.

Van Raamsdonk, L.W.D. and Van Der Maesen, L.J.G. (1996) Crop–weed complexes: the complex relationship between crop plants and their wild relatives. *Acta Botanica Neerlandica* 45, 135–155.

Wijnheijmer, E.H.M., Brandenburg, W.A. and Ter Borg, S.J. (1989) Interactions between wild and cultivated carrots (*Daucus carota* L.) in The Netherlands. *Euphytica* 40, 147–154.

5

Gene Exchange Between Wild and Crop in *Beta vulgaris*: How Easy is Hybridization and What Will Happen in Later Generations?

HENK VAN DIJK

Laboratoire de Génétique et Evolution des Populations Végétales, UMR CNRS 8016, Université de Lille, Bâtiment SN2, 59655 Villeneuve d'Ascq, France, E-mail: henk.van-dijk@univ-lille1.fr

Abstract

Four conditions are needed for the introgression of crop genes into wild relatives: (i) pollination is possible; (ii) zygotes are formed (compatibility of pollen and ovules); (iii) F_1 hybrids and later hybrid generations are viable and fertile; and (iv) there is no selection against the introduced gene in backcross generations.

In our study area, France, wild (*Beta vulgaris* ssp. *maritima*) and cultivated (ssp. *vulgaris*) beets have several zones of contact where pollination is possible: in the seed production areas with the surrounding wild populations, and in the sugarbeet fields with weedy forms or with wild coastal populations. Limiting factors, apart from distance, are male sterility and, in the sugarbeet fields, the low percentage of flowering crop plants. The difference in ploidy level may diminish the success of hybridization. The resulting F_1 hybrids, unless they have a triploid cultivated parent, are vigorous plants, very well adapted to the agroecosystem. The cultivated characteristics, however, may be a handicap in natural ecosystems.

Further generations are also known to be very successful in the agroecosystem, albeit only in the sugarbeet crop: these are the weed beets, which behave like annual weeds with a long-lived soil seed bank. We did not find substantial evidence for introgression of cultivated traits into wild populations outside the agroecosystem, either in the wild populations near the seed production areas in southwest France, or in coastal populations. The fate of transgenes, however, is not necessarily exposed to the same selection pressures as genes for classical cultivated traits. Herbicide tolerance and virus or nematode resistance may be selected for in situations where the corresponding herbicide is used or where plants live in contact with the enemy in question.

Introduction

Gene exchange between crops and wild plants has gained considerable interest now that transgenic crops are ready for the market or have already been introduced. At first sight, root crops such as beet, carrot or chicory do not constitute a substantial problem, but, since they are grown from seed, there is an important possibility of gene flow in seed production areas, in conditions where the wild species is sympatric. This is true for beet (and also for carrot; see Hauser *et al.*, Chapter 4, this volume), and creates a situation where gene flow in both directions has to be taken into account: from crop to wild, but also from wild to crop, introducing weedy genes into the crop. The major gene exchange possibilities will be described, together with a discussion of the possible consequences.

The Study Species: Distribution and Characteristics

Beta vulgaris ssp. *maritima* (L.) Arcang. is considered as the wild ancestor of all cultivated forms of *B. vulgaris* ssp. *vulgaris* (sugarbeet, fodderbeet, red table beet and leaf beets). The wild subspecies grows almost exclusively near the coasts of Europe (Mediterranean, Atlantic, Channel, Irish Sea, North Sea, western Baltic), northern Africa and the Middle East. Non-coastal populations are only known in southern regions (Spain and southern France, among others). Our study area is France, with wild populations growing all along the Channel, Atlantic and Mediterranean coasts, but also inland from Narbonne (Mediterranean) up to Nérac (halfway between Bordeaux and Toulouse). These inland populations form a continuum, but there is no connection with the Atlantic coast near Bordeaux (see Fig. 5.1).

The major crop is sugarbeet, which is cultivated in our study area in northern France (Fig. 5.1). Sugarbeet seeds were formerly (before 1960) produced in several regions rather close to the sugar production area, but later moved on to southwest France (between Bordeaux and Toulouse, and therefore within the region of the inland wild beets, see Fig. 5.1). Other such seed production areas in Europe are situated in Italy (the Po delta; Bartsch and Schmidt, 1997) and in the Ukraine (the Crimea peninsula; see Slyvchenko and Bartsch, Chapter 14, this volume).

The wild subspecies *maritima*, the seabeet, is diploid ($2n = 18$), perennial, wind-pollinated, self-incompatible and gynodioecious. The species needs long days for flowering induction after a variable exposure time to low temperatures (vernalization), but there is a major gene *B* that suppresses the vernalization requirement completely (Boudry *et al.*, 1994). Northern populations (North Sea and English Channel) are 100% *bb* and have a rather strong quantitative vernalization requirement (Boudry *et al.*, 2002), whereas in the Mediterranean area the frequency of the *B* allele is intermediate and the quantitative vernalization requirement among *bb* plants is low. The inland populations near the present seed production areas are almost fixed for the *B* allele (see Fig. 5.1).

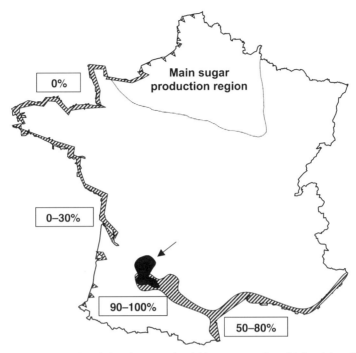

Fig. 5.1. The geographical distribution of wild beet (coastal and inland; hatched), the main sugarbeet seed production area (black, see arrow) and the main sugar production region in France. Wild beets have variable percentages of *B-* (which means all plants without a requirement for vernalization, whether *BB* or *Bb*), while cultivated forms do have a requirement for vernalization and are *bb* (diploid) or *bbb* (triploid).

Sugarbeet is supposed to remain vegetative and is harvested in autumn but, in virtually all sugarbeet fields, a few 'bolters' can be found (plants with stem elongation and subsequent flowering, having been subjected to more cold than the others or having a lower vernalization requirement). They may flower and produce seeds when not destroyed. Classic varieties in Europe are triploid, made by pollination of male-sterile seed-bearer plants by tetraploid pollinator plants, but, gradually, more diploid varieties are being developed. Plants in the seed production area used to be destroyed after reproduction and are therefore enforced biennials, but even without destruction they normally die after reproduction. Sugarbeet is able to survive mild winters (Pohl-Orff *et al.*, 1999; own observations near Lille), but we only very exceptionally observed survival after reproduction followed by a second flowering period in the next summer.

An Overview of Gene Exchange Between Crop and Wild

Wild beet and sugarbeet are sympatric at several places in the study area. In northern France, sugarbeet fields near the coast are sometimes located at short distances from wild populations, although the latter are not very abundant along this part of the coast and are mostly limited to estuaries (Fig. 5.1). Sugarbeet is grown for its roots and there has been intensive artificial selection to avoid flowering, especially by increasing the vernalization requirement. Before 1960, farmers had to remove excess seedlings, because the sown seedballs consisted of several fruits grown together, with one seed per fruit. They also followed their crop with more attention than at present and assiduously removed their bolters. Gene flow from crop to wild was therefore rather hypothetical at that time. After the introduction of mono-germ seed (one seed per propagule) and a higher level of mechanization, less attention was paid to bolters, and a few bolters per hectare can now be found in virtually all sugarbeet fields (Desplanque *et al.*, 2002).

However, even if these bolters are not kept from flowering, their male fertility is limited for two reasons. First, a part of the sugarbeet plants is fully or partially male sterile, which is a direct consequence of the way in which sugarbeet seed is produced: by pollination of male-sterile seed-bearer plants by male-fertile pollinator plants. Male sterility can be restored, which we could verify at places where sugarbeets were not harvested at the edges of fields destined for the construction of buildings in the following year. In some cases, they succeeded in surviving and flowering; the majority turned out to produce considerable amounts of pollen, but male steriles were also observed.

A second factor with a negative influence on gene flow is the ploidy level of sugarbeet. Traditionally, the varieties used in our study area are triploid. Although the abandoned sugarbeet plants mentioned above produced large quantities of seeds, we observed that the majority of the plants grown from these seeds were severely deformed or died in the early stages of development, undoubtedly due to problems at the chromosome level.

A second possibility of contact between wild and crop is in the areas where sugarbeet seed is produced. Before 1960, these areas where sometimes located near the coast, e.g. in northern Brittany. Pollen flow from the pollinator plants to seabeet populations may have taken place, as well as a contamination of the seed lots by hybrids formed by the arrival of wild pollen on the seed-bearer plants, but this did not lead to detectable problems of seed quality (M. Desprez, Cappelle-en-Pévèle, personal communication). In the present major seed production area in southern France, inland wild beets are present in the surroundings but there is a striking difference between these wild beets and the seabeets in northern Brittany. The latter all show the genotype *bb*, just like the sugarbeet crop, whereas the wild inland beets of southern France are almost all *BB*, which means that the contamination by wild pollen here leads to *Bb* heterozygotes. The *B* allele is dominant over *b*, at least under optimal conditions (Boudry *et al.*, 1994), which means that the hybrid plants will bolt, flower and produce seeds unless they are eradicated. The change to the new seed production areas coincided with the new agricultural practice, and in many sugarbeet fields these F_1 hybrid bolters were neglected. Their seeds stayed in the soil and formed a long-lived seed bank (with a lifetime of at least several decades). Thus surviving easily to the next sugarbeet culture, the phenomenon 'weed beet' was born: their progeny maintained themselves, gaining in weedy performance. Weed beets are abundant in northern France, but also in some other European countries (e.g. see Soukup and Holec, Chapter 16, this volume).

In the seed production areas, gene flow is possible in both directions; the consequences, if any, of the arrival of cultivated genes in the wild inland populations will be discussed later, but the appearance of weed beets in the sugarbeet crop greatly enhanced the possibilities of gene flow, as will be explained in the next section.

Weed Beets: a Bridge Between Crop and Wild

The genetic information that makes a beet weedy is first of all the *B* allele. In the crop–wild F_1 hybrid, the strong vernalization requirement of sugarbeet is thus completely annihilated, which guarantees a rapid bolting and therefore enough time for seed ripening, if the farmer takes no preventive measures. In the F_2 and later generations, *bb* genotypes will segregate, and will be counter-selected unless they have a low quantitative vernalization requirement. The selection pressure on low or absence of vernalization requirement is almost absolute, because non-flowering plants will be destroyed at the end of the crop season without leaving any progeny. We measured *B* allele frequencies in older weed beet populations and found values between 60 and 80%. This is significantly higher than the initial 50% in the F_1 hybrids, but lower than expected when the selection pressure is solely acting on the *B/b* locus. The quantitative vernalization requirement was about as low as in the wild inland populations (for data on the latter populations, see Boudry *et al.*, 2002) and much lower than in sugarbeet, which suggests that an important part of the natural selection has been acting on this character. Farmers sow their sugarbeet seed as early in spring as possible but have to avoid a too intensive exposure to cold. Weed beets, which have a lower cold requirement and germinate together with the crop, will therefore easily exceed their vernalization threshold.

Once established, weed beet populations are difficult to eliminate if sugarbeet is regularly grown on these fields. The specific herbicides for sugarbeet do not destroy them because they belong to the same species as the crop. Even on severely infested fields, farmers continue growing sugarbeet, in order to maintain their quota, and also because growing sugarbeet is economically profitable even if there is loss of biomass due to the weed beets. As a result, on many sugarbeet fields, considerable quantities of beet pollen and seed are produced. The very limited possibilities of gene flow from the agroecosystem towards the wild seabeets as they existed before the introduction of weed beets are no longer valid (see Cuguen *et al.*, Chapter 15, this volume, for more details). Figure 5.2 shows the major gene exchange possibilities in the present situation.

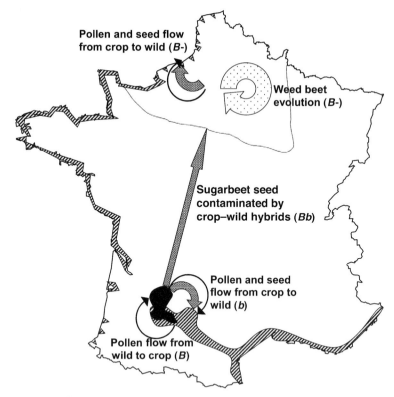

Fig. 5.2. Gene flow by seed (thick arrows) and pollen (thin arrows) between crop and wild. Weed beets occur in the sugarbeet fields and are the progeny of crop–wild hybrids formed in the seed production area.

Gene Flow from Crop to Wild in the Seed Production Region

In the seed production areas, massive amounts of pollen are produced each year. It can therefore be expected that the wild plants found nearby will have many crop–wild hybrids in their progeny. Although wild beets in these areas are systematically eliminated in order to minimize contamination, we nevertheless succeeded in 1995 in localizing a wild population in Montmaur (east of Toulouse) at a distance of 550 m from a sugarbeet seed production field. We harvested seeds and analysed their progeny. Crop–wild hybrids are morphologically very different from pure wild plants (thick roots, different growth form and leaves of different shape and colour), but we did not detect any cultivated traits like these. In 63 other populations in this region, situated at larger distances from the seed production fields than the Montmaur population, we found that 1–2% of the plants had progeny with cultivated morphological traits, but in the large majority of cases there was a clear resemblance to leaf beet (Swiss chard). Also a few hybrids with red table beet were found, and in only two populations were sugarbeet traits clearly present in a few individuals.

Because the pollinator plants used for seed production are for the greater part tetraploid, one should also expect triploids in the progeny of the surrounding wild beets. Desplanque *et al.* (1996) examined 81 progeny of wild plants from the sugarbeet seed production area, including several from the Montmaur population, but did not find any triploids, using a method based on counting chloroplasts in stoma guard cells (Desplanque *et al.*, 2002). Both results suggest that gene flow from sugarbeet to wild by pollination of wild plants with cultivated pollen is very limited. A plausible reason is that diploid pollen (from tetraploid pollen donors) is far less competitive compared with the local haploid wild pollen (Scott and Longden, 1970; Hecker, 1988). This will change when diploid pollinators are used, but we do not have data for that situation.

A different way in which gene flow from crop to wild in the seed production area may occur is by loss of seeds during the seed production process. In 1995, we observed in one of the fields large amounts of viable seeds on the cut pollinator plants which apparently were not destroyed early enough after flowering. Commercial seed may be lost during harvest and transport and, in that case, when they leave descendants, their specific male-sterile cytoplasm can be used to check whether there has been effective seed flow in the past followed by crosses with wild individuals. Desplanque (1999) found this cytoplasm in five out of 86 plants, from five different populations out of 23. These plants were morphologically indistinguishable from the other wild plants, which suggests an introgression of the cytoplasmic genetic information in a nuclear genetic wild background. In conclusion, gene flow from sugarbeet to wild in the past could have been more important by seed loss than by direct pollen flow.

Scenarios for the Introgression of Transgenes

Gene flow between crop and wild is of particular importance now that transgenic varieties could possibly be authorized in Europe. Because of the key role of weed beets, a major question is whether their formation can happen again, but now with transgenic sugarbeet as the cultivated parent (Desplanque *et al.*, 2002). One category of transgenes plays an important role: the tolerance for total herbicides, because the introduction of varieties with such a tolerance may be the only economically attractive solution for the weed beet problem. All non-tolerant plants, even the members of the same species, are destroyed by the herbicide. In the case of the appearance of transgenic weed beets, however, this solution will be only temporary! Desplanque *et al.* (2002) argue that the incorporation of the transgene into tetraploid pollinators is the best scenario to reduce the probability of tolerant weed beets, because the wild pollen then forms hybrids with non-tolerant seed-bearer plants leading to non-tolerant F_1 hybrids. A second advantage is that the diploid pollen of the tetraploid pollinators will have less chance to transmit their transgenes to the wild beets near the seed production fields.

If, however, the transgene escapes to the weed beets (an additional possibility is by crosses between variety bolters and existing non-tolerant weed beets in neighbouring fields), they may create a problem in all cultures that are treated with the corresponding herbicide, and not only in sugarbeet. It is to be expected that in the agroecosystem, such a weedy form will maintain itself as long as the herbicide is being used. Outside the agroecosystem, there is no particular reason for a successful introgression of the transgene. The fact that we found very few hybrids, even in the inland wild beet populations, although the conditions are often quite similar to the agroecosystem (edges of sunflower fields, roadsides), suggests that the domestication traits prevent a massive transmission of cultivated genes to the wild. Only if the herbicide in question is applied, for example, on roadsides, will the selection pressure in favour of plants possessing the tolerance gene be so strong that a successful introgression will be realized. In the more natural coastal populations, this is rather unlikely.

Other important candidates of transgenes for beet are virus resistance and nematode resistance. Such traits, when they escape to, for instance, weed beets, pose less of a problem when compared with herbicide tolerance. The fitness advantage is quantitative rather then absolute (plants are not prevented from being killed, but may produce more seeds; e.g. see Snow *et al.*, 2003), and it may even be beneficial that the weeds cannot form a refuge for these enemies. In natural populations, all depends on the presence of, and the role played by, the corresponding enemy. Even in the case of a successful introgression, the ecological consequences may be limited: we did not observe substantial damage by viruses and nematodes in seabeet, although juvenile mortality may easily be missed, and further obervations are therefore required.

Conclusions

Hybrids between crop and wild can easily be formed in beet, in spite of the supposed absence of flowering of the cultivated forms. On the one hand, gene exchange is possible in the seed production area, where wild inland beets are growing at distances from the crop which are relatively short for wind pollination. On the other hand, wild coastal populations in the sugar production region may receive pollen and seed from the sugarbeet fields, especially when these are infested by weed beets.

According to our knowledge of weedy and wild populations, the only introgression of transgenes with significant negative consequences is herbicide tolerance in weed beets, but it should be emphasized that this is an economic and not an ecological problem. The herbicide and the crop varieties tolerant to that herbicide will become useless if such an introgression is successful. The decision to introduce such varieties has to be taken while paying attention to the level of gene flow.

References

Bartsch, D. and Schmidt, M. (1997) Influence of sugar beet breeding on populations of *Beta vulgaris* ssp. *maritima* in Italy. *Journal of Vegetation Science* 8, 81–84.

Boudry, P., Wieber, R., Saumitou-Laprade, P., Pillen, K., Vernet, P., Van Dijk, H. and Jung, C. (1994) Identification of RFLP markers closely linked to the bolting gene B and their significance for the study of the annual habit in beets (*Beta vulgaris* L.). *Theoretical and Applied Genetics* 88, 852–858.

Boudry, P., McCombie, H. and Van Dijk, H. (2002) Vernalization requirement of wild beet *Beta vulgaris* ssp. *maritima*: among population variation and its adaptive significance. *Journal of Ecology* 90, 693–703.

Desplanque, B. (1999) Betteraves mauvaises herbes et rudérales: diversité génétique, traits d'histoire de vie et flux de gènes au sein du complexe d'espèces cultivées–sauvages *Beta vulgaris* ssp. PhD thesis, Université de Lille 1.

Desplanque, B., Boudry, P., Broomberg, K. and Van Dijk, H. (1996) Flux géniques entre betteraves sauvages et cultivées: des populations de betteraves mauvaises herbes transgéniques peuvent-elles s'installer rapidement? *Xème Colloque International sur la Biologie des Mauvaises Herbes*, Dijon, pp. 239–245.

Desplanque, B., Hautekèete, N. and Van Dijk, H. (2002) Transgenic weed beets: possible, probable, avoidable? *Journal of Applied Ecology* 39, 561–571.

Hecker, R.J. (1988) Pollen characteristics of diploid and tetraploid sugarbeet. *Journal of Sugar Beet Research* 25, 55–62.

Pohl-Orf, M., Brand, U., Driessen, S., Hesse, P., Lehnen, M., Morak, C., Mücher, T., Saeglitz, C., Von Soosten, C. and Bartsch, D. (1999) Overwintering of genetically modified sugar-beet, *Beta vulgaris* L. subsp. *vulgaris*, as a source for dispersal of transgenic pollen. *Euphytica* 108, 181–186.

Scott, R.K. and Longden, P.C. (1970) Pollen release by diploid and tetraploid sugar beet plants. *Annals of Applied Biology* 66, 129–135.

Snow, A.A., Pilson, D., Rieseberg, L.H., Paulsen, M.J., Pleskac, N., Reagon, M.R., Wolf, D.E. and Selbo, S.M. (2003) A Bt transgene reduces herbivory and enhances fecundity in wild sunflowers. *Ecological Applications* 13, 279–286.

6 Hybridization Between Wheat and Wild Relatives, a European Union Research Programme

YOLANDE JACOT[1], KLAUS AMMANN[1], PIA RUFENER
AL MAZYAD[1], CRISTINA CHUECA[2], JACQUES DAVID[3],
JONATHAN GRESSEL[4], IÑIGO LOUREIRO[2], HAIBO WANG[5]
AND ELENA BENAVENTE[6]

[1]Botanical Garden, University of Bern, Altenbergrain 21, CH-3013, Bern,
Switzerland; [2]INIA, Madrid, Spain; [3]INRA, Mauguio, France; [4]Weizmann
Institute of Science, Rehovot, Israel; [5]Heibei Academy of Agricultural
and Forestry Sciences, Shijiahuang, China; [6]ETSIA, Madrid, Spain;
E-mail: yolande.jacot@ips.unibe.ch

Abstract

This contribution is based mainly on ongoing research in an EU Framework 5
research programme. Wheat is a self-compatible, wind-pollinated species whose
flowers are often cleistogamous. Crop to crop gene flow is very limited. Field
experiments demonstrate that the distance of wheat pollen dispersal resulting in
hybrids is usually a few metres. The safe distance to keep intermixing of traits below
0.5% does not seem to exceed 1 m. Crop to wild gene flow has been detected between
wheat and *Aegilops ovata* in the field or in herbarium specimens. Backcrosses and F_1
and F_2 progeny are known from experimental gardens. However, there is no evidence
that hybrids are stabilized in any of the populations observed. Spontaneous
amphiploidy could be the main avenue for transferring genes from cultivated durum
wheat to wild *A. ovata*. Artificial hybridization with wheat used as pollen donor can
reach higher levels for some species such as *A. cylindrica*, *A. ovata* or *A. biuncialis*, but
is generally very low with other wild relatives. Risk assessment of gene escape from
wheat should be based on these hybridization limits.

Introduction

Wheat is one of the main food grains consumed directly by humans and
it is grown in all temperate zones of the world. The origin of the tribe
Triticeae is Transcaucasia from where the species diversified by crossing
and polyploidization. Of particular importance in the evolution of wheat

were Mediterranean and Irano-Turanian regions that are considered as centres of diversification. These circumstances led to the situation of wheat as a highly compatible complex of diploid and polyploid species that forms a network of outcrossing and self-fertilization, permanently influencing cultivated wheat, especially in Mediterranean areas where wheat and related species have overlapping distribution areas and can cross and transfer genes.

With the introduction of transgenic wheat, attention has to be paid to potential gene flow and to learn how to manage transgenes in future agricultural practice, learning also from gene flow of traditional traits. The aim of this contribution is to summarize the present day knowledge of gene flow from wheat to its wild relatives. We specifically base our information on work carried out under the framework of an EU project (reference number INCO No. ERBIC18 CT98 0391). This project focuses on gene flow of *Triticum* and *Setaria* and their wild relatives. Here, we concentrate on the analysis of field experiments and herbarium analysis on *Triticum* sp. in Spain, France, Switzerland and China.

Transgene location on the chromosome can influence the likelihood of gene escape. Wheat is an allopolyploid with three genomes: *A*, *B* and *D*. Because the *D* genome is shared with some *Aegilops* species, gene flow to this species is expected to be more likely if genes are inserted in this part of the genome. However, there are also wild species that have partial homology to the *A*, *B* or *D* genome of wheat.

Wheat is a self-compatible, wind-pollinated species whose flowers are often cleistogamous. *Triticum aestivum* L. (bread wheat) is a hexaploid, and *T. durum* Desf. (durum wheat) is a tetraploid species. Both species are allopolyploid. Professional and industrial *Triticum* seed growers keep to failsafe rules to maintain the purity of the genomes of their produce: usually a distance of 0.5 m is sufficient in order to avoid mixing during the harvest, and 3 m are necessary to avoid significant gene flow ($< 0.5\%$, based on common practice). Nevertheless, references of crosses at greater distances were reported by some authors (Waines and Hegde, 2003). Still, gene flow towards the wild relatives of wheat also needs consideration, since transgenes could persist in wild populations. Some wild relatives are suspected to hybridize with wheat in Europe (Armstrong, 1945; Petrova, 1962; Sharma and Gill, 1983; Ainsworth *et al.*, 1987; Maan, 1987; Fedak, 1991; van Slageren, 1994). Most of them belong to the genus *Aegilops*. They are not known as weeds, except for *Aegilops cylindrica*, a species that originated in northern Turkey, a pernicious weed in the northern Great Plains of the USA. It is rarely found in Central Europe, although it is found more frequently in southern Europe. In this contribution, we address the following questions:

1.   What is the potential of hybridization between wheat and its wild relatives in both artificial and natural conditions?
2.   How far can the wheat pollen move and at which distance from the pollen source can wheat pollen fertilize wheat relatives?

Hybridization Potential

Artificial hybridization

Artificial hybridization rates have been studied by means of cultivation experiments in the greenhouse (Zhao *et al.*, 2000; Loureiro *et al.*, 2001; Lu *et al.*, 2002a). Fourteen species were studied. Bread wheat has been crossed as pollinator with *Agropyron repens, Aegilops bicornis, Aegilops biuncialis, Aegilops cylindrica, Aegilops neglecta, Aegilops ovata, Aegilops peregrina, Aegilops speltoides, Aegilops squarrosa, Aegilops tauschii, Aegilops triuncialis, Elytrigia elongata, Roegneria ciliaris, Secale cereale, T. durum, T. monococcum, T. timopheevii* ssp. *armeniacum* and *T. turgidum* ssp. *dicoccoides. A. neglecta, A. ovata* and *A. triuncialis* were also crossed with durum wheat.

In a first set of experiments, several individuals of wild relatives were emasculated (glumes not cut). The first third was hand pollinated with wheat pollen. The green seeds formed were collected and their embryos were cultivated *in vitro*. The second third was hand pollinated and seeds were maintained on the plants until maturation. The last third was surrounded by flowering plants of wheat, and seeds were collected at maturity. Those experiments were repeated twice, in 2000 and 2001.

In a second series of experiments, wheat relatives were castrated before flowering and artificially pollinated twice in the following 1–2 days with mixed pollen of *T. aestivum*. One drop (10 µl) of GA_3 (50 mg/l) or 2.4-D (100 mg/l) was applied to each castrated floret during the second and third day after pollination. After 12–17 days, the hybrid spikes were collected and the rate of seed set was determined.

Artificial hybridization between wheat used as a pollen donor and wild relatives is likely to occur at low rates. After hand pollination, hybrids were observed between wheat and 12 out of 17 related taxa (see Table 6.1). Pollen of *T. durum* was able to sire viable seeds with all three *Aegilops* species used in the experiment (Table 6.2). Hybrids appear to have intermediate characters, but most of them are at least partially sterile.

Natural hybridization

In a third series of experiments, hybridization rates were studied in field experiments with mixed plantings in 2001 (Zhao *et al.*, 2000; Lu *et al.*, 2002a). Eight species were cultivated in alternate rows with bread wheat (*A. bicornis, A. biuncialis, A. cylindrica, A. ovata, A. speltoides, A. repens, T. durum* and *T. monococcum*). *A. ovata* has also been tested for hybridization with durum wheat. Plants were cultivated in the greenhouse in pots in order to closely match the flowering period of wheat in the field. Before flowering, they were placed in the wheat field. Seeds were harvested from the wild species and grown in the greenhouse for the purpose of identifying the hybrids between wheat and its relatives based on morphological and cytological study. Hybrids were obtained between bread wheat and *T. durum* and three wild

Table 6.1. Summary of the crosses between *T. aestivum* and its wild relatives.

Taxa studied	Manual crossing	Field conditions	Spontaneous hybridization
Agropyron repens (L.) Beauv.	0	0	–
Aegilops bicornis (Forsskal) Jaub. & Spach	0	0	–
Aegilops biuncialis Vis.	x	x	–
Aegilops cylindrica Host	x	x*	0
Aegilops neglecta Req. *ex* Bertol.	0	–	–
Aegilops ovata L.	x	x	x
Aegilops peregrina (Hack. in J. Fraser) Marie & Weiller	x (F_1 + BC_1)	–	–
Aegilops speltoides Tausch.	x*	0	0
Aegilops tauschii Coss.	x*	–	–
Aegilops triuncialis L.	0	–	–
Aegilops squarrosa L.	–	–	0
Elytrigia elongata (Host.) Nevski	x*	–	–
Roegneria ciliaris (Trin.) Nevski	x*	–	–
Secale cereale L.	0	–	–
Triticum durum Desf.	x	x	–
Triticum monococcum L.	x	No overlap of flowering period	
Triticum timopheevii (Zhuk.) Zhuk. ssp. *armeniacum* × (Jacubz.) van Slageren	(F_1)	–	–
Triticum turgidum (L.) Zhuk. ssp. *dicoccoides* (Körn *ex* Asch. & Graebn.) Thell.	x (F_1 + BC_1)	–	–

Based on data from David *et al.* (2002), Loureiro *et al.* (2001), Lu *et al.* (2002b), Wang (2001) and Zhao *et al.* (2000).
0, no hybrids obtained experimentally or found spontaneously; x, hybrids obtained experimentally or found spontaneously; –, no data; *, with use of embryo rescue.

Table 6.2. Summary of the crosses between *T. durum* and wild relatives.

	Manual crossing	Field conditions
Aegilops neglecta	x	–
Aegilops ovata	x	x
Aegilops triuncialis	x	–

Based on data from David *et al.* (2002) and Zhao *et al.* (2000).
x, hybrids obtained experimentally or found spontaneously; –, no data.

relatives (*A. ovata*, *A. cylindrica* and *A. biuncialis*). *T. durum* crosses freely with *A. ovata* (see Tables 6.1 and 6.2). Although hybridization is experimentally possible, the flowering period of wheat and wild species such as *A. biuncialis*, *A. cylindrica* and *A. ovata* does not overlap in countries such as China.

Spontaneously occurring hybridization on field margins in the natural presence of the wild relatives has been analysed in southern France (David *et al.*, 2002). In 1999 and 2000, sympatric populations of *Aegilops* and wheat were screened for the presence of hybrid plants. *Aegilops* seeds were collected, sown in the greenhouse and transplanted to the field plots when sufficiently developed. Hybrid and non-hybrid *Aegilops* plants were examined morphometrically. Nine hybrid sites, with mostly the pair *A. ovata/ T. durum* co-occurring, were studied in 1999. The hybrid morphology was intermediate between the two parental species and corresponded to the description of *Triticum triticoides*. Only one hybrid bore viable seeds. These results demonstrate that the production of non-reduced gametes, which could allow a bridging for genes between species, is rare.

Herbarium analysis

In addition, hybridization records from Switzerland, southern France and northern Spain, present in herbaria, and covering a time span of several decades have been collected and analysed morphologically and statistically (Ammann and Jacot, 2003). Biogeographical studies and morphological character analyses were performed on herbarium specimens of wheat and related wild species (*A. cylindrica*, *A. ovata*, *A. speltoides* and *A. squarrosa*) in order to identify the occurrence and frequency of spontaneous hybrids in nature. We used the 'biogeographical assay' method (Kjellson *et al.*, 1997), which provides information about differences in potential gene flow between given regions. This method is a combination of two methods, i.e. 'herbarium specimen survey' and 'morphological character analysis'. A set of 75 different metric, semi-quantitative and nominal characters that allow differentiation among the taxa studied were determined. Inflorescence and flower characters are essential differential characters for species and hybrid identification. Ideally, samples to be measured should include 50 individuals in order to use multivariate statistics. However, for some species, we were limited by the number of herbarium sheets available in the collections. The biometrical data were analysed by principal coordinate (PCO) analysis using weighted Gower coefficient metrics with the R package for Multivariate and Spatial Analysis, version 4.0d5 (Casgrain and Legendre, http://www.fas.umontreal.ca/BIOL/legendre/). We compared each *Aegilops* species with hexaploid bread wheat in pairs in order to search for intermediate forms.

On the set of 75 measured parameters, about 45 parameters distinguishing wheat from the wild species or hybrid forms were chosen. They concern the stem, the leaf (form and size) and the spike (form, size, density, awn number, awn length, rachis, spikelet, glume, lemma and palea). Fruit shape and fertility characters were not included in the statistics because many pressed plants were collected in the flowering state or before fruit maturity. All the scattergrams show a good differentiation of the two pools: *Aegilops* and *Triticum* are separated for all the species studied (see Fig. 6.1). The *T. aestivum* group has wider scatter diagrams in all cases (it is morphologically less uniform) than the *Aegilops* group. This may be due to the fact that many different varieties of bread wheat, including very old ones, are included in the analysis. Hybrid F_1 plants (*A. triticoides* Req. *ex* Bertol.), F_2 as well as backcrosses of *A. triticoides* × bread wheat (*A. speltaeforme* Jord.) were found

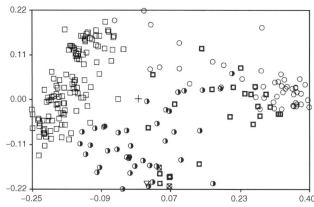

Fig. 6.1.   Scattergram *Aegilops ovata* and *Triticum aestivum*, based on unpublished data of Y. Jacot and P. Rufener Al Mazyad. *Aegilops ovata*, □; *Triticum aestivum*, ○; *Aegilops triticoides* (F$_1$), ◑; *Aegilops speltaeforme* (BC$_1$), ◘; *Aegilops triticoides* (F$_2$), ⊠.

between *A. ovata* and *T. aestivum*. The scattergram (Fig. 6.1) shows the well separated parental groups and the hybrid forms in between. The F$_1$ form *A. triticoides* is well distributed between the two parental species, and the F$_2$ is well distributed in the middle of the hybrid area. On the other hand, the backcross from *A. speltaeforme* is closer to the wheat parent.

The herbarium data survey demonstrates that *A. ovata* can form F$_1$ hybrids in the wild with *T. aestivum*. Backcrosses of F$_1$ hybrids with the wheat parent and F$_2$ are only known from botanical or private gardens. The hybrid plants found in the herbaria came from the south of France, Spain and the north of Italy. Those plants were found in wheat field borders or in disturbed areas. It is known that the distribution of several *Aegilops* × *Triticum* hybrids is concentrated in the south of France. In our herbaria studies, × *A. triticoides* Req. ex. Bertol., the primary hybrid between bread wheat (*T. aestivum* L. ssp. *aestivum*) and *A. geniculata* Roth., a subspecies of *A. ovata*, comes from the south of France, the north of Spain and the north of Africa. Further to this, × *A. speltaeforme* Jord., the backcross form of *A. triticoides* ((= *A. geniculata* Roth. × *T. aestivum* L. ssp. *aestivum*) × *T. aestivum* L. ssp. *aestivum*), originates from botanical gardens and was cultivated in private gardens.

Pollen Dispersal

In Spain, a field assay of pollen flow from wheat (cv. Chinese Spring) to wheat (cv. Pavon), *T. turgidum* (cv. Nita), *A. repens* and *A. speltoides* was conducted three times. Receptors (emasculated plants) were located at a regular distance (0–14 m) from the pollen source (2 m^2) (Loureiro *et al.*, 2001). In a second field trial, castrated wheat plants were arranged around a wheat field (100 m^2) at various distances (6–110 m) along eight lines radiating from the pollen donor. The proportion of hybrid plants was measured by morphological analyses of F_1 seedlings.

To estimate the distance of pollen spread and the effective cross-pollination under natural conditions, in 2000, an experiment was set up with blue-grained wheat (cv. *Zénia*) planted in a central plot (16 m²) surrounded by normal wheat. Blocks of 1 m² each of the normal wheat were harvested at regular distances (1–3 m) along eight lines radiating from the pollen donor. In 2001, the experiment was repeated: then, the central plot measured 100 m² and wheat spikes were harvested at regular distances (0–40 m). The results demonstrate that pollen was capable of fertilizing emasculated spikes of wheat up to 80 m away from the pollen source. However, the pollen movement was low after 50 m. A dispersal curve can be drawn by pooling all raw data coming from different locations and different years (see Fig. 6.2).

From these field experiments, we can determine that pollen is able to fertilize emasculated spikes at appreciable rates only at a short distance (< 4 m). At distances greater than 4 m, hybridization can still occur at low rates, and hybrids were found even at 14 m from the pollen source. These data are mean values of four localities all around the pollen source. They show large differences in relation to wind direction, temperature, and in general to environmental conditions during the flowering period (Loureiro *et al.*, 2001). Experiments with the blue grain variety (not castrated) show that after 1 m, the likelihood of pollination decreases distinctly and is nearly nil beyond 20 m (Fig. 6.3). The percentage again depends on the environmental conditions. The dispersal of pollen is even lower (4 m) for durum wheat (data not shown).

Outlook

The distance of pollen dispersal under natural conditions resulting in hybrids found in the experiments is limited to a few metres. However, pollen

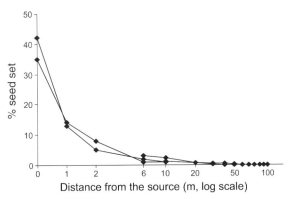

Fig. 6.2. Percentage of flowers producing seeds in castrated wheat plants at different distances from the pollen source (unpublished data of C. Chueca and H. Wang).

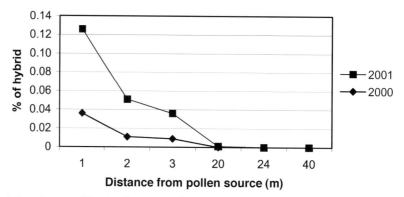

Fig. 6.3.   Cross-pollination between wheat with blue grain and normal wheat under natural conditions (unpublished data of H. Wang).

dispersal is an imprecise and uncertain process that is influenced by genotypes of varieties (Sharma and Gill, 1983) and (as a rule strongly) stochastic environmental conditions (Gusta and Chen, 1987; Waines and Hegde, 2003). The safe distance to keep intermixing thresholds of auto-gamous traits below agriculturally relevant amounts of 0.5% does not seem to exceed 1 m. However, studies carried out in similar environmental conditions and with similar crop genotypes are needed in order to obtain a more precise and generalized picture. If hybrid wheat with protruding anthers is commercialized, a re-evaluation of effective pollen dispersal according to cultivar characteristics will be required.

Artificial hybridization rates between wheat used as pollen donor and wild relatives are generally very low, but could reach more than 10% for some species such as *A. cylindrica* (Seefeldt *et al.*, 1999), *A. ovata* or *A. biuncialis* (Loureiro *et al.*, unpublished data), and even 20% under greenhouses conditions (J. David, personal communication). The differences between the years of the field experiments are huge due to all kinds of parameters, such as temperature and humidity.

The herbarium data surveys show that *A. ovata* can form F_1 hybrids with bread wheat in the wild. Hybrid plants detected in the herbaria were found to occur in wheat field borders or in disturbed areas. Backcrosses of F_1 hybrids with the wheat parent and F_2 are only known from botanical or private gardens. Data analysis shows that F_1 hybrids have overall characteristics closer to wheat than F_1 or F_2 progeny. However, it is not clear if further selfing or backcross generations could be distinguished from the wild relatives or the wheat species.

Hybrids appear to have intermediate characters, but most of them are at least partially sterile. This suggests that the likelihood of further introgression will be low and, therefore, gene flow from cultivated wheat to wild relatives is unlikely to be important under natural conditions in Europe and China, unless some particularly crossable species were present in the cropping area. There is no evidence that hybrids were stabilized in any population, but transgenes might stabilize in wild populations in circumstances of repeated backcrossing with transgenic wheat. Wheat genes can be detected in species with a considerable phylogenetic distance such as *Elymus caninus* (Bodrov, 1962; Guadagnuolo *et al.*, 2001). Traits such as herbicide resistance might have a selective advantage in agroecosystems under herbicide application, and hybrids may stabilize under this constant selection pressure.

Although the reproductive output of those F_1 hybrids is low, some seeds have been obtained, including amphiploids (colchicine-induced or natural) with a higher fertility (David *et al.*, 2002). This shows that gene flow is likely to occur at a low rate. Even if the fertility of the hybrids is very low, it is possible: (i) to backcross the hybrids between bread wheat and *A. ovata*, *A. cylindrica* and *A. tauschii* with both parents; and (ii) to self interspecific hybrids of the hybrids between *A. ovata* and *T. aestivum*.

However, for those genera related to wheat that are very rich in polyploids, the ploidy level can have an effect on fertility. Hybrids between durum wheat and *A. ovata* did not produce seeds, but doubling the hybrid chromosome number allowed one amphiploid hybrid to be obtained that was highly fertile, producing both pollen and seeds (David *et al.*, 2002). Rare spontaneous amphiploids were also observed originating from hybrids producing unreduced gametes. Spontaneous amphiploidy could be the main avenue for transferring genetic information from cultivated durum wheat to wild *A. ovata*. Unreduced gametes play a central role in this scenario as they permit the formation of amphiploids. Nevertheless, no permanent populations of wild amphiploids have been described so far, so it is necessary to screen more thoroughly for such events. *A. cylindrica*, *A. ovata* and *A. biuncialis* could be interesting targets for environmental monitoring. The choice of the species depends on the country: *A. cylindrica* is of particular interest in the USA where it is known as an invasive weed, and *A. ovata* is of interest in Mediterranean regions where it grows in the field margins.

Weed specialists (Wang *et al.*, 2001) have some doubts about the possibility that hybrids between wheat and those grasses might become noxious weeds in Europe, as long as no selection advantage is introduced with specific (trans)genes. Still, many wild *Triticum* species cannot be controlled by the herbicides applied in weed control, due to their close relatedness to wheat. Crosses of wheat with its wild relatives have been reported to be unlikely (with the exception of *A. cylindrica*) and, when they do occur, the hybrids are said to be sterile (Maan, 1987; Fedak, 1991). However, hybrids between *Aegilops* species and wheat were described extensively by botanists during the 19th century, and that with *A. cylindrica* is a weed, producing some fertile offspring capable of further backcrossing (Seefeldt *et al.*, 1998; Wang *et al.*, 2001, 2002). The identification of wild species which can cross naturally with wheat is therefore a key issue for future monitoring of transgene dispersal. As well as results gained with the help of laboratory techniques, we need to know more about what happens under field conditions.

Acknowledgements

We thank the European Union for supporting this project (INCO No. ERBIC18 CT98 0391).

References

Ainsworth, C.C., Miller, T.E. and Gale, M.D. (1987) α-Amylase and β-amylase homoeoloci in species related to wheat. *Genetical Research* 49, 93–103.

Ammann, K. and Jacot, Y. (2003) Vertical gene flow. In: Ammann, K., Jacot, Y. and Braun, R. (eds) *Methods for Risk Assessment of Transgenic Plants.* Birkhäuser, Basel, pp. 19–33.

Armstrong, J.M. (1945) Investigations in *Triticum Agropyron* hybridisation. *Empirical Journal of Experimental Agriculture,* 13, 41–53.

Bodrov, M.S. (1962) Hybridisation between wheat and *Elymus.* In: Tsitsin, N.V. (ed.) *Wide Hybridisation of Plants.* Translated from Russian, Israel Program for Scientific Translation, Jerusalem, pp. 238–241.

David, J., Dusautoir, J.C., Patry-Brès, C. and Benavente, E. (2002) Spontaneous formation of amphiploids between durum wheat and *Aegilops ovata* L. through unreduced gametes. Polyploidy Meeting, INA-PG, Paris, 24 January 2002.

Fedak, G. (1991) Intergeneric hybrids involving the genus *Hordeum.* In: Gupta, P.K. and Tsuchiya, T. (eds) *Chromosome Engineering in Plants. Genetics, Breeding, Evolution. Part A.* Elsevier, Amsterdam, pp. 433–448.

Guadagnuolo, R., Savova-Bianchi, D. and Felber, F. (2001) Gene flow from wheat (*Triticum aestivum*) to jointed goatgrass (*Aegilops cylindrica* Host) as revealed by RAPD and microsatellite markers. *Theoretical and Applied Genetics* 103, 191–196.

Gusta, L.V. and Chen, T.H.H. (1987) The physiology of water and temperature stress. In: Heyne, E.G. (ed.) *Wheat and Wheat Improvement.* Agronomy Monograph No. 13 ASSSSSA-CSSA-SSSA publishers, Madison, Wisconsin, pp. 115–150.

Kjellson, G., Simonsen, V. and Ammann, K. (1997) *Methods for Risk Assessment of Transgenic Plants, 2. Pollination, Gene Transfer and Population Impacts.* Birkhauser Verlag, Basel.

Loureiro, I., Escorial, M.C., Garcia-Baudin, J.M. and Chueca, M.C. (2001) Capacidad potencial de diffusion de genes de trigo a especxies afines a traves del polen. *Actas Congreso SEMH,* pp. 129–133.

Lu, A.Z., Zhao, H., Wang, T.Y. and Wang, H.B. (2002a) Possibility of target gene introgresssion from transgenic wheat into non-transgenic plants through pollen. *Acta Agriculturae Boreali-Sinica* 17, 1–6.

Lu, A.Z., Zhao, H., Wang, T.Y. and Wang, H.B. (2002b) Possibility of target gene introgression from transgenic wheat into non-transgenic plants through pollen. *Acta Agriculturae Boreali-Sinica* 17, 1–6.

Maan, S.S. (1987) Interspecific and intergeneric hybridisation in wheat. In: Heyne, E.G. (ed.) *Wheat and Wheat Improvement,* 2nd edn. Agronomy Series No. 13, American Society of Agronomy, Crop Science Society of America, Soil Science Society of America, Madison, Wisconsin, pp. 453–461.

Petrova, K.A. (1962) Hybridisation between wheat and *Elymus.* In: Tsitsin, N.V. (ed.) *Wide Hybridisation of Plants.* Translated from Russian, Israel Program for Scientific Translation, Jerusalem, pp. 226–237.

Seefeldt, S., Zemetra, R., Young, F.L. and Jones, S. (1998) Production of herbicide-resistant jointed goatgrass (*Aegilops cylindrica*) × wheat hybrids in the field by natural hybridisation. *Weed Science* 46, 632–634.

Seefeldt, S., Young, F.L., Zemetra, R. and Jones, S. (1999) The production of herbicide-resistant jointed goatgrass (*Aegilops cylindrica*) × wheat hybrids in the field by natural hybridization and management strategies to reduce their occurrence. *BCPC Symposium Proceeding* 72, 159–164.

Sharma, H.C. and Gill, B.S. (1983) Current status of wide hybridisation in wheat. *Euphytica* 32, 17–31.

van Slageren, M.W. (1994) *Wild Wheats: a Monograph of* Aegilops L. *and* Amblyopyrum *(Jaub & Spach) Eig (Poaceae)*. Wageningen Agricultural University Papers 94 7.

Waines, J.G. and Hegde, S.G. (2003) Intraspecific gene flow in bread wheat as affected by reproductive biology and pollination ecology of wheat flowers. *Crop Science* 43, 451–463.

Wang, Z.N., Zemetra, R.S., Hansen, J. and Mallory-Smith, C.A. (2001) The fertility of wheat × jointed goatgrass hybrid and its backcross progenies. *Weed Science* 49, 340–345.

Wang, Z.N., Zemetra, R.S., Hansen, J., Hang, A., Mallory-Smith, C.A. and Burton, C. (2002) Determination of the paternity of wheat (*Triticum aestivum* L.) × jointed goatgrass (*Aegilops cylindrica* host) BC1 plants by using genomic *in situ* hybridization (GISH) technique. *Crop Science* 42, 939–943.

Zhao, H., Wu, Z., Wu, W., Xie, X., Ma, M., Godovikova, V.A., Lu, M. and Wang, H. (2000) Ecological safety assessment of herbicide resistant transgenic wheat–II. Studies on interspecific and intergeneric hybridisation between common wheat and its wild relatives. *Journal of Hebei Agricultural Sciences* 4, 6–9.

7

Molecular Genetic Assessment of the Potential for Gene Escape in Strawberry, a Model Perennial Study Crop

ANNE L. WESTMAN, SANDY MEDEL, TIMOTHY P. SPIRA, SYRIANI RAJAPAKSE, DAVID W. TONKYN AND ALBERT G. ABBOTT

Department of Biological Sciences, 132 Long Hall, Clemson University, Clemson, SC 29634, USA; E-mail: aalbert@clemson.edu

Abstract

The risk of gene escape from crops to closely related wild species is a major concern, but can be difficult to assess when crop and wild species are genetically similar. Molecular markers with high resolving power can be informative tools for measuring crop–wild gene flow. As a first step towards assessing the potential for gene escape from cultivated strawberry (*Fragaria × ananassa*) to its close relative wild strawberry (*Fragaria virginiana*) in south-eastern USA, we examined the extent and patterns of amplified fragment length polymorphism (AFLP) marker variation in *Fragaria* species and identified markers characteristic of each species. An array of *Fragaria* entries, including *F. × ananassa* cultivars grown in the South-east and *F. virginiana* populations far from cultivation, was evaluated using four AFLP primer pairs. Marker variation was moderate to high (254 fragments, 50–75% polymorphic within species), and patterns of variation were consistent with the species' taxonomy. In *F. virginiana*, marker-based distances and geographical distances between entries were not correlated. In *F. × ananassa*, marker variation was consistent with cultivar pedigrees. Twenty-six markers (10% of the total) were present in cultivars and absent or rare in wild *F. virginiana* entries, including markers for modern cultivars and markers for cultivars no longer widely grown. Sixty-six markers were present in wild *F. virginiana* and absent from cultivars. The cultivar markers were used to estimate past and present gene flow from cultivated to wild strawberry. Frequencies of the 26 markers characteristic of strawberry cultivars were scored in wild populations in the vicinity of cultivated fields. Seventeen of the markers were present in two distant populations, including markers for previously grown cultivars and markers for cultivars currently grown. Within and between populations, plants had different combinations of cultivar markers. One population included plants with markers from more than one cultivar, and these plants were classified as second- or later generation hybrids. Cultivar markers were present at widely varying frequencies and their distribution between populations was non-random, suggesting differences in marker persistence. These results are consistent with past and present gene flow, from strawberry cultivars to local *F. virginiana* populations and between hybrid and non-hybrid plants.

Introduction

The risk of gene escape from crops to closely related wild species is a major concern in both agricultural and natural environments. Many crops potentially can hybridize with closely related wild species, or crop plants themselves may become established in the wild. Depending on their fitness, crop–wild hybrids and escapes could potentially out-compete wild relatives, become weedy pests and could serve to promote the persistence and spread of particular genetic associations, thus potentially decreasing germplasm variability. As the development of transgenic crops has increased, concern about these risks has multiplied (Wolfenbarger and Phifer, 2001). The potential for gene escape from a crop to wild relatives is increased when the crop and wild relative: are sexually compatible; have high outcrossing rates; have the same pollinators; and overlap in flowering time and geographical ranges. The risk of spread and persistence of introgressed characters is even greater when plants are perennial, propagate by both seed and vegetative clone, and produce edible fruits dispersed by animals.

One such perennial crop species is strawberry (*Fragaria* × *ananassa*). Strawberry is a hybrid of two closely related species native to the Americas (*F. chiloensis* and *F. virginiana*; $2n = 8x = 56$ for all species). The cultivated and wild species exhibit all the traits that increase the potential for gene escape, spread and persistence: all three species are sexually compatible, are outcrossing perennials that produce edible fruit and spread by both seed and vegetative runners (Darrow, 1966). In addition, the wild species have wide overlapping geographical ranges in the USA, including regions where strawberries are cultivated (Hancock *et al.*, 1990). Strawberry is a significant crop worldwide, with the USA and Spain as the leading producers (Fig. 7.1), and field trials of transgenic strawberry are underway in many countries

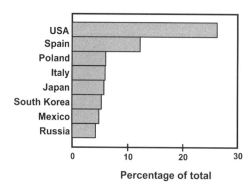

Fig. 7.1. World strawberry production in 1999 (total production, in metric tonnes) (FAO, 1999).

(Fig. 7.2). Thus, we are investigating strawberry as an excellent model for studies of gene flow from perennial crop to wild species as it occurs in field situations.

For this purpose, we are using molecular markers to examine the extent of past and present crop–wild gene flow around strawberry fields in the Piedmont region of the south-eastern USA. In this chapter, we describe our approach to development and application of amplified fragment length polymorphism (AFLP) markers suitable for evaluating gene flow. We present evidence that: (i) AFLP marker technologies are highly suitable for assessing gene flow from cultivated to wild strawberry populations; (ii) significant past and present introgression of cultivated markers has occurred in wild strawberry populations surrounding the cultivated fields in the Piedmont region; (iii) there appears to be potential long-term persistence of introgressed markers in wild populations near cultivated fields; and (iv) marker flow can occur at significant distances from the fields.

Materials and Methods

Plant material

For marker development, 52 single-plant entries were evaluated. These included: 13 *F.* × *ananassa* entries comprising all major commercial cultivars historically (and currently) grown in the region (Caldwell, 1989; Galletta, 1997; Poling, 1998); 31 *F. virginiana* comprising genebank accessions from ten sites (one sampled plant/site) and population samples from eight sites (one to five plants/population); a set of *F. chiloensis* genotypes, in order to assess whether fragment markers present in the hybrid *F.* × *ananassa* were shared with one or both of its parent species and, finally, a representative genotype of *F. vesca* (a diploid putative ancestor of the octoploid species (Darrow, 1966) was included as an outgroup comparison (Table 7.1)).

For field studies, single-plant entries were sampled from *F. virginiana* populations in upstate South Carolina. Plants from varying distances (50–500 m) around the Robertson Farm, South Carolina (designated RF) were initially sampled, and later two populations (designated Populations 1 and 2) from more distant locations were examined. Population 1 is ≥ 5 km from large (> 1 ha) commercial strawberry farms that have grown modern cultivars for 10–15 years; Population 2 is ≥ 7 km from small farms (< 1 ha)

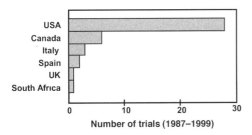

Fig. 7.2. Transgenic strawberry field trials for 1987–1999 and participating countries. Transgenes were for disease resistance, herbicide tolerance, fruit quality and flowering time.

Table 7.1.   Test array of entries from four *Fragaria* species.

ID[a]	Location	County	Comments	Collected	Source[b]
F. vesca			Cultivar Mignonette	1998	NC
Fchil38	Coastal N. CA		Del Norte; aphid resistance	1981	USDA
Fchil358	Coastal N. CA		Cape Mendocino	1984	USDA
Fchil1377	Los Lagos, Chile		Valcano Michimahuida	1992/95	USDA
Fchil372	Cuzco, Peru		Selection from cultivated type	1965	USDA
Fchil31	Coastal CA		Dry Lagoon	1981	USDA
Fchil1330	Coastal OR	Lane	OR Dunes Natl Rec Area	1994	USDA
Fchil33	Coastal OR	Lincoln	Yachats State Park	1977	USDA
F. virginiana					
959	Central-west SC	Greenwood	Sumter Natl Forest, ditchbank	1995	USDA
9511	Central-west SC	Abbeville	Sumter Natl Forest	1995	USDA
Ocon (3)	North-west SC	Oconee	Oconee Station, forest edge	1998	Clemson
FHD (5)	North-west SC	Oconee	Sumter Natl Forest, forest edge	1999	Clemson
CHD3	North-west SC	Greenville	Roadside near Caesar's Head	1999	Clemson
9613	North-east NC	Bertie	Woods near stream	1996	USDA
9617	North-east NC	Pasquotank	Roadside near woods	1996	USDA
1627	North-east NC	Jones	Woodland edge	1996	USDA
Math (4)	North-west NC	Jackson	Woodland edge	1997	Clemson
BSO (4)	North-west NC	Transylvania	Blue Ridge Pkwy, forest edge	1998	Clemson
DC4	North-west NC	Jackson	Blue Ridge Pkwy, forest edge	1998	Clemson
RTF1	North-west NC	Haywood	Blue Ridge Pkwy, forest edge	1998	Clemson
Fv2 (2)	North-west NC	Transylvania	Blue Ridge Pkwy, forest edge	1997	Clemson
9619	South-east AL	Lee	Open woodland + roadside	1996	USDA
9625	South-east AL	Tallapoosa	Roadside	1996	USDA
9627	South-east AL	Tallapoosa	Roadside	1996	USDA
9633	South central AL	Dallas	Roadside + limestone cedar glade	1996	USDA
9510	GA			1995	USDA

F. × ananassa cultivars

ID	Cultivar	Cultivated at present?	Original maternal parent[d]	Date introduced	Breeding programme
SwtCh[c]	Sweet Charlie	Moderate	FL 80–456 (breeding line)	1994	FL
Cam[c]	Camarosa	Moderate	Nich Ohmer	1992	CA
Chand[c]	Chandler	Widespread	Nich Ohmer	1983	CA
Card	Cardinal	Seldom	The Native Iowa (*F. virginiana*)	1975	AK
Eglow	Earliglow	Seldom	Aberdeen	1975	MD
Titan	Titan	Seldom	Chesapeake	1971	NC
Apollo	Apollo	Seldom	Chesapeake	1970	NC
Atlas	Atlas	Seldom	Chesapeake	1970	NC
Seq[c]	Sequoia	None	Hudson Bay (*F. virginiana*)	1968	CA
Ebelle	Earlibelle	None	The Native Iowa (*F. virginiana*)	1964	NC
Sun	Sunrise	None	Missionary	1964	MD
Alb	Albritton	None	The Native Iowa (*F. virginiana*)	1951	NC
Blake	Blakemore	None	Missionary	1929	MD

[a]Number in parentheses indicates number of entries evaluated from a population.
[b]NC, James Ballington, North Carolina State University, Raleigh, North Carolina; USDA, USDA-ARS National Clonal Germplasm Repository, Corvallis, Oregon; Clemson, collected by researchers at Clemson University, Clemson, South Carolina.
[c]Modern cultivars developed in California or Florida; other cultivars are traditional to the South-east.
[d]From Dale and Sjulin (1990).

where traditional and modern cultivars are currently grown and > 22 km from large farms of modern cultivars. Each population comprised ± 100 plants. Within each population, the plants were sampled at least 3 m apart, to minimize the chances of sampling multiple clones of an established plant.

For wild plants, sites were selected (by the authors and regional strawberry expert J. Ballington, North Carolina State University) as remote as possible from past and present cultivation (most samples were collected in the National forests of the Blue Ridge Mountains where there was no evidence of prior cultivation, and cultivation would have been unlikely due to the park designation of the region). Multiple plants at sampling sites were selected at least 3 m apart, to minimize the number of clonal propagants in the study.

DNA isolation

For each entry, leaf tissue (100 mg) was immersed in liquid N_2 and finely ground, then genomic DNA was extracted and treated with ribonuclease A (Promega). Samples were extracted in DIECA (diethyldithiocarbamic acid, sodium salt) buffer, using the protocol in Lamboy and Alpha (1998).

AFLP analysis

Protocols were modified from Vos et al. (1995), using oligonucleotide primers and polymerase chain reaction (PCR) amplification cycles described therein. For our analyses, four primer pairs were used: E-ACC/M-CAG, E-ACT/M-CAA, E-ACT/M-CAC and E-AGG/M-CAT. Reaction products were evaluated by vertical electrophoresis on 6.0% denaturing gels (20:1 acrylamide–bisacrylamide, 7.5 M urea) in 1 × TBE buffer (50 mM Tris, 50 mM boric acid, 1 mM EDTA, pH 8.0) at constant power (70 W, 40–50 V/cm). After electrophoresis, gels were transferred to filter paper and dried. X-ray film (Kodak) was exposed to dried gels. Each primer pair × entry was evaluated on at least two gels.

Data analysis

For each primer pair, presence/absence of each fragment was recorded for each entry. Fragments of the same size in different entries were considered homologous. Pairwise distances between entries were computed using distance metric D_D, the complement of Dice's (1945) similarity coefficient, defined as $1 - [2n_{xy}/(n_x + n_y)]$, where n_{xy} = number of fragments common to plants x and y; and n_x and n_y = total number of fragments in x and y. Distances were calculated for each primer pair separately and for combined data of all pairs (Westman et al., unpublished results). Correlations between distance matrices for the separate pairs were evaluated with Mantel (1967) matrix correspondence tests, using the R package (Casgrain *et al.*, 1999; Westman *et al.*, unpublished results). Subsequent analyses were conducted with the combined data. The Dice distances were used to conduct principal coordinate analysis (PCA), with SPLUS5 (MathSoft), and neighbour-joining (NJ) analysis, with PHYLIP version 3.5c (Felsenstein, 1995). Consensus analysis was conducted from ten replicate NJ analyses, with data entered in a different order each time. Geographical distances between the *F. virginiana* entries' collection sites were computed from latitude and longitude, using RAPDISTANCE version 1.04 (Armstrong *et al.*, 1996). Correlation between geographical and marker distances was evaluated with the Mantel test. For nine cultivars, the relative contributions of 29 parental clones (Sjulin and Dale, 1987) were used to calculate pairwise distances, with $D_{ij} = \Sigma_k |(x_{ik} - x_{jk})| / 100$ (x_{ik} and x_{jk} = contribution of parental clone k to cultivars i and j) (Graham *et al.*, 1996). Correlation between pedigree and marker distances was evaluated with the Mantel test.

　　　Fragments were selected as 'cultivar markers' for detecting crop–wild gene flow, based on stringent criteria (fragments present in at least one *F. × ananassa* cultivar and absent from all 31 *F. virginiana* entries) or on frequency (present in at least three different cultivars and fewer than four *F. virginiana* entries). Fragments were selected as wild *F. virginiana* markers based on stringent criteria (present in at least one *F. virginiana* entry and absent from *F. × ananassa* and *F. chiloensis*).

Results

Fragment polymorphism and distribution

The four primer pairs generated 254 fragments (53–71 fragments/pair). Overall, 89.4% of the fragments were polymorphic (84.5–92.3% for each primer pair) and 41.3% were unique to entries from one of the species.

Variation between and within species

Average marker distances were higher between entries from different species than between entries of the same species (Westman *et al.*, unpublished results). In cluster analyses, entries from the same species were grouped together (Fig. 7.3). Between species, *F. × ananassa* and *F. virginiana* were grouped together most closely. The *F. vesca* entry was distinct from the octoploids, but was most similar to *F. chiloensis*. The range of intraspecific distances was highest for *F. virginiana* and lowest for *F. × ananassa*. Within four of the five subsampled *F. virginiana* populations (FHD, Fv, Math and Ocon), most or all subsampled entries for a population were grouped together (Fig. 7.3). Over all *F. virginiana* entries, however, geographical and marker distances were not correlated as judged by the Mantel test ($r = 0.123$,

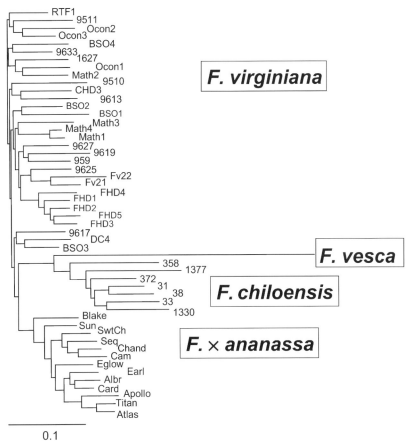

Fig. 7.3. Neighbour-joining analysis of variation in *Fragaria* species. Each node was supported in ten out of ten dendrograms, with data entered in random order each time.

$P < 0.154$) (Westman *et al.*, unpublished results). In general, cultivars with similar pedigrees were clustered together.

Fragaria virginiana **markers**

Sixty-six fragments were present only in the entries of *F. virginiana* (Fig. 7.4). Most were at low frequency, each marker present in from one entry (18 markers) to 12 entries (one marker); however, one marker was in 78% of the entries.

To test the ability of the chosen AFLP markers to track gene flow, we examined a panel of cultivars of known pedigree for their genetic related-ness. Cluster analyses (Fig. 7.4) of the cultivars based on these markers closely mirrored the known pedigrees for these cultivars and substantiated the use of this technology for introgression studies.

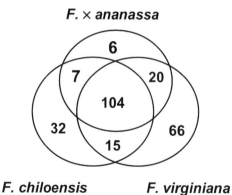

F. × ananassa

F. chiloensis   **F. virginiana**

Fig. 7.4. Distribution of AFLP fragments
in the octoploid *Fragaria* species (numbers
of fragments present in each species or
combination of species).

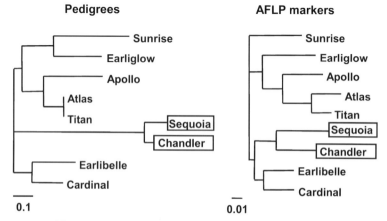

Fig. 7.5. Neighbour-joining analysis of pedigree-based distances and AFLP
marker-based distances between *F. × ananassa* cultivars.

Cultivar markers

Twenty-six fragments were selected as cultivar markers (Fig. 7.5). Thirteen
fragments were selected based on stringent criteria (present in *F. × ananassa*,
absent from all *F. virginiana* entries). Thirteen additional fragments were
selected based on their frequencies (present in at least three *F. × ananassa* and
fewer than four *F. virginiana* entries). Among the 26 markers, nine were only
in traditional cultivars and two were only in modern cultivars. Eighteen of
the cultivar markers (including six traditional and two modern cultivar
markers) were also present in *F. chiloensis* entries.

Our test population of plants was chosen from three field sites. At one of
the sites, samples were taken within 100 m of one farm and in excess of
500 m. At the two remaining sites, Population 1 was approximately 5 km
from a series of surrounding farms; Population 2 was from a site approximately 7 km from a number of farms; however, these farms did not surround
the site but were located adjacent to it.

AFLP results using our test panel of markers revealed a significant number of markers present in these wild populations in the vicinity of the cultivated fields (Fig. 7.6). In summary, 20 of 24 cultivar markers were present in one or more populations including markers for cultivars no longer grown in the Piedmont region and markers present in cultivars currently in propagation. The number of markers ranged from zero to eight (Fig. 7.7). Within and between populations, the plants had different combinations of markers. Some plants had markers from more than one cultivar and thus appeared to be later generation hybrids. We have evaluated the field hybrids further by comparison with marker frequencies determined in controlled crosses between wild and cultivar plants, and the data suggest that most if not all field plants studied at 5–7 km from the fields are not first-generation

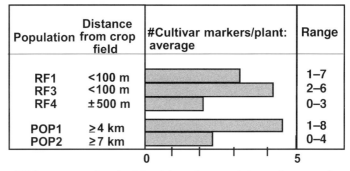

Fig. 7.6. AFLP summary data for 26 cultivar characteristic markers tested on *F. virginiana* populations from sites adjacent to strawberry farms in the South Carolina Piedmont region. Population size estimated at 50 for RF1, 300 for RF4, and 100 for all others. Number of plants sampled per population ranged from six (RF3) to 18 (POP1).

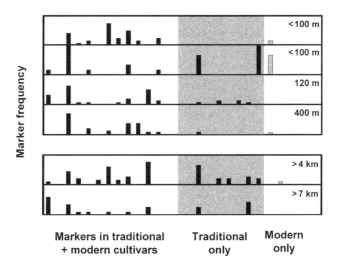

Fig. 7.7. Frequency of cultivar characteristic markers in *F. virginiana* populations from sites adjacent to strawberry farms in the South Carolina Piedmont region.

hybrids (data not shown). In addition, cultivar markers were present at widely varying frequencies, suggesting differences in persistence (Fig. 7.8).

Finally, for the populations distant from the fields, the number of plants having multiple cultivar markers is quite high for Population 1 (surrounded by cultivated fields) in comparison with Population 2 (adjacent to farms on one side) and the truly wild accessions in our study (Figs 7.8 and 7.9). This further underscores the likelihood of introgression from crop to wild plants as a function of proximity to the cultivated fields.

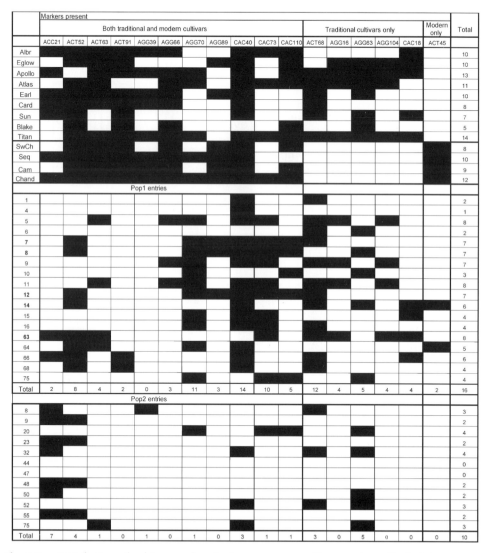

Fig. 7.8. Distribution of cultivar markers in *F.* × *ananassa* cultivars and two *F. virginiana* populations (selection of 18 and 12 individuals). Entries 7, 8, 12, 14 and 63 in Population 1 could not be first-generation hybrid progeny of a cross between wild *F. virginiana* (containing no cultivar markers) and one of the listed cultivars.

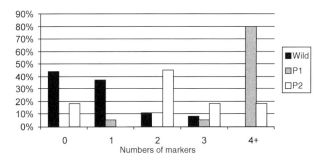

Fig. 7.9. Evaluation of the percentage of plants in field populations and truly wild plant accessions having multiple cultivar characteristic markers.

Conclusion

AFLP analysis was an appropriate method for describing variation in cultivated strawberry (*F. × ananassa*) and its wild parent species. Reproducible fragments were easily identified and scored, and patterns of variation generated by the four primer pairs were correlated. Levels of polymorphism in our study were comparable with randomly amplified polymorphic DNA (RAPD) analyses of *Fragaria* spp. (Hancock *et al.*, 1994; Graham *et al.*, 1996; Harrison *et al.*, 1997; Porebski and Catling, 1998), but discrimination was greater: only six fragments from one AFLP primer pair were needed to uniquely identify 13 cultivars.

The validity of marker analyses can be assessed by comparing marker variation with known relationships between taxa. In our array of *Fragaria* entries, AFLP variation was consistent with the known taxonomy: *F. vesca* was distinct from the octoploids, *F. × ananassa* was intermediate to its parent species *F. chiloensis* and *F. virginiana*, and more than 95% of the fragments in *F. × ananassa* were shared with its parents. The AFLP marker distances between cultivars were generally correlated with their pedigrees, and inconsistencies may be related to selection or drift during cultivar development. Marker–pedigree correlation was higher in our study than in most RAPD analyses of strawberry cultivars (Gidoni *et al.*, 1994; Graham *et al.*, 1996), but Degani *et al.* (2001) found that coefficients of co-ancestry for 19 cultivars were more closely correlated with RAPDs than AFLPs. Differences between studies may reflect the use of different cultivars, primers and measures of marker distance.

Only 10% of the AFLP fragments were identified as 'cultivar markers' appropriate for detecting crop–wild gene flow. This is not surprising, since *F. × ananassa* was developed relatively recently and its breeding has involved backcrossing to both parent species (Hancock and Luby, 1993). The most reliable markers in *F. × ananassa* for gene flow to *F. virginiana* may be those that are derived from the *F. chiloensis* parent. In our study, the majority of fragments identified as cultivar markers were also present in *F. chiloensis*. Half of the markers we selected were rare, not absent, in the wild *F. virginiana* entries. However, as long as criteria for marker selection are clearly stated, such 'imperfect' cultivar markers can generate useful information about crop–wild introgression (Bartsch *et al.*, 1999). Utilization of these markers in field studies reported here demonstrated significant gene flow from field to wild populations. Additionally, hybrid plants at fairly great distances from the fields displayed significant numbers of cultivar markers.

Markers characteristic of cultivars from different time periods can provide a picture of both present gene flow and the persistence of cultivar markers from past gene flow. Such a comparison is possible in the southeastern USA, where a progression of strawberry cultivars has been grown. Traditional cultivars developed in the eastern and central USA for matted row cultivation were widely grown until the late 1980s (Darrow, 1966; Caldwell, 1989; Galletta, 1997). Cultivars from California and Florida, developed for annual plasticulture, were first introduced in the late 1970s but initially were not widely grown (Walker Miller, Clemson, 1999, personal communication). In the late 1980s, they rapidly became the major cultivars: in North Carolina, they increased from 9% of commercial acreage in 1988 to more than 77% in 1995, and a similar increase occurred in South Carolina (Caldwell, 1989; Poling, 1998; unpublished results). The modern cultivars have pedigrees quite different from the traditional cultivars (Sjulin and Dale, 1987), and markers for each cultivar type were identified in our study. These cultivar-specific markers used in this study enabled us to assess the persistence of older cultivar markers in current hybrid populations. The results suggest that there is persistence of older cultivar markers in hybrids around the cultivated fields where these traditional cultivars are no longer grown.

One potential result of crop–wild hybrid persistence in wild populations is the loss of genetic variants unique to the wild species. Genetic assimilation by hybridization with crops has been implicated in the extinction of several wild plant species (Ellstrand, 1992), but many reports are based on observation of morphological traits. In this study, the markers we identified as characteristic of wild and cultivated strawberry species can be used to examine rates at which wild genetic variation is lost after hybridization, by evaluating their relative frequencies in *F. virginiana* populations near strawberry farms. Such analyses are seldomly conducted, but may be critical for assessing the impact of crop–wild hybrids in natural ecosystems.

References

Armstrong, J., Gibbs, A., Peakall, R. and Weller, G. (1996) The RAPDistance package. Version 1.04. Available at: ftp://life.anu.edu.au/pub/RAPDistance

Bartsch, D., Lehnen, M., Clegg, J., Pohl-Orf, M., Schuphan, I. and Ellstrand, N.C. (1999) Impact of gene flow from cultivated beet on genetic diversity of wild sea beet populations. *Molecular Ecology* 8, 1733–1741.

Caldwell, J.D. (1989) Southern United States strawberry cultivars. *Fruit Varieties Journal* 43, 33–37.

Casgrain, P., Legendre, P. and Vaudor, A. (1999) The R package for multidimensional and spatial analysis. Version 4.0 d1. Available at: http://alize.ere.umontreal.ca/~casgrain/R/

Crawley, M.J., Hails, R.S., Rees, M., Kohn, D. and Buxton, J. (1993) Ecology of transgenic oilseed rape in natural habitats. *Nature* 363, 620–623.

Dale, A. and Sjulin, T.M. (1990) Few cytoplasms contribute to North American strawberry cultivars. *HortScience* 25, 1341–1342.

Darrow, G.M. (1966) *The Strawberry: History, Breeding and Physiology*. Holt, Rinehart and Winston, New York.

Degani, C., Rowland, L.J., Saunders, J.A., Hokanson, S.C., Ogden, E.L., Golan-Goldhirsh, A. and Galletta, G.J. (2001) A comparison of genetic relationship measures in strawberry (*Fragaria* × *ananassa* Duch.) based on AFLPs, RAPDs, and pedigree data. *Euphytica* 117, 1–12.

Dice, L.R. (1945) Measures of the amount of ecologic association between species. *Ecology* 26, 297–302.

Ellstrand, N.C. (1992) Gene flow by pollen: implications for plant conservation genetics. *Oikos* 63, 77–86.

FAO (1999) FAOSTAT Database. Available at: www.fao.org

Felsenstein, J. (1995) PHYLIP (Phylogeny Inference Package) version 3.5c. University of Washington, Seattle. Available at: http://evolution.genetics.washington.edu/phylip.html

Galletta, G.J. (1997) Small fruit breeding for the southern United States: strawberries. *Fruit Varieties Journal* 51, 138–143.

Gidoni, D., Rom, M., Kunik, T., Zur, M., Izsak, E., Izhar, S. and Firon, N. (1994) Strawberry cultivar identification using randomly amplified polymorphic DNA (RAPD) markers. *Plant Breeding* 113, 339–342.

Graham, J., McNicol, R.J. and McNicol, J.W. (1996) A comparison of methods for the estimation of genetic diversity in strawberry cultivars. *Theoretical and Applied Genetics* 93, 402–406.

Hancock, J.F. and Luby, J.J. (1993) Genetic resources at our doorstep: the wild strawberries. *Bioscience* 43, 141–147.

Hancock, J.F., Maas, J.L., Shanks, C.H., Breen, P.J. and Luby, J.J. (1990) Strawberries (*Fragaria* spp.). In: Moore, J.N. and Ballington, J.R. (eds) *Genetic Resources in Temperate Fruit and Nut Crop*. International Society of Horticultural Sciences, Wageningen, The Netherlands, pp. 489–546.

Hancock, J.F., Callow, P.A. and Shaw, D.V. (1994) Randomly amplified polymorphic DNAs in the cultivated strawberry, *Fragaria* × *ananassa*. *Journal of the American Society for Horticultural Science* 119, 862–864.

Harrison, R.E., Luby, J.J., Furnier, G.R. and Hancock, J.F. (1997) Morphological and molecular variation among populations of octoploid *Fragaria virginiana* and *F. chiloensis* (Rosaceae) from North America. *American Journal of Botany* 84, 612–620.

Lamboy, W.F. and Alpha, C.G. (1998) Using simple sequence repeats (SSRs) for DNA fingerprinting germplasm accessions of grape (*Vitis* L.) species. *Journal of the American Society for Horticultural Science* 123, 182–188.

Mantel, N. (1967) The detection of disease clustering and a generalized regression approach. *Cancer Research* 27, 209–220.

Poling, E.B. (1998) Optimal growing regions for Chandler plasticulture. In: *Strawberry Plasticulture Notebook*. The North Carolina Strawberry Association. Available at: http://www.smallfruits.org/Notebook/noteb.htm

Porebski, S. and Catling, P.M. (1998) RAPD analysis of the relationship of North and South American subspecies of *Fragaria chiloensis*. *Canadian Journal of Botany* 76, 1812–1817.

Sjulin, T.M. and Dale, A. (1987) Genetic diversity of North American strawberry cultivars. *Journal of the American Society for Horticultural Science* 112, 375–385.

Vos, P., Hogers, R., Bleeker, M., Reijans, M., van de Lee, T., Horenes, M., Frijters, A., Pot, J., Peeman, J., Kuiper, M. and Zabeau, M. (1995) AFLP: a new technique for DNA fingerprinting. *Nucleic Acids Research* 23, 4407–4414.

Wolfenbarger, L.L. and Phifer, P.R. (2001) The ecological risks and benefits of genetically engineered plants. *Science* 290, 2088–2093.

8 Gene Flow in Forest Trees: Gene Migration Patterns and Landscape Modelling of Transgene Dispersal in Hybrid Poplar

GANCHO T. SLAVOV[1], STEPHEN P. DIFAZIO[2] AND STEVEN H. STRAUSS[1]

[1]Oregon State University, Corvallis, OR 97331, USA; [2]Oak Ridge National Laboratory, Oak Ridge, TN 37830, USA; E-mail: Steve.Strauss@oregonstate.edu

Abstract

The extent to which transgene flow from plantations can be effectively predicted, managed and monitored will be a critical factor influencing the adoption of transgenic plantations. Studies of historical and contemporary gene flow levels, via genetic structure surveys and parentage analyses, demonstrate that gene flow is generally extensive in both wind- and animal-pollinated forest tree species. Marker studies, however, give little insight into the extent and consequences of gene flow in actual deployment scenarios, because realized gene flow depends on a large number of interacting ecological and genetic factors. For example, despite the potential for extensive gene flow, there appear to be extremely low levels of introgression into wild poplar stands from plantations of hybrids. We introduce a spatial simulation model called STEVE that we have used to integrate a variety of ecological and genetic processes that affect gene flow from poplar plantations in the US Pacific Northwest. The model enables simulation of gene flow on a large landscape over long time frames (e.g. 50–100 years), and permits virtual experiments that investigate how diverse genetic, ecological and management factors might influence the magnitude and variance of gene flow. Similar approaches could be used elsewhere to help identify and prioritize research needs, suggest the strength of mitigation measures such as engineered sterility where they are warranted, and to aid in design of monitoring programmes for large-scale research and commercial applications. We present simulation results which suggest that long-distance gene flow dynamics and creation of a suitable habitat for tree establishment are of considerably more importance to predicting transgene spread than knowledge of intrapopulation mating patterns. The model also predicts that sterility mechanisms, even if incomplete, can greatly impede transgene spread.

Introduction

Early studies of gene flow in forest trees focused on the physical dispersal potential of pollen and seeds. Direct tracking of propagule dispersal and retrospective analyses of population structure have revealed the astonishing mobility of tree pollen (Lanner, 1965; Levin and Kerster, 1974; Di-Giovanni *et al.*, 1996) and seeds (Clark *et al.*, 1998; Cain *et al.*, 2000). The last few decades have seen extraordinary growth in knowledge of gene flow patterns as a result of extensive biochemical and molecular marker studies (e.g. Loveless and Hamrick, 1984). Most of these studies have employed allozymes, which provide a low cost means for studying patterns of genetic differentiation and, by inference, gene flow. However, recent studies, particularly using highly polymorphic DNA markers, have enabled more precise and more direct inferences, particularly of within-population mating patterns and immigration (reviewed below).

Until recently, interest in gene flow was largely restricted to a modest subset of evolutionary and population biologists, and to plant breeders, for whom it was important to interpret patterns of phylogeny, adaptation and degree of contamination in seed production populations. With the advent of genetic engineering, and the biopolitical and legal controversies it has generated, gene flow has moved on to the public stage (e.g. Ellstrand, 2001). Because of the strong sentiments against genetic engineering of some sectors of the public, even small amounts of contamination appear capable of generating great concern (Thompson, 2001). The extent to which we understand, can predict and can efficiently monitor gene flow may therefore be an important biological determinant of whether transgenic crops are adopted and publicly accepted.

Because of the extensive gene flow possible from trees, and their very limited history of domestication, it has long been known that genes from intensively bred or engineered trees are highly likely to enter wild populations unless very special measures – such as use of sterile trees – are taken to avoid it (Strauss *et al.*, 1995). In the southern hemisphere, where some of the most intensive plantation forestry in the world occurs, most of the trees grown are exotics. It has been suggested that there will be less ecological and social concern over dispersal of transgenic exotics compared with that of native tree species (Burdon and Walter, 2001). However, in many of these places, the planted trees have feral, 'naturalized' forms, and in some cases these have given rise to invasive weeds that, because of their size, can have highly deleterious impacts on local ecosystems (Richardson, 1998; Ledgard, 2001). In contrast, transgenic forms of native species are expected to occupy the same niches as the native species, and thus should not displace other species (Strauss, 1999) or radically alter ecosystem processes. Dispersal of some kinds of transgenes in exotic species may therefore present more of an ecological hazard – if it promotes spread or makes control more difficult – than that from native trees.

The frequency and consequences of gene flow may therefore be globally important issues for the future of plantation forestry. Unfortunately, we have a limited knowledge base for risk assessments. There have been very few studies of gene flow from conventional plantations to wild populations. Due to the long lifespans and generation times of trees, predictions of impacts must consider very large temporal and spatial scales, and thus embrace high levels of uncertainty about future ecological conditions (James *et al.*, 1998). New studies, and new tools, will be useful for making rational inferences about possible impacts of transgenic plantations.

We review in detail what is known about gene flow in forest trees, and then consider landscape simulation methods that might be employed for analysing, predicting and monitoring gene flow from tree plantations. As a case study, we focus on a recent analysis we undertook for poplars (genus *Populus*) in the Pacific Northwest USA. We suggest that simulation models can help to guide research and to aid in prediction and monitoring during commercial development, helping to inform environmental assessments of transgenic forestry.

Measuring Historical Levels of Gene Flow

The advent of genetic markers revolutionized methods of measuring gene flow. Over the last three decades, rates of migration typically have been inferred from the degree of genetic differentiation among populations as measured by the fixation index F_{ST} (Wright, 1931), or its many extensions and analogues (e.g. G_{ST}, representing the interpopulation component of total gene diversity (Nei, 1973)). Other approaches for measuring historical gene flow include the 'rare allele method' (Slatkin, 1985) and coalescent methods (Beerli and Felsenstein, 1999). All such methods of measuring gene flow are 'indirect' because they apply genetic models (and underlying assumptions) to infer long-term levels of gene flow (Neigel, 1997; Sork *et al.*, 1999). Indirect measures reflect the complex interactions of all demographic parameters and evolutionary forces acting on a population, and the resulting gene flow estimates should be taken as long-term averages estimated over a large number of populations (Sork *et al.*, 1999).

Although inapplicable for estimating contemporary gene flow, indirect approaches have provided a number of valuable insights about historical forces shaping forest tree genetic structure. There have been a large number of studies of allozyme gene diversity, geographical structure and gene flow among populations of forest trees (reviewed in Govindaraju, 1989; El-Kassaby, 1991; Hamrick *et al.*, 1992; Müller-Starck *et al.*, 1992; Hamrick and Nason, 2000), and a few generalizations have emerged.

1. *Trees are characterized by higher genetic diversity and lower levels of differentiation compared with other plant groups* (Hamrick *et al.*, 1992). The interpopulation component of total gene diversity (based on G_{ST}) of woody species rarely exceeds 10–15% (Table 8.1). This low differentiation suggests extensive gene flow among tree populations. However, some authors have hypothesized that the observed patterns of genetic variation may

Table 8.1. Neighbourhood sizes and per generation migration events estimated by two indirect methods for some tree genera.

Genus	G_{ST}	N_em	N_em*	N_a
Abies	0.063	3.72	8.71	260
Picea	0.055	4.30	–	133–260
Pseudotsuga	0.074	3.13	–	–
Pinus	0.065	3.60	1.51–21.83	158–534
Quercus	0.107	2.09	2.24–6.74	–
Populus	0.041	5.85	–	–
Eucalyptus	0.169	1.23	0.76–6.51	–

G_{ST}, averages from Hamrick *et al.* (1992); N_em, per generation migration events estimated based on the G_{ST} values; N_em*, estimated for some species using the method of Slatkin (1985), from Govindaraju (1989); N_a, neighbourhood area in m^2 (area from which the parents of some central individual may be treated as randomly drawn) estimated for some species, from Govindaraju (1988).

also be caused by other aspects of tree life history. For example, long pre-reproductive phases tend to mitigate founder effects, and long generation times delay differentiation through genetic drift (e.g. Kremer, 1994; Austerlitz *et al.*, 2000).

2. *Wind-pollinated tree species typically have interpopulation differentiation levels of less than 10%, which translates to more than two successfully established migrants per population in each generation in ideal populations.* Hamrick *et al.* (1992) reported an average interpopulation differentiation of 8% for wind-pollinated trees based on 146 data sets. In species with large and continuous ranges, interpopulation differentiation is often below 3% (e.g. *Pinus ponderosa* and *P. contorta* (El-Kassaby, 1991); *P. sylvestris, Picea abies, Quercus petraea* and *Fagus sylvatica* (Müller-Starck *et al.*, 1992)). At the opposite extreme, species with small and fragmented populations can have interpopulation differentiation in the range of 15–30% (e.g. *P. cembra, P. halepensis, P. nigra* and *Castanea sativa* (Müller-Starck *et al.*, 1992); *P. torreyana* and *P. muricata* (Hamrick *et al.*, 1992)). This observation suggests that even though long-distance pollen dispersal is possible, its effect may be insufficient for genetic homogenization of spatially isolated populations.

3. *Outcrossed animal-pollinated tree species have a detectably (but not significantly) higher degree of interpopulation genetic differentiation compared with wind-pollinated trees.* The average interpopulation differentiation for animal-pollinated trees with mixed mating systems was 10% based on 37 data sets (Hamrick *et al.*, 1992). As in wind-pollinated species, spatial distribution seemed to be a good predictor of the degree of differentiation. Moran (1992) reviewed interpopulation differentiation in Australian eucalypts and cited values in the range of 8–12% for widespread species (e.g. *Eucalyptus saligna, E. cloeziana* and *E. delegatensis*); 30% for the highly disjunct *E. nitens* (widespread, but with a highly discontinuous distribution); and 61% for isolated populations of *E. caesia*. However, 67 scattered and putatively isolated low

density populations of the insect-pollinated, and presumably bird-dispersed, *Sorbus torminalis* had interpopulation differentiation of only 15% (Demesure *et al.*, 2000), suggesting that gene flow rates can, in some cases, be high among spatially isolated populations (but see Austerlitz *et al.*, 2000).

Direct Methods of Measuring Gene Flow Through Parentage Analysis

Direct observations of gene dispersal obviate the need for tenuous assumptions about historical conditions. Instead, they provide short-term 'immigration snapshots'. Generally, direct methods require the genotyping of all potential parents in a population and estimation of the proportion of progeny that could *not* have been produced by within-population mating. One approach employs simple paternity exclusion, feasible where the maternal genotype can be readily determined (Smith and Adams, 1983; Devlin and Ellstrand, 1990). However, the low variability of allozyme markers greatly limits the ability to distinguish between local and immigrant genotypes (Adams, 1992). To help overcome this problem, a number of methods employ maximum likelihood either to assign parentage (Meagher, 1986; Adams *et al.*, 1992; Smouse and Meagher, 1994; Kaufman *et al.*, 1998) or to estimate mating parameters that provide the best fit to observed progeny genotypes (Devlin *et al.*, 1988; Roeder *et al.*, 1989; Adams and Birkes, 1991).

In the early 1990s, highly variable DNA-based markers began to become affordable for parentage analyses. The high genetic resolution provided by microsatellite and amplfed fragment length polymorphism (AFLP) data allowed scientists to conduct paternity analyses based on genotypic exclusion with acceptable levels of discrimination (Dow and Ashley, 1998; Streiff *et al.*, 1999; Lian *et al.*, 2001), as well as to apply maximum likelihood assignments with greater confidence (Gerber *et al.*, 2000; Kameyama *et al.*, 2000).

Seed orchards

There have been a number of studies of genetic contamination of forestry 'seed orchards'. Seed orchards are plantations in which selected genotypes are placed to allow cross-pollination for production of seeds to be used for reforestation. Most of these orchards are within the range of native or planted populations, and distinct orchard blocks (that service distinct ecogeographical regions) are, in many cases, planted close together for management efficiency. Thus, there are often high levels of unwanted pollen immigration into the orchard blocks. This can result in substantial loss of genetic gain compared with expectations based on selection theory, and can compromise adaptability of seed orchard crops to their intended plantation environments (Adams and Burczyk, 2000).

Pollen contamination in seed orchards is usually estimated via simple paternity exclusion, adjusting the observed proportion of immigrants by the probability that an immigrant gamete will be distinguishable from the potential orchard gamete pool (Adams and Burczyk, 2000). Other statistical procedures have also been implemented (El-Kassaby and Ritland, 1986; Plomion *et al.*, 2001). These studies have revealed great variation in pollen contamination, even when the same analytical approaches and tree species are considered (Table 8.2; Adams and Burczyk, 2000). None the less, it is clear that pollen contamination is often very large, commonly exceeding 40%, even when the closest stands of the same species are several hundred metres away.

Commercial plantations

We know of very few studies of the effects of gene flow from forest plantations on wild populations. The best-studied cases are from interspecific hybrids in the genus *Populus*, which are often planted in proximity to wild populations. In Europe, most of the planted cottonwood hybrids include the native *Populus nigra* as a parent. These hybrids are interfertile with wild populations of *P. nigra*, whose populations are greatly reduced in extent due to destruction of riparian habitat by farming and human habitations. None the less, several studies have reported low levels of gene flow into wild *P. nigra* stands (Legionnet and Lefèvre, 1996; Benetka *et al.*, 1999, 2002; Lefèvre *et al.*, 2001; Heinze and Lickl, 2002). Likewise, in the Pacific Northwest USA, gene flow from hybrid poplar plantations into wild American black cottonwood (*Populus trichocarpa*) populations was extremely low, despite the presence of

Table 8.2. Estimated pollen contamination in clonal seed orchards of forest trees.

Tree species	S	D_i	b	m
Picea abies	13.2	None	–	0.55–0.81[a]
	–	–	0.10–0.17	–
Picea glauca	–	1000	–	0.01
Pseudotsuga menziesii	1.8–3.3	None	–	0.29–0.91
	20	500	–	0.11
Pinus sylvestris	22.9	2000	–	0.48
	3	1000	0.15	–
	6	500	0.38	–
	12.5	> 100	–	0.72
Pinus taeda	2	100	–	0.36
Pinus pinaster	–	None	0.36[b]	–
Pinus contorta	4.4	200	0.05[c]	0.08[c]
Pinus thungbergii	0.5	> 500	0.02[d]	–

S, area in which all potential parents have been genotyped (ha); D_i, distance (m) to the nearest population or tree of the same species; b, observed proportion of immigrant pollen gametes; m, pollen contamination adjusted for the probability to distinguish local and migrant gametes.
References from Adams and Burczyk (2000), unless otherwise indicated: [a]Pakkanen *et al.* (2000); [b]Plomion *et al.* (2001); [c]Stoehr and Newton (2002); [d]Goto *et al.* (2002).

large male plantations in close proximity to native female trees and the large-scale clearing of wild populations for agriculture (DiFazio, 2002).

Wild populations

Due to restrictions on marker resolution and the high cost of genotyping, spatially distinct forest stands are usually chosen for study, and parentage analysis is then applied to a sample of progeny. Typically, seeds are collected from mother trees of known genotype, and paternity estimated by comparing inferred paternal genotypes with those of all potential fathers in the analysed population (e.g. Schnabel and Hamrick, 1995; Dow and Ashley, 1998; Kaufman *et al.*, 1998). Some researchers have also sampled seedlings and/or saplings, attempting to estimate both pollen- and seed-mediated gene flow (Dow and Ashley, 1996; Isagi *et al.*, 2000; Konuma *et al.*, 2000). However, such analyses require considerably more exclusion power than paternity analysis (Marshall *et al.*, 1998).

Results from paternity analyses in wind-pollinated species generally agree with predictions from studies of genetic structure (Table 8.3). In most cases, the frequency of immigrant pollinations was over 30% (e.g. 31% in *Pinus densiflora* (Lian *et al.*, 2001); 57% in *Quercus macrocarpa* (Dow and Ashley, 1998); 65% in *Q. robur* and 69% in *Q. petraea* (Streiff *et al.*, 1999)), but appeared rather low (6.5%) in a spatially isolated population of *Pinus flexilis* (Schuster and Mitton, 2000). Remarkably, Kaufman *et al.* (1998) reported extensive pollen-mediated gene flow (a minimum of 37%) in a population of the tropical pioneer *Cecropia obtusifolia*, even though the closest population of the same species was at least 1 km away. Kaufman *et al.* (1998) suggested that successful pollen travelled as far as 10 km. Similarly, DiFazio *et al.* (unpublished results) observed extensive pollen immigration into stands of *P. trichocarpa* that were isolated by up to 16 km from the nearest ungenotyped pollen source (Table 8.4).

As with indirect methods, gene flow estimates based on parentage analysis in animal-pollinated tree species were somewhat lower, but generally similar to immigration rates in wind-pollinated trees (e.g. 17–30% in *Gleditsia triacanthos* (Schnabel and Hamrick, 1995); 20–30% in *Rhododendron metternichii* (Kameyama *et al.*, 2000); 36–68% in *Swietenia humilis* (White *et al.*, 2002); and 74% in *Magnolia obovata* (Isagi *et al.*, 2000)). These results support the notion that animals can be effective agents of long-distance pollen and seed dispersal. However, there is also considerable variance in pollen immigration between species, even when isolation distances are similar (Stacy *et al.*, 1996), cautioning against broad generalizations.

Although paternity analysis based on highly variable markers appears to be the most effective current method for measuring gene dispersal in ecological time, estimates can be greatly affected by the presence of null alleles, and the misgenotyping of complex microsatellite and AFLP phenotypes. Both of these types of errors will cause overestimates of gene flow (Marshall *et al.*, 1998). Despite high reported exclusion probabilities, it is

Table 8.3. Mean pollination distance and immigration estimates from parentage analyses in wind- and animal-pollinated trees.

Tree species	S	D_i	d_{wp}	d_{op}	m	Reference
Wind-pollinated						
Pinus flexilis	–	> 2000	133–140	155–265	0.07	Schuster and Mitton (2000)
Pinus densiflora	9.1	–	68	–	0.31	Lian et al. (2001)
Pinus pinaster	0.8	–	–	–	0.15	González-Martínez et al. (2002)
Quercus macrocarpa	5	> 100	75	–	0.57	Dow and Ashley (1998)
Quercus robur	5.8	> 100	22–58	333	0.65	Streiff et al. (1999)
Quercus petraea	5.8	> 100	18–65	287	0.69	Streiff et al. (1999)
Cecropia obtusifolia	8.6	> 1000	–	–	0.37	Kaufman et al. (1998)
Animal-pollinated						
Gleditsia triacanthos	3	> 85	–	–	0.17–0.30	Schnabel and Hamrick (1995)
Ficus (from three different species)	–	> 1000	–	–	> 0.90	Hamrick and Nason (2000)
Rhododendron metternichii	1	> 50	–	–	0.20–0.30	Kameyama et al. (2000)
Magnolia obovata	69	–	131	–	0.74[a]	Isagi et al. (2000)
Neobalanocarpus heimii	42	–	188–196	–	0.21–0.69[a]	Konuma et al. (2000)
Enterolobium cyclocarpum	9.8	300	–	–	0.69–0.74	Apsit et al. (2001)
Swietenia humilis	–	> 1000	–	3100	0.47	White et al. (2002)[b]
Eucalyptus regnans[c]	0.5	40	16–27	–	0.49–0.51	Burczyk et al. (2002)

S, area in which all potential parents have been genotyped (ha); D_i, distance to the nearest population (tree) of the same species (m); d_{wp}, average pollination distance within the reference stand (m); d_{op}, mean pollination distance from assumed dispersal curve (m); m, proportion of offspring with immigrant paternal gametes.
[a]Offspring having one or both parents located outside the reference stand.
[b]Estimates for population Butus/Jicarito.
[c]Seed orchard (values for neighbourhoods with radius 40 m are shown).

therefore important to treat many of the published estimates with caution; their accuracy will improve over the next several years as applications of molecular technology and statistical methods mature.

Spatial Simulation Modelling

Extrapolation of short-term or historical gene flow observations to spatial and temporal scales that are relevant for management and ecological policy remains a major challenge (Levin, 1992; Turner et al., 2001). Because of the time and expense required for a typical parentage analysis study, only a

Table 8.4. Population and gene flow statistics from three microsatellite-based studies of pollen dispersal in *Populus trichocarpa* in Oregon.

Site	r^a	Mothers[b]	Fathers[c]	n^d	D_i^e	d_{wp}^e	d_{op}^e	P^f	M^g	G^h
Willamette	0.25	5	221	255	100–300	138	847	89	55	44
Luckiamute	1	5	57	423	1000–1100	128	–	98	5	76
Vinson	10	28	54	712	2680–9760	1093	1270	273	197	34

From DiFazio *et al.* (unpublished results).
[a]Radius of sampled area (km).
[b]Number of trees from which seeds were collected.
[c]Number of reproductively mature male trees within sampled area.
[d]Number of progeny genotyped.
[e]Defined in Table 8.3.
[f]Number of seeds for which a single putative father was compatible within the sampled area.
[g]Number of seeds for which multiple putative fathers were compatible within the sampled area.
[h]Percentage of seeds for which no compatible fathers were identified within the sampled area (i.e. minimum estimates of gene flow).

limited number of populations and years can be examined (Ouborg *et al.*, 1999; Cain *et al.*, 2000). However, ecologically significant levels of establishment may occur only once or twice per generation (James *et al.*, 1998), and in particular habitats. An emerging solution is the use of spatial simulation models to extrapolate results of short-term gene flow studies with knowledge of ecological processes (King, 1991; Dunning *et al.*, 1995). Such models provide an extensible framework for integrating data from disparate demographic and genetic field studies with landscape-scale analyses of ecosystem dynamics (Sork *et al.*, 1999; Higgins *et al.*, 2000). In addition, these models allow 'virtual experiments' through sensitivity analyses in which selected components of the system are manipulated to determine their importance in determining long-term outcomes (Turner *et al.*, 2001).

Case Study: Gene Flow in Poplar

We analysed gene flow in wild black cottonwood populations, and from hybrid poplar plantations, in the northwestern USA. The primary objective of these studies was to provide data for assessing the extent of transgene dispersal that is likely to occur should transgenic hybrid poplars be cultivated in the region. We studied gene flow using parentage analysis in three wild populations with contrasting ecological characteristics (DiFazio *et al.*, unpublished results), and gathered data on seedling establishment and survival in experimental plots and in the wild (DiFazio *et al.*, 1999). We also inferred landscape-level spatial and temporal dynamics of black cottonwood establishment from a chronosequence of geographical information system (GIS) layers encompassing some of the same populations included in the field studies (Fig. 8.1).

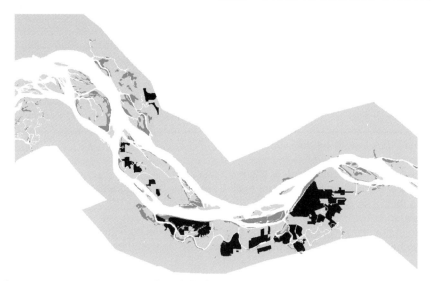

Fig. 8.1.   GIS representation of modelled area in northwestern Oregon (23 km ×
237 km: 846 km²). White shows main channels of the Columbia River; black, areas
of hybrid cottonwood plantations; dark grey, wild black cottonwood stands; light grey
are non-poplar lands (mostly farms, coniferous uplands and wetlands; DiFazio, 2002).

The model, called STEVE (simulation of transgene effects in a variable
environment – but also named to reflect that four Steves contributed to
its development!), provides a spatially explicit representation of gene flow
(DiFazio, 2002; Fig. 8.2). It operates on a landscape grid (23 km × 37 km,
100 m² cells) containing information about elevation, habitat type and poplar
populations. The simulation has an annual time step, with modules to
simulate creation and conversion of poplar patches, growth, reproduction,
dispersal and competition within poplar cohorts. The simulations track
two genotypes, transgenic and conventional. Transgenic trees originate
in plantations and may spread to the wild through pollen, seed and/or
vegetative propagules. The relative amounts of propagules produced in
each location are proportional to basal area (i.e. trunk cross-sectional area)
of each genotype, modulated by a fecundity factor.

   We structured and parameterized the model based on results of our field
studies of gene flow. They indicated that long-distance dispersal is considerable for *Populus* (Table 8.4; DiFazio *et al.*, unpublished results), with the tail
of the distribution quite 'fat' (*sensu* Kot *et al.*, 1996). We therefore chose to
model gene dispersal as a two-stage process, with local dispersal modelled
explicitly by a negative exponential distribution, and long-distance dispersal
modelled as if a portion of the pollen and seeds were panmictic at the landscape scale. This is analogous to a mixed model approach (Clark *et al.*, 1998).
Seed dispersal was modelled in the same way, though based on more limited
field studies of movement, and assumed much less local and long-distance
movement than for pollen. *Populus* seeds are very light and contain cotton

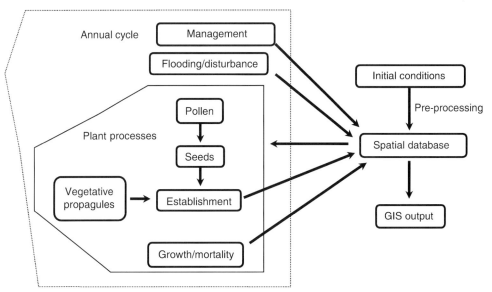

Fig. 8.2. The STEVE model. The model begins with pre-processing of GIS layers representing initial simulation conditions. Data are stored in a spatial database containing information about elevation, cover type, poplar populations, plantations and agricultural fields. Simulation begins with management activities such as plantation harvesting and herbicide spraying. Poplar establishment and mortality is simulated in the disturbance function. Seed, pollen and vegetative propagules are produced proportional to basal area of each genotype, followed by dispersal, establishment, growth and mortality. Outputs are text files and spatial data layers.

appendages that facilitate wind and water dispersal. Therefore, a portion of the seeds is expected to attain stochastic long-distance dispersal (Wright, 1952).

This method of modelling pollen and seed dispersal had major implications for gene flow from transgenic plantations. Modelled gene flow was highly sensitive to changes in the proportion of pollen and seed dispersed over long-distances (Fig. 8.3A and B), but relatively insensitive to the slope of local dispersal curves (Fig. 8.3C and D). This was primarily because poplars require very intense disturbance, abundant moisture and freedom from most competition by other plants for successful establishment. These conditions are rarely met in space and time. The majority of establishment sites therefore occurred beyond the local seed and pollen shadows of the plantations (Fig. 8.3F). Also, because long-distance dispersal was insensitive to wind in this model (pollen and seed were assumed to be panmictic at the landscape scale), wind speed had no detectable effect on gene flow from plantations (Fig. 8.3E). Long-distance dispersal ensured that a proportion of plantation-derived propagules would encounter stochastic establishment sites regardless of distance from plantations, which explains why this portion of the dispersal function was overwhelmingly important in determining gene flow. One implication of this result is that future research on gene flow in *Populus* would benefit most from better definition of the dynamics of long-distance

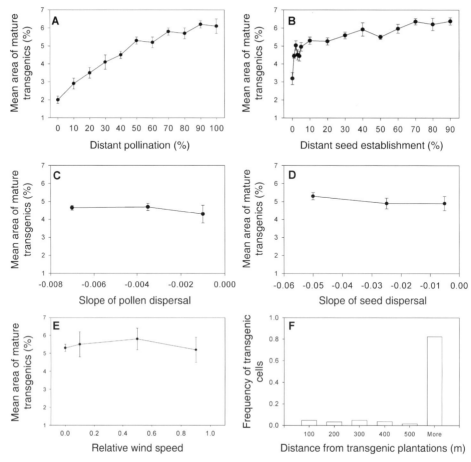

Fig. 8.3.   Effects of dispersal and wind on simulated gene flow. Error bars are 1 SE from ten repetitions with each set of parameter values. (A) Effects of distant pollination on transgene flow. Distant pollination is the proportion of seeds that are fathered by trees that do not occur in the local population. This parameter has a strong effect on transgene flow, reflecting the importance of long-distance pollen dispersal. (B) Effects of distant seed establishment on transgene flow. Distant seed establishment had minor effects except at very low levels. (C and D) Effects of varying the slope of the negative exponential distributions depicting local pollen and seed dispersal, respectively. Varying this slope had little effect on gene flow. (E) Effect of relative wind speed, with wind direction set at 90°. (F) Distance of portions of transgenic cohorts (cells) from mature transgenic plantations. The local pollen and seed shadows end at 440 and 220 m, respectively.

dispersal, rather than from studies of local pollen movement and mating between trees within stands.

Sensitivity analyses allowed us to study the consequences for gene flow of many ecological conditions and transgenic deployment scenarios over a 50–100 year time frame (DiFazio, 2002). For example, we studied the consequences of:

- Transgenes that imparted herbicide resistance with respect to various scenarios of herbicide use and disturbance on the landscape;
- Transgenic trees with insect resistance, with varying levels of insect attack in wild populations;
- Reductions in fertility due to transgenic or other sources, and implications of various levels of efficiency and stability;
- The effects of transgenes with positive or negative effects under natural selection; and
- Effects of transgenic versus non-transgenic plantation area, plantation gender and rotation length (time to harvest).

Most of these simulations also included stochastic variation, so that natural environmental variances and uncertainty in parameter estimates could be reflected in model outputs. They also included a number of conservative assumptions, such that the estimates of transgene dispersal are expected to be worst-case estimates. Ideally, the model structure and parameters would be continually revised based on research results, and by data from monitoring programmes during commercial deployment. Some key results from sensitivity studies (DiFazio, 2002) were that:

- Vegetative propagation appeared to be much less important than sexual propagules in determining extent of migration under virtually all scenarios;
- An imperfect and tightly linked sterility gene could dramatically slow the spread of a gene that provided a strong selective advantage; and
- The spread of neutral genes could be greatly attenuated by sterility levels whose effectiveness (stability) was in the vicinity of 95%.

The most important contribution of spatial simulation models such as STEVE is that they provide a comprehensive, explicit logical framework for thinking about the long-term consequences of different options for deploying transgenic, as well as conventionally bred, plants. Models help to reduce the immense ecological complexity of tree gene flow to a set of specific, testable predictions that can guide further research, and inform business plans, regulatory decisions and, ultimately, public views about transgenic technology in plantation forestry.

References

Adams, W.T. (1992) Gene dispersal within forest tree populations. *New Forests* 6, 217–240.

Adams, W.T. and Birkes, D.S. (1991) Estimating mating patterns in forest tree populations. In: Fineschi, S., Malvolti, M.E., Cannata, F. and Hattemer, H.H. (eds) *Biochemical Markers in the Population Genetics of Forest Trees*. S.P.B. Academic Publishing bv, The Hague, The Netherlands, pp. 157–172.

Adams, W.T. and Burczyk, J. (2000) Magnitude and implications of gene flow in gene conservation reserves. In: Young, A., Boshier, D. and Boyle, T. (eds) *Forest*

Conservation Genetics: Principles and Practice. CSIRO Publishing, Collingwood, Victoria, Australia, pp. 215–224.

Adams, W.T., Griffin, A.R. and Moran, G.F. (1992) Using paternity analysis to measure effective pollen dispersal in plant populations. *American Naturalist* 140, 762–780.

Apsit, V.J., Hamrick, J.L. and Nason, J.D. (2001) Breeding population size of a fragmented population of a Costa Rican dry forest tree species. *Journal of Heredity* 92, 415–420.

Austerlitz, F., Mariette, S., Machon, N., Gouyon, P.H. and Godelle, B. (2000) Effects of colonization processes on genetic diversity: differences between annual plants and tree species. *Genetics* 154, 1309–1321.

Beerli, P. and Felsenstein, J. (1999) Maximum-likelihood estimation of migration rates and effective population numbers in two populations using a coalescent approach. *Genetics* 152, 763–773.

Benetka, V., Mottl, J., Vacková, I., Pospíšková, M. and Dubsky, M. (1999) Estimation of the introgression level in *Populus nigra* L. populations by means of isozyme gene markers. *Silvae Genetica* 48, 218–223.

Benetka, V., Vacková, I., Bartáková, I., Pospíšková, M. and Rasl, M. (2002) Introgression in black poplar (*Populus nigra* L. ssp. *nigra*) and its transmission. *Journal of Forest Science* 48, 115–120.

Burczyk, J., Adams, W.T., Moran, G.F. and Griffin, A.R. (2002) Complex patterns of mating revealed in a *Eucalyptus regnans* seed orchard using allozyme markers and the neighbourhood model. *Molecular Ecology* 11, 2379–2391.

Burdon, R.D. and Walter, C. (2001) Perspectives on risk in transgenic forest plantations in relation to conventional breeding and use of exotic pines and eucalypts. *Symposium Proceedings: Tree Biotechnology in the New Millennium.* Oregon State University, Corvallis, Oregon. Available at: http://www.fsl.orst.edu/tgerc/iufro2001/eprocd.htm

Cain, M.L., Milligan, B.G. and Strand, A.E. (2000) Long-distance seed dispersal in plant populations. *American Journal of Botany* 87, 1217–1227.

Clark, J.S., Fastie, C., Hurtt, G.S., Jackson, T., Johnson, C., King, A.G., Lewis, M., Lynch, J., Pacala, S., Prentice, C., Schupp, E.W., Webb, T. III and Wyckoff, P. (1998) Reid's paradox of rapid plant migration. *Bioscience* 48, 13–24.

Demesure, B., Le Guerroué, B., Lucchi, G., Prat, D. and Petit, R.J. (2000) Genetic variability of a scattered temperate forest tree: *Sorbus torminalis* L. (Crantz). *Annals of Forest Science* 57, 63–71.

Devlin, B. and Ellstrand, N.C. (1990) The development and application of a refined method for estimating gene flow from angiosperm paternity analysis. *Evolution* 44, 248–259.

Devlin, B., Roeder, K. and Ellstrand, N.C. (1988) Fractional paternity assignment: theoretical development and comparison to other methods. *Theoretical and Applied Genetics* 76, 369–380.

DiFazio, S.P. (2002) Measuring and Modeling Gene Flow from Hybrid Poplar Plantations: Implications for Transgenic Risk Assessment. PhD Dissertation, Oregon State University, Corvallis, Oregon. Available at: http://www.fsl.orst.edu/tgerc/dif_thesis/difaz_thesis.pdf

DiFazio., S.P., Leonardi, S., Cheng, S. and Strauss, S.H. (1999) Assessing potential risks of transgene escape from fiber plantations. In: Lutman, P.W. (ed.) *Gene Flow and Agriculture: Relevance for Transgenic Crops.* Symposium Proceedings No. 72. British Crop Protection Concil, Farnham, UK, pp. 171–176.

Di-Giovanni, F., Kevan, P.G. and Arnold, J. (1996) Lower planetary boundary layer profiles of atmospheric conifer pollen above a seed orchard in northern Ontario, Canada. *Forest Ecology and Management* 83, 87–97.

Dow, B.D. and Ashley, M.V. (1996) Microsatellite analysis of seed dispersal and parentage of saplings of bur oak, *Quercus macrocarpa*. *Molecular Ecology* 5, 61–627.

Dow, B.D. and Ashley, M.V. (1998) High levels of gene flow in Bur oak revealed by paternity analysis using microsatellites. *Journal of Heredity* 89, 62–70.

Dunning, J.B., Stewart, D.J., Danielson, B.J., Noon, B.R., Root, T.L., Lamberson, R.H. and Stevens, E.E. (1995) Spatially explicit population models: current forms and future uses. *Ecological Applications* 5, 3–11.

El-Kassaby, Y.A. (1991) Genetic variation within and among conifer populations: review and evaluation of methods. In: Fineschi, S., Malvolti, M.E., Cannata, F. and Hattemer, H.H. (eds) *Biochemical Markers in the Population Genetics of Forest Trees*. S.P.B. Academic Publishing bv, The Hague, The Netherlands, pp. 61–76.

El-Kassaby, Y.A. and Ritland, K. (1986) Low levels of pollen contamination in a Douglas-fir seed orchard as detected by allozyme markers. *Silvae Genetica* 35, 224–229.

Ellstrand, N.C. (2001) When transgenes wander, should we worry? *Plant Physiology* 125, 1543–1545.

Gerber, S., Mariette, S., Streiff, R., Bodénès, C. and Kremer, A. (2000) Comparison of microsatellites and amplified fragment length polymorphism markers for parentage analysis. *Molecular Ecology* 9, 1037–1048.

González-Martínez, S.C., Gerber, S., Cervera, M.T., Martínez-Zapater, J.M., Gil, L. and Alia, R. (2002) Seed gene flow and fine-scale structure in a Mediterranean pine (*Pinus pinaster* Ait.) using nuclear microsatellite markers. *Theoretical and Applied Genetics* 104, 1290–1297.

Goto, S., Miyahara, F. and Ide, Y. (2002) Monitoring male reproductive success in a Japanese black pine clonal seed orchard with RAPD markers. *Canadian Journal of Forest Research* 32, 983–988.

Govindaraju, D.R. (1988) Life histories, neighborhood sizes, and variance structure in some North American conifers. *Biological Journal of the Linnean Society* 35, 69–78.

Govindaraju, D.R. (1989) Estimates of gene flow in forest trees. *Biological Journal of the Linnean Society* 37, 345–357.

Hamrick, J.L. and Nason, J.D. (2000) Gene flow in forest trees. In: Young, A., Boshier, D. and Boyle, T. (eds) *Forest Conservation Genetics: Principles and Practice*. CSIRO Publishing, Collingwood, Victoria, Australia, pp. 81–90.

Hamrick, J.L., Godt, M.J.W. and Sherman-Broyles, S.L. (1992) Factors influencing levels of genetic diversity in woody plant species. *New Forests* 6, 95–124.

Heinze, B. and Lickl, E. (2002) Rare, but steady, introgression in Austrian black poplar as a long-term risk? In: Van Dam, B. and Bordács, S. (eds) *Genetic Diversity in River Populations of European Black Poplar – Implications for Riparian Eco-system Management*. Proceedings of an international symposium. Csiszár Nyomda, Budapest, Hungary, Symposium held 16–20 May 2001, Szekszárd, pp. 169–175.

Higgins, S.I., Richardson, D.M. and Cowling, R.M. (2000) Using a dynamic landscape model for planning the management of alien plant invasions. *Ecological Applications* 10, 1833–1848.

Isagi, Y., Kanazashi, T., Suzuki, W., Tanaka, H. and Abe, T. (2000) Microsatellite analysis of the regeneration process of *Magnolia obovata* Thunb. *Heredity* 84, 143–151.

James, R.R., DiFazio, S.P., Brunner, A.M. and Strauss, S.H. (1998) Environmental effects of genetically engineered woody biomass crops. *Biomass and Bioenergy* 14, 403–414.

Kameyama, Y., Isagi, Y., Naito, K. and Nakagoshi, N. (2000) Microsatellite analysis of pollen flow in *Rhododendron metternichii* var. *hondoense*. *Ecological Research* 15, 263–269.

Kaufman, S.R., Smouse, P.E. and Alvarez-Buylla, E.R. (1998) Pollen-mediated gene flow and differential male reproductive success in a tropical pioneer tree, *Cecropia obtusifolia* Bertol. (Moraceae): a paternity analysis. *Heredity* 81, 164–173.

King, A.W. (1991) Translating models across scales in the landscape. In: Turner, M.G. and Gardner, R. (eds) *Quantitative Methods in Landscape Ecology*. Springer-Verlag, New York, pp. 479–517.

Konuma, A., Tsumura, Y., Lee, C.T., Lee, S.L. and Okuda, T. (2000) Estimation of gene flow in the tropical-rainforest tree *Neobalanocarpus heimii* (Dipterocarpaceae), inferred from paternity analysis. *Molecular Ecology* 9, 1843–1852.

Kot, M., Lewis, M. and van den Driessche, P. (1996) Dispersal data and the spread of invading organisms. *Ecology* 77, 2027–2042.

Kremer, A. (1994) Diversité génétique et variabilité des characters phénotypiques chez les arbres forestiers. *Génétique, Sélection, Évolution* 26 (Supplement 1), 105–123.

Lanner, R.M. (1965) Needed: a new approach to the study of pollen dispersion. *Silvae Genetica* 15, 50–52.

Ledgard, N. (2001) The spread of lodgepole pine (*Pinus contorta*, Dougl.) in New Zealand. *Forest Ecology and Management* 141, 43–57.

Lefèvre, F., Kajba, D., Heinze, B., Rotach, P., de Vries, S.M.G. and Turok, J. (2001) Black poplar: a model for gene resource conservation in forest ecosystems. *Forestry Chronicle* 77, 239–244.

Legionnet, A. and Lefèvre, F. (1996) Genetic variation of the riparian pioneer tree species *Populus nigra* L. I. Study of population structure based on isozymes. *Heredity* 77, 629–637.

Levin, D.A. and Kerster, H.W. (1974) Gene flow in seed plants. In: Dobzhansky, T., Hecht, M.T. and Steere, W.T. (eds) *Evolutionary Biology 7*. Plenum Press, New York, pp. 139–220.

Levin, S.A. (1992) The problem of pattern and scale in ecology. *Ecology* 73, 1943–1967.

Lian, C., Miwa, M. and Hogetsu, T. (2001) Outcrossing and paternity analysis of *Pinus densiflora* (Japanese red pine) by microsatellite polymorphism. *Heredity* 87, 88–98.

Loveless, M.D. and Hamrick, J.L. (1984) Ecological determinants of genetic structure in plant populations. *Annual Review of Ecology and Systematics* 15, 65–95.

Marshall, T.C., Slate, J., Kruuk, L.E.B. and Pemberton, J.M. (1998) Statistical confidence for likelihood-based paternity inference in natural populations. *Molecular Ecology* 7, 639–655.

Meagher, T.R. (1986) Analysis of paternity within a natural population of *Chamaelirium luteum*. 1. Identification of most-likely male parents. *American Naturalist* 128, 199–215.

Moran, G.F. (1992) Patterns of genetic diversity in Australian tree species. *New Forests* 6, 49–66.

Müller-Starck, G., Baradat, P. and Bergmann, F. (1992) Genetic variation within European tree species. *New Forests* 6, 23–47.

Nei, M. (1973) Analysis of gene diversity in subdivided populations. *Proceedings of the National Academy of Sciences USA* 70, 3321–3323.

Neigel, J.E. (1997) A comparison of alternative strategies for estimating gene flow from genetic markers. *Annual Review of Ecology and Systematics* 28, 105–128.

Ouborg, N.J., Piquot, Y. and Van Groenendael, J.M. (1999) Population genetics, molecular markers, and the study of dispersal in plants. *Journal of Ecology* 87, 551–568.

Pakkanen, A., Nikkanen, T. and Pulkkinen, P. (2000) Annual variation in pollen contamination and outcrossing in a *Picea abies* seed orchard. *Scandinavian Journal of Forest Research* 15, 399–404.

Plomion, C., LeProvost, G., Pot, D., Vendramin, G., Gerber, S., Decroocq, S., Brach, J., Raffin, A. and Pastuszka, P. (2001) Pollen contamination in a maritime pine polycross seed orchard and certification of improved seeds using chloroplast microsatellites. *Canadian Journal of Forest Research* 31, 1816–1825.

Richardson, D.M. (1998) Forestry trees as invasive aliens. *Conservation Biology* 12, 18–26.

Roeder, K., Devlin, B. and Lindsay, B.G. (1989) Application of maximum likelihood methods to population genetic data for the estimation of individual fertilities. *Biometrics* 45, 363–379.

Schnabel, A. and Hamrick, J.L. (1995) Understanding the population genetic structure of *Gleditsia triacanthos* L.: the scale and pattern of pollen gene flow. *Evolution* 49, 921–931.

Schuster, W.S.F. and Mitton, J.B. (2000) Paternity and gene dispersal in limber pine (*Pinus flexilis* James). *Heredity* 84, 348–361.

Slatkin, M. (1985) Rare alleles as indicators of gene flow. *Evolution* 39, 53–65.

Smith, D.B. and Adams, W.T. (1983) Measuring pollen contamination in clonal seed orchards with the aid of genetic markers. In: *Proceedings of the 17th Southern Forest Tree Improvement Conference*, Athens, Georgia, pp. 64–73.

Smouse, P.E. and Meagher, T.R. (1994) Genetic analysis of male reproductive contributions in *Chamaelirium luteum* (L.) Gray (Liliaceae). *Genetics* 136, 313–322.

Sork, V.L., Nason, J., Campbell, D.R. and Fernandez, J.F. (1999) Landscape approaches to contemporary gene flow in plants. *Trends in Ecology and Evolution* 14, 219–224.

Stacy, E.A., Hamrick, J.L., Nason, J.D., Hubbell, S.P., Foster, R.B. and Condit, R. (1996) Pollen dispersal in low-density populations of three neotropical tree species. *American Naturalist* 148, 275–298.

Stoehr, M.U. and Newton, C.R. (2002) Evaluation of mating dynamics in a lodgepole pine seed orchard using chloroplast DNA markers. *Canadian Journal of Forest Research* 32, 469–476.

Strauss, S.H. (1999) Report of the poplar working group. In: Traynor, P.L. and Westwood, J. (eds) *Ecological Effects of Pest Resistance Genes in Managed Ecosystems*. Virginia Polytechnic Institute, Blacksburg, Virginia, pp. 105–112.

Strauss, S.H., Rottmann, W.H., Brunner, A.M. and Sheppard, L.A. (1995) Genetic engineering of reproductive sterility in forest trees. *Molecular Breeding* 1, 5–26.

Streiff, R., Ducousso, A., Lexer, C., Steinkellner, H., Gloessl, J. and Kremer, A. (1999) Pollen dispersal inferred from paternity analysis in a mixed stand of *Quercus robur* L. and *Q. petraea* (Matt.) Liebl. *Molecular Ecology* 8, 831–841.

Thompson, P.B. (2001) The ethics of molecular silviculture. In: Strauss, S.H. and Bradshaw, H.D. Jr (eds) *Proceedings of the First International Symposium on Ecological and Societal Aspects of Transgenic Plantations*. College of Forestry, Oregon State University, Corvallis, Oregon, pp. 85–91. Available at: http://www.fsl.orst.edu/tgerc/iufro2001/eprocd.pdf

Turner, M.G., Gardner, R. and O'Neill, P. (2001) *Landscape Ecology in Theory and Practice: Pattern and Process.* Springer-Verlag, New York.

White, G.M., Boshier, D.H. and Powell, W. (2002) Increased pollen flow counteracts fragmentation in a tropical dry forest: an example from *Swietenia humilis* Zuccarini. *Proceedings of the National Academy of Sciences USA* 99, 2038–2042.

Wright, J.W. (1952) *Pollen Dispersion of Some Forest Trees.* Northeastern Forest Experiment Station Paper 46.

Wright, S. (1931) Evolution in Mendelian populations. *Genetics* 16, 97–159.

9

Implications for Hybridization and Introgression Between Oilseed Rape (*Brassica napus*) and Wild Turnip (*B. rapa*) from an Agricultural Perspective

CAROL NORRIS[1], JEREMY SWEET[1], JOHN PARKER[2] AND JOHN LAW[1]

[1]NIAB, Huntingdon Road, Cambridge CB3 0LE, UK; [2]Cambridge University Botanic Garden, Cory Lodge, Bateman Street, Cambridge CB2 1JF, UK; E-mail: carol.norris@niab.com

Abstract

Brassica rapa ssp. *rapa* (turnip) and *B. rapa* ssp. *oleifera* (turnip rape) are both grown as crop plants in many parts of the world. In addition, *B. rapa* ssp. *sylvestris* (wild turnip, bargeman's cabbage or navew) is found both in agricultural and non-agricultural environments as a weed and as feral populations, and is the most likely of the cruciferous species found in the UK to form hybrids with oilseed rape. It is also the species on which the largest number of interspecific hybridization studies have been conducted.

Extensive hybridization between *B. napus* and *B. rapa* was found under field conditions in the Humberside area of England. Several experiments were conducted on a field trial of genetically modified (GM), herbicide-tolerant (HT) oilseed rape in a field where *B. rapa* grew as a weed, to examine the extent of hybridization and possible introgression. Examination of seed harvested from *B. rapa* individuals growing as a weed in this field showed that between 0 and 43% of the progeny were hybrids. Soil samples also showed an established seed bank of hybrids. Preliminary investigations using amplified fragment length polymorphism (AFLP) analyses of seed bank and field samples indicated possible evidence of historical introgression. Results from these experiments and observation are used to consider the implications of growing GM HT oilseed rape and the way in which farming methods and environmental conditions could influence hybridization and introgression.

Introduction

Brassica rapa (wild turnip) is known to be a self-incompatible annual or biennial plant that is locally abundant on roadsides, in arable fields, on waste ground and particularly along riverbanks in Britain. It has a chromosome number of $2n = 20$ and is thought to be one of the progenitor species (*AA*) of oilseed rape (*Brassica napus AACC*) along with *B. oleracea* (*CC*). *B. napus* and *B. rapa* are sexually compatible and hybridize readily under certain conditions.

Hybridization frequencies between oilseed rape and wild/weedy *B. rapa* vary between studies and are much higher where *B. rapa* occurs as a weed in oilseed rape crops, rather than in its 'wild' habitat along riverbanks (Jørgensen and Andersen, 1994; Jørgensen *et al.*, 1996; Landbo *et al.*, 1996; Scott and Wilkinson, 1998). Studies of frequencies of spontaneous hybridization between oilseed rape and *B. rapa* in experimental fields in Denmark showed that the frequency of interspecific hybridization varied significantly with experimental design (Jørgensen *et al.*, 1996). Frequencies of hybrids in progeny ranged from 9% in a 1:1 mix of *B. rapa* and *B. napus* with oilseed rape as the mother plant, to 93% where the mother plants were single isolated *B. rapa* plants surrounded by oilseed rape. *B. rapa* is highly self-incompatible, and a single plant will accept pollen from other plants and even pollen from other closely related species more easily than its own. Therefore, interspecific pollination is more likely to occur where *B. rapa* is well spaced out in the field. Where *B. rapa* occurs in dense patches, hybrids are less likely to occur because of pollen competition.

Scott and Wilkinson (1998) measured gene flow from oilseed rape into natural populations of *B. rapa* growing outside field boundaries along the riverbanks of the Thames. A low level of hybridization was found (between 0.4 and 1.5%) in populations growing adjacent to oilseed rape fields. However, they found that less than 2% of the hybrid seedlings survived and no hybrid plants were found in the wild. Scott and Wilkinson (1998) therefore suggested that establishment of GM *B. napus* × *B. rapa* plants would be at very low frequencies and that introgression of transgenes from *B. napus* into wild *B. rapa* populations would be very unlikely or at most would be very slow.

Gene flow from oilseed rape into weedy *B. rapa* had not been examined in the UK until now. In other countries, hybridization studies between oilseed rape and weedy *B. rapa* have focused on the hybridization and introgression of transgenes into weedy *B. rapa* under controlled conditions, but have not examined the extent of historical introgression with conventional oilseed rape varieties under field conditions. However, some studies in Denmark recently have examined introgression from weedy oilseed rape into weedy *B. rapa* using amplified fragment length polymorphism (AFLP) markers (Hansen *et al.*, 2001). In this study, they screened weedy *B. rapa* and volunteer oilseed rape growing in an organic cereal crop with 24 species-specific AFLP markers. From 102 plants screened, 44 appeared to be introgressed beyond the F_1 generation.

Hybridization between oilseed rape and weedy *B. rapa* and the introgression of *B. napus* genes into *B. rapa* under agricultural conditions have been studied at a field site in Humberside in the UK. In addition, the individual varieties of oilseed rape grown in the field in the last 10 years have also been examined in preliminary molecular tests in order to give an indication of the likelihood of the long-term persistence of oilseed rape genes in weedy *B. rapa* populations.

Materials and Methods

Site description

B. rapa has occurred as a weed in the Humberside area of Yorkshire in the UK for over 30 years and, according to local farmers, has been growing in oilseed rape fields for at least 10 years. The site used in this study was an arable field at Patrington, North Humberside. The farm principally grows oilseed rape and cereal crops. Records have been kept of all varieties of oilseed rape grown in this field since 1988.

B. rapa plants at the site grow only within the agricultural field, not outside it. *B. rapa* is poorly controlled by the standard herbicides used on oilseed rape and appears in patches in the fields that remain in much the same place from year to year (B. Beeney, NIAB, 1998, personal communication). *B. rapa* is generally well controlled in cereal crops, which are usually grown for 2 years after a rape crop. Therefore, the weed only has the opportunity to reach maturity and produce seeds 1 year in every 4.

A trial area of glufosinate ammonium-tolerant transgenic winter oilseed rape was drilled in autumn 1998 in an area of the field where *B. rapa* is known to have been particularly abundant over the last 10 years.

Ploidy level testing by flow cytometry

Ploidy level testing by flow cytometry was used to confirm triploid hybrid ($2n = 29$), backcross to *B. napus* ($2n = 29–38$) and backcross to *B. rapa* ($2n = 20–29$) status of individual plants. Samples from the site were tested for DNA amount by Plant Cytometry Services, The Netherlands, according to the methods of Brown *et al.* (1991).

Flow cytometry results were presented as DNA ratios derived by the *Brassica* sample by the peak median of the standard peak (*Lactuca sativa* leaf material). The DNA ratio of a *B. rapa* plant gave values between 0.17 and 0.20, *B. napus* between 0.39 and 0.44, and hybrids between 0.29 and 0.33. This enabled *B. rapa*, *B. napus* and hybrid types to be easily distinguished, and the method was used as confirmation of both morphological identification and the identification of hybrids by herbicide spot testing (as in Norris and Sweet, 2002).

AFLP analysis

The AFLP technique (as in Norris and Sweet, 2002) was used to try to establish differences in band patterns between individual oilseed rape cultivars, and to identify whether specific varietal markers were present in the weedy *B. rapa* population, or in hybrids with *B. napus*.

Reference samples were: cultivars of *B. napus* which had previously been grown at the trial site (Apex, Cobra and Rocket), *B. rapa* (turnip rape variety Kova) and *B. oleracea* (kale, Brussels sprouts and cabbages). Field samples were taken from both mature plants and plants grown in the glasshouse from seeds derived from the soil seed bank. For the seed bank samples, 50 cores were taken from across the field and seeds grown as described. These were additional samples to those used in the seed bank analyses described below. Samples used for AFLP analyses from the soil cores were selected by their *B. rapa* morphology and by flow cytometry.

Soil sampling

Soil cores were taken using a soil gauge auger to investigate proportions of oilseed rape, *B. rapa* and hybrid seeds in the soil seed bank. Fifty soil cores (2.5 cm diameter, 35 cm depth) were taken from one fixed quadrat within each of six plots across the trial. Fixed quadrats were 3 m^2. The 50 cores from each plot were placed in plastic bags and weighed.

The bags of soil were mixed well and divided into trays. The soil was spread out, watered and placed in the glasshouse under clear plastic to prevent dehydration of the soil. Trays were kept moist for 10 days to allow seeds to germinate. Emerging seedlings were transplanted to pots, and the soil was mixed again to allow more seeds to germinate. Proportions of oilseed rape, *B. rapa* and hybrid seeds in the soil seed bank were then estimated from germinated seeds. Samples were taken on four occasions: February 1999 (before harvest of the GM crop), November 1999 (immediately after harvest of the GM crop), February 2000 (in the following wheat crop) and November 2000 (after harvest of the wheat crop). These sampling dates were chosen to establish any changes in the proportions of *B. rapa*, *B. napus* and hybrid seeds in the soil seed bank over time and in different crops. Estimates of numbers of seeds per m^2 were carried out according to methods used by P. Lutman (NIAB, 2000, personal communication).

Seed sampling from *B. napus* and *B. rapa*

B. rapa plants growing in the oilseed rape crop were selected by morphology and flow cytometry and tagged early in the growing season in November or December so that seed samples could be collected from them at maturity. Notes were made of the spatial distribution of the tagged plants in relation to other *B. rapa* plants and to the oilseed rape crop. All the mature seedpods were harvested from each selected plant. The seeds from each plant were germinated and the progeny examined for herbicide-tolerant hybrids by spraying with glufosinate ammonium solution.

All the seedpods from four *B. napus* plants growing in an area of the field densely infested by *B. rapa* were also sampled to look for reciprocal hybridization. The seeds from the individual plants were mixed together and 7512 seeds were germinated in the glasshouse. The seedlings were examined morphologically for hybrids as the maternal parent was glufosinate tolerant so hybrids could not be screened for by this method. Hybrids were identified as being a slightly paler green, less glaucous and hairier (both leaf and stem) than oilseed rape seedlings. One hundred oilseed rape seedlings from amongst these were randomly tested by flow cytometry to check that hybrids were not being missed by using morphology alone.

Postharvest sampling in 1999

The crop following the trial of GM oilseed rape was wheat. When the wheat had started to establish and before herbicide application, ten 1 m^2 quadrats were placed in areas where *B. rapa*, *B. napus* and putative hybrids were seen. Every *Brassica* plant was herbicide spot tested (as in Norris and Sweet, 2002) in each quadrat for tolerance to glufosinate. Leaf material from each plant was also sent for analysis by flow cytometry.

Ploidy analyses of the progeny of hybrid plants

Seed was collected from four susceptible *B. napus* × *B. rapa* hybrids whose identity had been checked by flow cytometry. This seed enabled an assessment to be made of the nature of the progeny produced by hybrids under natural conditions and the extent of backcrossing to *B. rapa*. The seeds were grown and spot tested for glufosinate tolerance to identify those pollinated by the GM *B. napus* (backcrosses to *B. napus*). Susceptible plants were then tested by flow cytometry to determine their DNA levels.

Results

Flow cytometry

This method was successful in confirming the identity of *B. rapa* individuals. However, backcross plants could not be identified easily by this method unless their DNA amount was different from that of either parent or the triploid hybrid. Triploid hybrids were easily distinguished by flow cytometry due to their DNA amount being intermediate between that of *B. rapa* and *B. napus*. Backcrosses with oilseed rape and hybrids resulting from unreduced gametes could also be identified by their DNA amount being higher than that of *B. napus*, although they could not be distinguished from each other.

AFLP

A preliminary study using the AFLP technique was used as a means of detecting possible introgression of *B. napus* genes into *B. rapa*. Different primer sets were examined to establish the most effective combination of primers for the species used. The primer set MSEI1 PSTI1 gave the most polymorphisms between oilseed rape varieties, although the three varieties examined had over 90% of bands in common of the 100 bands scored. Despite this, a small number of discriminatory bands between oilseed rape varieties were observed. In two of the oilseed rape varieties (Apex and Cobra), 5% of bands were polymorphic, while in Rocket 6% of bands differed from the other two oilseed rape varieties. Hybrids had bands in common with both parental species. In some cases, it was possible to identify oilseed rape-specific varietal bands in the hybrids. The *B. rapa* individuals were highly polymorphic, with 32% of bands polymorphic between only four individuals.

Two gels from sample sets using the primer set MSEI1 PSTI1 were informative. The first compared *B. rapa* (eight weedy samples and four turnip rape samples), *B. napus* (six samples), *B. oleracea* (six samples) and hybrids (two samples). When principal coordinate (PCO) plots were run using information from this gel, the weedy *B. rapa* and turnip rape formed two distinct but closely related groups. *B. napus* varieties grouped together while *B. rapa* × *B. napus* hybrids taken from the field fell between the parents, although closer to *B. napus* than to *B. rapa*. *B. oleracea* also gave a distinct group.

The second sample set was a mixture of weedy *B. rapa* field samples (15), *B. rapa* samples from the soil seed bank (ten), hybrids (two samples) and two backcrosses to *B. napus* identified by flow cytometry. Again, the species separated distinctly when analysed. However, two of the apparent weedy *B. rapa* plants from the seed bank fell close to the F_1 hybrid and backcross hybrids, which may indicate past hybridity. These two plants initially were identified by flow cytometry as having DNA amounts slightly higher than that of *B. rapa* (0.21).

Soil sampling

The 190 plants germinated from seed in soil cores from six plots appeared, from flow cytometry, to represent *B. napus*, *B. rapa* and their F_1 hybrid. No plants were identified as aneuploids from their DNA amounts in these samples, unlike the samples grown from soil cores used in the AFLP analyses where two plants were found to have DNA amounts higher than normal *B. rapa* but lower than a triploid hybrid. The number of oilseed rape seeds greatly increased in all cores following the oilseed rape harvest of 1999 as expected, but fell sharply by November 2000. The numbers of *B. rapa* seeds in the cores remained constant during the sampling period and were generally lower than oilseed rape. Hybrids occurred at a low frequency but apparently increased a little over time (Table 9.1). The proportion of hybrids ranged from 5.5 to 27.3, and apparently showed a marked increase over the sampling period. However, this is a reflection of the crash in oilseed rape abundance in November 2000.

Frequency of hybridization between *B. rapa* and *B. napus*

The number of seeds collected and tested from individual plants of *B. rapa* and *B. napus* varied according to maturity at the time of seed sampling. Hybridization frequencies in the field obtained from analysis of the progeny of individual *B. rapa* plants are shown in Table 9.2. The proximity of *B. rapa* plants to other plants did not appear to affect the extent of hybridization. Plants 1 and 2 produced no hybrids. Plant 3 had only one hybrid in a total of over 328 seeds tested. In contrast, almost half the seeds taken from plants 4 and 5 were hybrid (93/195).

Table 9.1. Numbers of *Brassica* seeds found in 50 soil cores in plots at the Patrington site sampled at four times: February 1999 (F 99); November 1999 (N 99); February 2000 (F 00) and November 2000 (N 00).

Plot	B. rapa				B. napus				F_1 hybrid			
	F 99	N 99	F 00	N 00	F 99	N 99	F 00	N 00	F 99	N 99	F 00	N 00
1	1	6	4	0	2	13	14	1	1	1	1	0
2	2	1	3	0	6	21	10	1	0	1	2	1
3	2	0	1	1	2	7	3	0	0	0	0	2
4	0	1	0	1	2	8	14	2	0	1	0	1
5	3	4	2	3	3	7	5	2	0	1	2	1
6	3	0	1	5	5	1	0	0	1	0	0	1
Totals	11	12	11	10	20	57	46	6	2	4	5	6

Table 9.2. Numbers of hybrid seedlings from seeds harvested from individual *B. rapa* plants in the field at Patrington.

Plant number	No. of progeny	No. of hybrids[a]	Proximity to other B. rapa plants
1	291	0	< 1 m
2	16	0	> 4 m
3	328	1	> 1 m
4	28	12	> 2 m
5	167	81	> 2 m

[a]Hybrids detected by their glufosinate tolerance.

A total of 7512 seeds were germinated from *B. napus* plants growing adjacent to *B. rapa*. No hybrids were identified on morphological criteria. Flow cytometry of 100 plants randomly chosen from amongst them similarly indicated no hybrids.

Post-GM harvest quadrat counts and hybridization

A total of 67 plants in ten quadrats germinated after the harvest of the GM oilseed rape crop, and after the following wheat crop was sown (Table 9.3). They consisted of 23 *B. napus*, 28 *B. rapa* and 16 hybrids as determined by flow cytometry. Glufosinate tolerance was detected in six *B. napus* plants and four hybrids.

Determination of DNA levels of hybrid progeny

The numbers of tolerant offspring amongst the progeny of susceptible F_1 hybrids varied between plants (Table 9.4). The proximity of the sampled F_1 hybrid plant to *B. rapa* individuals did not appear to influence the level of its pollination by transgenic *B. napus*. Indeed, the hybrid that was furthest away from a *B. rapa* plant (> 10 m) carried a lower proportion of tolerant offspring than the other hybrids. The highest proportion of tolerant progeny (52%) was produced by a hybrid only 2 m away from the nearest *B. rapa* plant. A range of DNA ratios was found amongst 13 glufosinate-susceptible offspring of a *B. napus* × *B. rapa* hybrid, indicating backcrossing to both parents (Fig. 9.1). Eleven of the 13 fell between *B. rapa* and *B. napus*. The remaining two had DNA ratios higher than *B. napus* but not as high as would be expected from unreduced gametes on the F_1 hybrid pollinated by reduced *B. napus*.

Table 9.3. Quadrat counts and herbicide testing on postharvest emerging *Brassica* plants at Patrington, November 1999.

Quadrat (1 m²)	No. of plants tested	No. of *B. napus*	No. of *B. rapa*	No. of hybrids
A	6	2	4	0
B	8	6 (2)	1	1
C	4	0	3	1 (1)
D	7	7 (3)	0	0
E	7	2 (1)	1	4
F	6	0	3	3
G	8	4	4	0
H	4	1	2	1
I	9	1	5	3 (2)
J	8	0	5	3 (1)
Total	67	23 (6)	28 (0)	16 (4)

Numbers in parentheses are glufosinate tolerant.

Table 9.4.   Frequencies of glufosinate-tolerant seeds harvested from hybrid plants in the field.

Plant number	Proximity to other B. rapa plants (m)	No. of progeny tested	Numbers tolerant (%)
1	< 1	221	103 (46.6)
2	2	66	34 (51.5)
3	> 10	29	8 (27.5)
4	1	119	37 (31.1)

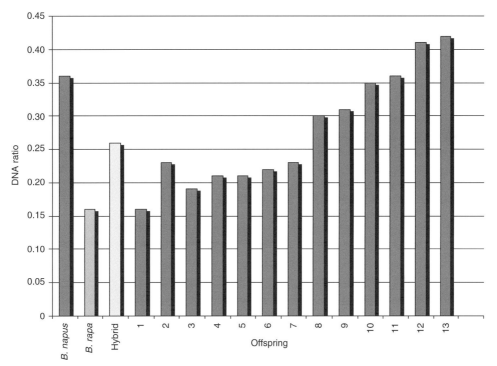

Fig. 9.1.   DNA ratios of glufosinate-susceptible offspring of a *B. napus* × *B. rapa* hybrid (plant 1) under field conditions from Patrington.

Offspring 1 had a DNA amount equal to that of *B. rapa*, while 10 and 11 had DNA ratios similar to *B. napus*. The plants varied greatly in morphology, and did not thrive under glasshouse conditions. Four survived to flowering (1, 2, 10 and 11), but the remainder died before reaching maturity.

Observations of the *B. rapa* population examined, and crop–weed interactions

The GM oilseed rape examined in this study was surrounded by 'set aside' land in the growing year. The weedy *B. rapa* population was unevenly distributed over the whole area, with only a few plants in the uncultivated set aside. The population was patchy, with densely grouped plants in a strip approximately 10 m × 50 m across the field, and was particularly concentrated in one area of the field. The population size was estimated at approximately 500 plants. In some parts of the field, there were isolated *B. rapa* individuals which were separated from others of the same species by more than 10 m.

Most of the *B. rapa* germinated at the same time as the oilseed rape. Germination tended to be quite even, with most plants germinating simultaneously, rather than being staggered over a period of months. However, a few individuals germinated later and were often at a disadvantage as they were by then over-shadowed by the developing crop canopy.

Flowering of *B. rapa* started before the stem extension of oilseed rape. The first individuals started to flower in early February, and flowering of some plants continued until the oilseed rape crop had finished flowering. Many *B. rapa* individuals had almost finished flowering when the rape started to flower in April/May. However, a large proportion of the *B. rapa* population had an overlap with the rape in their flowering duration.

A large proportion of the *B. rapa* individuals matured and shed seed 2–3 weeks earlier than the oilseed rape crop, so, by the time the crop was harvested, most *B. rapa* seed was already on the ground and escaped being harvested with the crop.

Interspecific hybrids also germinated at the same time as the oilseed rape. They were distinguishable by the two- or three-leaf stage, being intermediate in morphology between the two parental species, although more recently we have found hybrids with identical morphology to oilseed rape. Stem extension and flowering occurred later than in *B. rapa* and often at the same time as oilseed rape. The hybrid plants grew as tall as, or sometimes taller than, the rape. Anthers on hybrid plants were sometimes reduced in size or completely absent, and the quantities of pollen produced by these plants were visually lower than that of either parent. Most hybrids, however, were not generally completely male sterile. Seedpods on hybrids were often empty or contained very few seeds. Many of the seeds were aborted, shrivelled or malformed, although some filled, round seeds were formed and these were larger than those of either parent. Also, seeds produced on hybrid plants often germinated within the pod. A characteristic of hybrid plants at a late stage of development was their tendency to grow new flowering shoots after pods had been formed and the rest of the plant was starting to senesce. This did not occur in either parent, but is a common feature of near-sterile hybrids in many species (Stace, 1975).

Discussion

Extensive hybridization between oilseed rape and *B. rapa* occurs at the Patrington site. This finding supports previous work carried out on weedy

B. rapa in oilseed rape fields in Denmark (Jørgensen and Andersen, 1994; Jørgensen *et al.*, 1996; Landbo *et al.*, 1996).

The spatial distribution and growth habit of *B. rapa* at the Patrington site offers an explanation for the high frequency of hybridization. *B. rapa* is a mostly self-incompatible species, so very high proportions of interspecific hybrid seed are likely to be set on *B. rapa* mother plants when individuals are isolated from sources of intraspecific pollen and surrounded by oilseed rape. It has also been shown that when *B. rapa* and *B. napus* pollen are applied to *B. rapa* stigmas, they have equal fitness and so are equally likely to fertilize *B. rapa* (Hauser *et al.*, 1997). The spatial distribution of *B. rapa* was patchy at Patrington, with many isolated plants surrounded by *B. napus* pollinators. Thus, *B. rapa* pollen would have been under a great deal of competition from oilseed rape pollen in these fields.

Flowering time will have an influence on the probability of hybridization between *B. rapa* and *B. napus*. Observations of *B. rapa* populations in Humberside showed that when growing amongst winter oilseed rape, *B. rapa* started to flower earlier than the crop. Some individuals finished flowering before the oilseed rape started. Flowering of a proportion of the *B. rapa* then continued throughout the period of the crop flowering, so there was some overlap. At Patrington, the majority of the *B. rapa* population flowered synchronously with the oilseed rape.

It is likely then that the early flowerers maintain a discrete *B. rapa* population despite a high degree of hybridization in the population, and this would account for the appearance of the population in the same position in the field over many years.

Differences in dormancy and germination patterns between *B. rapa* and *B. napus* may have an effect on emergence and growth of interspecific hybrids. *B. rapa* produces seeds with many and varying germination requirements. Seeds from the same mother plant can have different levels of dormancy, as can seeds from different plants in the same population. This contributes to making *B. rapa* a persistent weed in disturbed habitats by increasing the probability of germination at a favourable time in an unpredictable environment (Rees and Long, 1992).

The ability of annual plants to survive in the environment is affected by certain environmental conditions and cues, and also by seed dormancy. Cold stratification in *B. rapa* has been found to induce light sensitivity and sensitize seeds to nutrient levels (Adler *et al.*, 1993). Adler *et al.* (1993) suggested that *B. rapa* seeds that have dispersed in the autumn will be subjected to cold stratification over the winter and are likely to germinate in the spring. This is in contrast to the findings at Patrington where most *B. rapa* germinated in the autumn at the same time as the crop. Late germinating plants were at a disadvantage due to light competition from the oilseed rape canopy.

B. rapa seeds are more likely to germinate at the same time as the crop seeds when the soil is fertile and cultivated, and crop cover is sparse. There may well be selection pressure for plants in the *B. rapa* population to germinate with the crop. This may be followed, however, by selection pressure against this type of germination, if herbicide is applied to the crop soon after emergence. Although conventional selective herbicides are not fully effective against weeds such as *B. rapa*, treated plants will nevertheless be affected in their development and may well be stunted.

B. napus seed at harvest does not show tendencies toward dormancy, and its seeds will germinate under most conditions. At harvest, however, certain environmental conditions may induce secondary dormancy. *B. napus* seeds regularly survive burial in the soil for 2–3 years (Pekrun and Lutman, 1998) and sometimes even up to 10 years, depending on the agricultural practice and whether they remain undisturbed. *B. napus* seeds tend to germinate after being brought to the surface by cultivation, but will remain dormant and viable if left buried.

Hybrids between *B. napus* and *B. rapa* have been shown to inherit the dormancy characteristics of the maternal parent. This could be due to seed coat influences that are maternally inherited (Mayer and Poljakoff-Mayber, 1989), but organelle inheritance and endosperm formation are also inherited unequally. However, there may also be some zygotic influence on dormancy (Adler *et al.*, 1993) as the hybrid dormancy characteristics are closer to those of the maternal parent, but are not identical to it. Adler *et al.* (1993) reported that the hybrid produced when *B. napus* is the maternal parent is slightly less dormant than the hybrid produced when *B. napus* is the pollen donor. The parentage of the hybrids found in the field at Patrington was unknown. The lower degree of dormancy of hybrid seeds compared with the weedy *B. rapa* parent could reduce the number of hybrid plants reaching the adult stage as they will predominantly be shed in disturbed habitats where the crop parent is found. The majority of hybrid seeds will germinate straight away; however, the staggered germination pattern of *B. rapa* allows it to germinate in years when conditions are optimum for its growth. This could allow repeated backcrossing to wild relatives many years after the initial introduction of a transgene (Linder and Schmitt, 1994, 1995).

If the hybrid seeds germinate immediately after harvest, as do a large proportion of shed *B. napus* seeds, then the normal agricultural practice of leaving seeds to chit for a minimum of 2 weeks before cultivation will ensure that the majority of hybrid plants do not survive. The uniformity of their germination requirement could reduce their fitness under field conditions. Alternatively, interspecific hybrids growing in agricultural conditions may maintain growth stages similar to the crop plant, thereby flowering and producing seed at similar times. F_1 interspecific hybrids vary considerably in their fitnesses, depending on parental genotypes (Hauser *et al.*, 1998a).

For long-term survival, hybrid seeds must remain dormant until suitable conditions for *B. rapa* and its hybrids occur in the agricultural field when *B. napus* is grown. In normal agricultural rotations in the UK, this is likely to be every 4–5 years. Hybrid plants emerging in interim years will usually be growing amongst cereal crops and exposed to herbicides used to eradicate them. Then few, if any, hybrids will reach maturity in years when cereals are grown. Seeds germinated from the soil seed bank when soil cores were taken at Patrington show that *B. napus* tends to germinate in a flush after the input of seeds from a crop. However, numbers of seeds in the soil then reduce dramatically in the following years until the next crop of oilseed rape. Numbers of *B. rapa* and hybrids, however, remained fairly constant over the sampling period, and no rapid reduction of their seeds in the seed bank was observed. Hence, the proportion of both hybrids and *B. rapa* in the total *Brassica* seed population increased with time.

The first backcross generation to *B. rapa* produces seeds with a more *rapa*-like germination pattern. Amongst the seeds produced, there will be differing germination requirements and some seeds will exhibit dormancy. Interspecific gene flow in the first backcross generation would therefore not be limited by seed germination characteristics to the same degree as that in F_1 hybrids.

In a study where reciprocal crosses were made (Hauser *et al.*, 1997), those with *B. napus* as the female parent were found to produce more viable seeds than when *B. rapa* was the female parent. This is contrary to the results found when 7512 *B. napus* seeds from plants at Patrington were screened for hybridity. No hybrids were identified amongst these *B. napus* parents growing adjacent to *B. rapa*. It is a possibility, however, that hybrids with morphology close to that of oilseed rape may not have been identified. The lack of hybrids detected on *B. napus* as the female parent is most easily explained by the extremely small amount of *B. rapa* pollen available locally compared with *B. napus* pollen.

The triploid (*AAC*) F_1 hybrid with $2n = 29$ often shows sterility or reduced fertility. Interestingly, it was found that at the Patrington site, hybrids formed from unreduced gametes of *B. rapa* are a common occurrence. These will have the genomic constitution *AAAC* ($2n = 39$). A selection of plants with DNA ratios between 0.36 and 0.39 were found from seeds taken from a *B. rapa* mother, which is higher than expected for a triploid hybrid. Some of these expressed tolerance to glufosinate, showing that they were pollinated by *B. napus* and were therefore a form of hybrid. These plants were not grown to maturity; therefore, there is no information about the comparative fertility of the two types of hybrid. It is possible, however, that some of the *AAAC* hybrids exist in the field as they appear to be commonly present as seed. However, they would be difficult to distinguish by flow cytometry, molecular markers or morphology from backcrosses to *B. napus*. Chromosome analysis would be required to assess genomic constitution. Plants that were thought to be backcrosses to *B. napus* by their flow cytometry peaks group together with the F_1 triploid hybrids when analysed by AFLP. This may indicate that these arose from F_1 hybrids with unreduced gametes.

Remarkably, F_1 hybrids were found to be fertile under field conditions, although seed production was low compared with that of either parent species. Thus, F_1 hybrids may provide an avenue for introgression to take place. Pollen production from hybrids was not measured, but is probably low due to the small anthers observed. The growth of new flowering shoots on hybrid plants after pod formation also indicates low fertility. In hybrids that are sterile or with low fertility, few seeds are produced, so energy resources are directed to continuing flower production to compensate. Many of the seeds produced on F_1 hybrid plants were inviable, had shrivelled or had germinated inside the pods. Hauser and Ostergard (1999) also reported germination of F_1 and F_2 seeds within pods, which may be an important loss of hybrid seeds. Hauser et al. (1998a) examined fitness of F_1 hybrids between weedy B. rapa and B. napus and remarkably found that seed set in hybrids was intermediate between its two parents, with B. rapa producing fewer seeds per plant than F_1 hybrids. This result is very different from that observed in the Patrington population where hybrids produced many fewer seeds per plant than B. rapa.

Backcrosses of F_1 hybrids to oilseed rape plants were also found in the field, but were less easy than the F_1 hybrids to identify by morphology alone as they appear very similar to oilseed rape. Backcrossing to B. rapa is likely to occur only in areas of the field where there are F_1 hybrids and are densely populated by B. rapa.

Flow cytometry testing of 12 susceptible offspring from a hybrid parent showed a remarkable range of DNA ratios. Flowers on the hybrid parent could have been pollinated by B. napus, B. rapa or another hybrid, or could have been selfed. Offspring with DNA ratios between those of B. rapa and the hybrid are likely to have been pollinated by B. rapa – the first backcross generation. Seven plants fell into this category. One plant had a DNA ratio indistinguishable from B. rapa and, in a field situation, would be impossible to identify as a backcross plant due to its B. rapa morphology.

Two plants grown from the soil samples used for AFLP analyses at the Patrington site, were shown by DNA ratio analysis to be backcrosses to B. rapa. However, plants with intermediate DNA levels between those of B. rapa and hybrids were not found growing in the field, which indicates that aneuploids generally do not survive to produce viable plants under natural conditions.

Backcrosses to B. rapa can have the same, or very similar, chromosome numbers as B. rapa, so these may not be identified by flow cytometry or by morphology. It is likely that individuals with a chromosome number close to that of B. rapa will be the most stable of the range of chromosome complements produced from backcrossing hybrids with B. rapa. They may occur at a higher frequency than other aneuploids due to selection taking place under field conditions. Hauser et al. (1998b) mentioned that backcross and F_2 plants are often aneuploid with unbalanced C chromosomes (backcrosses, $AA + 0–9$ C; F_2, $AA + 0–18$ C) and this may seriously reduce their fitness. This was supported by the fact that aneuploids produced from hybrid mother plants at Patrington grew poorly under glasshouse conditions.

The proposed reduced fitness of backcross plants is supported by the fact that no backcrosses to B. rapa individuals with DNA amounts between those of B. rapa and hybrid were found as plants growing in the field, but when progeny from known hybrids were grown under glasshouse conditions several were identified. For this reason, numbers of backcrosses to B. rapa plants identified in the field may be under-estimated, as those with DNA amounts very similar to B. rapa are indistinguishable from B. rapa. Some plants were identified by morphology and flow cytometry as being

B. rapa, but cluster together with F_1 hybrids. This suggests that these plants are either first-generation backcrosses to *B. rapa*, or later generation introgressed individuals. A larger sample size is needed to make firm conclusions about introgression taking place in this population. However, the evidence here from AFLP, hybrid progeny testing and field testing points towards the presence of an introgressing population.

The results of the field herbicide spot testing and the soil sampling show that the population found growing each year is only a snapshot of the true population in the seed bank. Only a proportion of the F_1 hybrids found were herbicide tolerant, suggesting that tolerant hybrids could still remain in the seed bank. Some of the non-tolerant hybrids may have arisen from hybridization in previous oilseed rape crops several years before. This means that introgression may occur over a long period of time as oilseed rape crops are generally only grown on the same field every 3 or 4 years. Hybridization and backcrossing between the two species can only take place in the years when oilseed rape is grown, as broad-leafed weeds such as *B. rapa* are easily killed by the herbicides applied to cereal crops. Thus, the long-term consequences of introducing a transgene into a weed population are as yet unknown, but these studies suggest they could be dramatic.

Familiarity with crop–weed interactions has enabled predictions to be made about the fate of the herbicide tolerance transgene under field conditions. Herbicide-tolerant hybrids have already been recorded post-harvest at the Patrington site, after a single GM crop. Will the next stage of introgression (backcrossing to the *B. rapa* parent) take place? The work discussed here took place mostly over one growing season, and the results have given an insight into the complex relationships between the *B. rapa* weed population and the oilseed rape crop. At the Patrington site, all possible combinations of crosses probably occurred in the field, whether at the seed or the mature plant stage. It has been impossible to distinguish the genomic make-up and parental origin of each individual without detailed molecular analyses and chromosome study.

The offspring from any of these F_1 and backcross hybrids could potentially cross with any of the others: the number of combinations of possible crosses is thus enormous. The results of this study have categorized the plants into major classes only – the two parental species, F_1 hybrids and backcrosses. What is more important than the exact identification of each cross, perhaps, is the discovery of extensive hybridization here, and the evidence of backcrossing in both directions. Thus a transgene is likely to persist past the initial F_1 hybridization and will move into further generations in an unpredictable way. In addition, the way in which the transgene will move between generations may also be determined by the fitness incurred on the plant by the trait it confers.

In an agricultural environment, the nature and extent of introgression will depend very much upon the type of cropping and the crop management. If weed management of cruciferous weeds is effective, there may remarkably be a higher potential for hybridization between *B. rapa* and oilseed rape, since individual *B. rapa* plants may escape herbicide treatment and thus become isolated within the crop. If weed management is poor, more *B. rapa* plants will be left in the field, and perhaps there will be less hybridization. For high levels of backcrossing to occur, the opposite scenario will apply. Backcrossing to *B. rapa* is more likely when hybrids are present in the field and there is an abundance of *B. rapa* plants (when weed management is poor).

The results presented here have shown that introgression between *B. napus* and *B. rapa* is likely to have already been occurring at the Patrington site for many years. The introduction of a transgenic oilseed rape to the site has allowed the extent of initial hybridization to be determined, and will enable further studies to be carried out on backcrossing events in the future. However, these data are from only one population of *B. rapa* and occupy only 2 years, so they cannot be deemed to be representative of every population. Further sites need to be studied over several years in order to gain more information about the nature and extent of introgression of oilseed rape genes into *B. rapa*.

References

Adler, L.S., Wikler, K., Wyndham, F.S., Linder, C.R. and Schmitt, J. (1993) Potential for persistence of genes escaped from canola: germination cues in crop, wild, and crop–wild hybrid *Brassica rapa*. *Functional Ecology* 7, 736–745.

Brown, S.C., Bergounioux, C., Tallet, S. and Marie, D. (1991) Flow cytometry of nuclei for ploidy and cell cycle analysis. In: Negruti, I. and Gharti-Chhetri, G. (eds) *A Laboratory Guide for Cellular and Molecular Plant Biology*. Springer-Verlag, Berlin, pp. 326–345.

Hansen, L.B., Siegismund, H.R. and Jørgensen, R.B. (2001) Introgression between oilseed rape (*Brassica napus*) and its weedy relative *B. rapa* in a natural population. *Genetic Resources and Crop Evolution* 48, 621–627.

Hauser, T.P. and Østergard, H. (1999) Precocious germination of *Brassica rapa* × *B. napus* seeds within pods. *Hereditas* 130, 89–93.

Hauser, T.P., Jørgensen, R.B. and Østergard, H. (1997) Preferential exclusion of hybrids in mixed pollinations between oilseed rape (*Brassica napus*) and weedy *B. campestris* (Brassicaceae). *American Journal of Botany* 84, 756–762.

Hauser, T.P., Jørgensen, R.B. and Østergard, H. (1998a) Fitness of F_1 hybrids between weedy *Brassica rapa* and oilseed rape (*B. napus*). *Heredity* 81, 429–435.

Hauser, T.P., Jørgensen, R.B. and Østergard, H. (1998b) Fitness of backcross and F_2 hybrids between weedy *Brassica rapa* and oilseed rape (*B. napus*). *Heredity* 81, 436–443.

Jørgensen, R.B. and Andersen, B. (1994) Spontaneous hybridisation between oilseed rape (*Brassica napus*) and weedy campestris (Brassicaceae): a risk of growing genetically modified oilseed rape. *American Journal of Botany* 81, 1620–1626.

Jørgensen, R.B., Andersen, B., Landbo, L. and Mikkelson, T.R. (1996) Spontaneous hybridisation between oilseed rape (*Brassica napus*) and weedy relatives. *Acta Horticulturae* 407, 193–199.

Landbo, L., Andersen, B. and Jørgensen, R.B. (1996) Natural hybridisation between oilseed rape and a wild relative: hybrids among seeds from weedy *B. campestris*. *Hereditas* 125, 89–91.

Linder, C.R. and Schmitt, J. (1994) Assessing the risks of transgene escape through time and crop–wild hybrid persistence. *Molecular Ecology* 3, 23–30.

Linder, C.R. and Schmitt, J. (1995) Potential persistence of escaped transgenes: performance of transgenic, oil-modified *Brassica* seeds and seedlings. *Ecological Applications* 5, 1056–1068.

Mayer, A.M. and Poljakoff-Mayber, A. (1989) *The Germination of Seeds*. Pergamon Press, Oxford.

Norris, C.E. and Sweet, J.B.S. (2002) Monitoring large scale releases of genetically modified crops (epg 1/5/84) incorporating report on project epg 1/5/30: monitoring releases of genetically modified crop plants. Available at: http://www.defra.gov.uk/environment/gm/research/epg-1-5-84.htm

Pekrun, C. and Lutman, P.J.W. (1998) The influence of post-harvest cultivation on the persistence of volunteer oilseed rape. *Aspects of Applied Biology* 51, 113–119.

Rees, M. and Long, M.J. (1992) Germination biology and the ecology of annual plants. *American Naturalist* 139, 484–508.

Scott, S.E. and Wilkinson, M.J. (1998) Transgene risk is low. *Nature* 393, 320.

Stace, C.A. (1975) *Hybridisation and the Flora of the British Isles*. Academic Press, London.

10 Asymmetric Gene Flow and Introgression Between Domesticated and Wild Populations

ROBERTO PAPA[1] AND PAUL GEPTS[2]

[1]Dipartimento di Biotecnologie Agrarie ed Ambientali, Università Politecnica delle Marche, Ancona, Italy; [2]Department of Agronomy and Range Science, University of California, Davis, California, USA; E-mail: rpapa@univpm.it

Abstract

Gene flow may have very different consequences for the structure of the genetic diversity of populations depending on its direction (symmetric versus asymmetric). If gene flow is symmetric between two or more populations, its effect will be the reduction of genetic diversity between the populations and an increase in the differences between individuals within each of the populations considered. Alternatively, if gene flow is much higher from one population (source) to another (recipient) than in the reverse direction, the long-term consequence will be the displacement of alleles of the recipient population with the replacement by alleles of the source population, unless there is strong selection against source alleles.

Recently, we have used amplified fragment length polymorphism (AFLP) markers to study introgression between wild and domesticated common beans (*Phaseolus vulgaris* L.) in Mesoamerica. We have shown by both phenetic and admixture population analysis that gene flow is about three- to fourfold higher from domesticated to wild populations than in the reverse direction.

In this work, we review the results obtained in *P. vulgaris* and we compare them with studies on other species, to understand to what extent asymmetric gene flow is a general phenomenon in the wild–domesticated context, and to investigate the main factors involved.

Even if the information available on other crops is insufficient to determine whether asymmetric gene flow and introgression from crops into their wild relatives is a general phenomenon, and whether it may also displace the genetic diversity of the wild population in specific genomic regions, as seems to be the case in the common bean, the data available on genetic differentiation between wild and domesticated populations, the inheritance of the domestication syndrome and the demography of the two types of populations generally suggest that asymmetric gene flow and introgression is the most likely hypothesis to explain the observed patterns of genetic diversity in wild and domesticated populations also in the other crops considered.

Introduction

Gene flow between wild and domesticated populations has been the subject of extensive investigations, particularly over the last decade, due to the release of transgenic varieties and their possible impact on genetic diversity of wild relatives (Hails, 2000; Gepts and Papa, 2003; Jenczewski *et al.*, 2003). Several studies have focused on the introgression of genes between wild and domesticated gene pools (for reviews, see Ellstrand *et al.*, 1999; Jarvis and Hodgkin, 1999). With very few exceptions, gene flow between wild and domesticated populations has been shown to be important in many different species, both allogamous (predominantly cross-pollinated) and autogamous (predominantly self-pollinated). Gene flow may have very different consequences for the structure of the genetic diversity of populations, depending on its direction (symmetric versus asymmetric). If gene flow is symmetric between two or more populations (i.e. the island model; Wright, 1931), its effect will be the reduction of genetic diversity between the populations and an increase in the differences between individuals within each of the populations considered. Alternatively, if gene flow is much higher from one population (source) to another (recipient) than in the reverse direction, the long-term consequence will be the displacement of alleles of the recipient population with the replacement by alleles of the source population, unless there is strong selection against source alleles (see below).

This model of migration is the so-called mainland–island model (Harrison, 1991), one-way migration model or unidirectional gene flow model. It originally was based on the presence of populations (source/mainland and sink/island) of different population sizes. In addition to demographic factors, one-way migration may be related to pre-zygotic (differential pollen migration; Taylor, 1995) and post-zygotic (e.g. epistatic hybrid inviability; Servedio and Kirkpatrick, 1997) unilateral reproductive barriers. The one-way migration model does not necessarily imply that migration is present just in one direction, but rather that one direction is largely prevalent over the other.

The directionality of introgression can also be promoted by selection when its intensity against exotic alleles is different in the two populations. In crop and wild populations, different types of selection are likely to be present for two different reasons.

1.   With few exceptions (Burke *et al.*, 2002), the genes of the domestication syndrome (Hammer, 1984) are characterized by complete or incomplete dominance (Doebley *et al.*, 1990; Koinange *et al.*, 1996; Xiong *et al.*, 1999; Poncet *et al.*, 2000), where the wild alleles are dominant while those domesticated are recessive. Thus, the first-generation hybrids will be much more similar to the wild than to the domesticated parent, leading to selection

being more effective against hybrids in the cultivated than in the natural environment.

2. In the cultivated environment, there is also a conscious, direct selection operated by farmers, in addition to that already imposed by the agroecosystem itself.

Recently, we have used amplified fragment length polymorphism (AFLP) markers to study introgression between wild and domesticated common beans (*Phaseolus vulgaris* L.) in Mesoamerica. Gene flow occurred principally in close-range sympatry, i.e. when two populations grew in close proximity. We have shown, by both phenetic and admixture population analysis, that gene flow is about three- to fourfold higher from domesticated to wild populations than in the reverse direction (Papa and Gepts, 2003).

The aim of this work is to review the results obtained in *P. vulgaris* in comparison with studies on other species, to understand to what extent asymmetric gene flow is a general phenomenon in the wild–domesticated context and to investigate the main factors involved.

Phaseolus vulgaris

In Papa and Gepts (2003), we described an analysis of gene flow between wild and domesticated populations of *P. vulgaris* ($2n = 2x = 22$) from Mesoamerica using AFLP markers. *P. vulgaris* is an annual autogamous species (Ibarra-Pérez *et al.*, 1997) and presents two centres of domestication (reviewed in Gepts, 1993), one in Mesoamerica and one in the Andes. In Mexico, wild and domesticated forms are sympatric in many locations of the country. Previous analyses have shown that the Mesoamerican domestication centre was most probably located in what is now the state of Jalisco and the neighbouring state of Guanajuato (Gepts, 1988). Thus, wild populations outside the area just mentioned do not constitute immediate progenitors of the sympatric domesticated populations. Landraces are still widespread in Mexico, and their differentiation is related mainly to seed colour and shape; for this reason, they represent the most important traits for which farmers select the seeds for the new planting season.

We analysed two samples. The first consisted of 24 domesticated accessions, 52 wild and ten weedy populations from Mesoamerica. The UPGMA (unweighted pair group method with arithmetic mean) dendrogram shows two main groups (Fig. 10.1), one with only wild accessions that appear to be distributed according to their geographical origins, and a second with all the domesticated types, all the weedy types, and 17 wild accessions that did not cluster according to their geographical origins. Moreover, the domesticated accessions are nested inside this second cluster, along with most of the weedy and four of the wild types. A similar structure was shown by the analysis of the second sample (overall, 382 individuals collected on a single plant basis from 31 populations), based on a recent collection of populations mainly in the state of Chiapas (Fig. 10.2). In this case, two groups are also

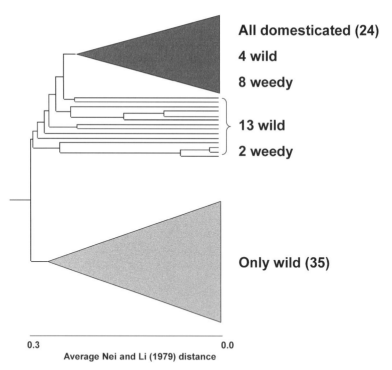

Fig. 10.1. UPGMA dendrogram of wild, weedy and domesticated accessions from Mesoamerica based on Nei and Li (1979) genetic distances between genotypes.

present, one with only wild populations, clustering according to their collection areas, and the other with all the domesticated and weedy populations. In both samples, we observe wild and weedy populations in the cluster of the domesticated, but never the opposite. This indicates that gene flow predominates from domesticated to wild populations. However, it does not appear to be very important in shaping the genetic structure of domesticated populations and may not be responsible for the racial differentiation observed in the domesticated Mesoamerican gene pool (Singh *et al.*, 1991). A possible alternative explanation is that the wild and weedy populations placed in the domesticated cluster are of direct descent from the wild populations from which domestication took place. However, this hypothesis is not consistent with the absence of any geographical structure in these groups of accessions. Earlier data have suggested that common bean domestication in Mesoamerica may have taken place in what is now the state of Jalisco and the neighbouring state of Guanajuato (Gepts, 1988).

From the analysis of the population structures of wild and domesticated populations, we have observed that gene flow between the wild and domesticated common bean is mainly present in sympatry, when the wild plants are also present within or immediately around the farmer's field. A lack of gene flow is apparent in allopatry and even in parapatry, here defined as a separation from just a few metres to 1 km in distinct spatial

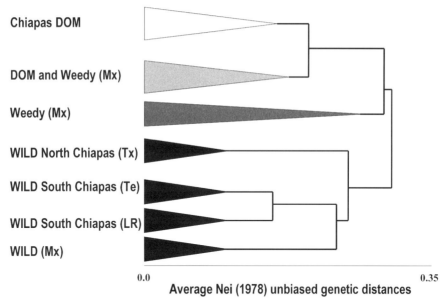

Chiapas DOM

DOM and Weedy (Mx)

Weedy (Mx)

WILD North Chiapas (Tx)

WILD South Chiapas (Te)

WILD South Chiapas (LR)

WILD (Mx)

0.0   0.35
Average Nei (1978) unbiased genetic distances

Fig. 10.2. UPGMA dendrogram of wild, weedy and domesticated (DOM) populations from Chiapas and others states of Mexico (Mx) based on Nei (1978) unbiased genetic distances between populations. Populations Te, LR and Tx indicate different collection areas from Chiapas, Mexico: Teopisca, Las Rosas and Tuxtla-Gutiérrez, respectively.

arrangements. For this reason, we assumed that sympatric populations are an admixture resulting from gene flow between two (one wild and one domesticated) isolated wild and domesticated populations. This assumption is also supported by the significantly higher genetic diversity present in the sympatric populations, as compared with both the allopatric and parapatric populations. Based on these considerations, we have used three admixture estimators on the assumption that the admixed population frequencies should be intermediate between the two parental populations (Cavalli-Sforza and Bodmer, 1971; Bertorelle and Excoffier, 1998): mR (Roberts and Hiorns, 1965), mC (Chakraborty *et al.*, 1992) and mY (Bertorelle and Excoffier, 1998). All the three estimators gave similar results, showing that the relative contribution of the domesticated parental population to the admixture wild population was three- to fourfold higher than the estimated contribution of the parental wild population. Using the same data, we show a genotype assignment analysis using a model-based clustering method implemented in the software STRUCTURE (Pritchard *et al.*, 2000) (Fig. 10.3). This analysis confirms that gene flow and introgression are asymmetric, being higher from the domesticated to the wild population than in the reverse direction. Some possible explanations of these observations are: (i) different dispersal ability between the domesticated and wild plants; (ii) different population sizes, as in the mainland–island model of migration; (iii) farmer selection against the first-generation hybrids, which are showing

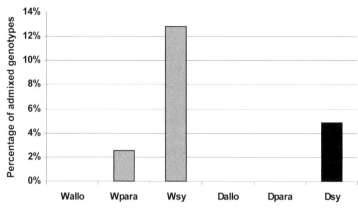

Fig. 10.3. Genotype assignment in Chiapas populations (sympatric (sy), parapatric (para) and allopatric (allo)) by a model-based clustering method implemented in the STRUCTURE software of Pritchard *et al.* (2000).Two populations were assumed: wild (W) and domesticated (D).

the wild phenotype; and (iv) different types of selection in the domesticated and wild environments.

Considering that there is no evidence that suggests a possibly higher pollen dispersal ability of domesticated versus wild populations, the results obtained for *P. vulgaris* may be explained by the combination of different population sizes and by farmer selection against F_1 hybrids. The observation that gene flow is mostly occurring only when wild plants are present within the field supports this hypothesis. Indeed, the number of domesticated plants that are present in a farmer's field is much higher than that of the wild plants. Therefore, the different population sizes are sufficient to explain the asymmetry of gene flow. The effects of different population sizes will also be enhanced by selection operated by the farmer for seed colour, shape and size. Because of the genetic control of these traits (wild alleles are dominant or partially dominant), the F_1 hybrids will be phenotypically closer to the wild parent rather than to the domesticated one and will thus be easily eliminated by the farmer. This type of selection is operating at the whole-genome level in the F_1 generation, like a post-zygotic reproductive barrier, thus preventing the recombination between gametes donated by the wild and domesticated parents. Considering only these hypotheses (population size, selection against F_1 hybrids in domesticated populations), we would expect that introgression will be homogeneous in the various parts of the genome and that its level will depend on the rate of migration. Heterogeneous introgression along the genome could be promoted only by the effect of selection. If selection favours the same exotic alleles in different populations (homogeneous selection), it will increase their frequency in these populations much more than would be expected by the sole effect of migration. Alternatively, as is the case of domestication genes distinguishing wild and domesticated populations, selection can favour alternative alleles in different environments (heterogeneous selection, e.g. wild versus culti-vated). In this case, the effect of heterogeneous selection will contrast with the effect of migration, reducing the introgression of exotic alleles from other populations. The level of differentiation between populations measured as F_{ST} is, under the island model, directly related to the average number of migrants per generation (N_m) by the relationship $F_{ST} = 1/(4N_m + 1)$. This relationship is valid if we consider neutral loci (absence of selection). In the case of genes that undergo homogeneous selection, we expect $F_{ST} < 1/(4N_m + 1)$, with $F_{ST} > 1/(4N_m + 1)$ if selection is heterogeneous. On the other hand, neutral loci could depart from this expectation because of linkage with loci under selection (hitchhiking, background selection). In this case, their frequencies, and the related parameters such as F_{ST}, will reflect the action of selection as a function of the effective recombination rate with loci under selection.

Thus, the hypothesis that asymmetric gene flow is due only to the effect of different population sizes and of the selection against F_1 hybrids in domesticated populations (post-zygotic reproductive barrier) is not consistent with our further data (R. Papa and P. Gepts, unpublished results), showing significantly different levels of introgression along the genome. Introgression (particularly from domesticated to wild populations) is very low in areas of the genome surrounding genes for the domestication syndrome, and much higher in the absence of linkage between the markers studied and domestication genes. In regions of the genome where domesti-cation genes and quantitative trait loci (QTLs) are located (D), the level of differentiation between wild and domesticated populations, measured as F_{ST}, was much higher than in other regions, where genes and QTLs are not involved in the domestication syndrome (ND) or no known genes and QTLs (UN) have been located. The average size of genomic regions affected by linkage disequilibrium, inferred from the average map distance of linked markers (7 cM), was quite large (14 cM), probably because of the auto-gamous breeding system of *P. vulgaris*. The level of diversity for UN markers was similar between wild and domesticated populations and, in the wild, much lower than for D and ND markers. In domesticated populations, in contrast, the level of diversity was low and almost constant among all three classes of markers (R. Papa and P. Gepts, unpublished results). This result could be due to the effect of purifying selection in UN genome areas or to the fact that genetic diversity in the wild populations was reduced by the effect of asymmetric gene flow, which takes place predominantly from domesti-cated to wild types (Papa and Gepts, 2003). These results clearly indicate that selection for adaptation to either wild or cultivated environments plays a major role in the asymmetry of introgression between the wild and domesticated common bean, in combination with other factors such as population size and farmer seed selection.

In Table 10.1, we present the expected action of selection in both wild and domesticated environments at one locus involved in the genetic control of the domestication syndrome with complete dominance (*AA* wild and *aa* domesticated genotype). According to this model, there will only be a slow

Table 10.1. Model of selection with complete dominance in the wild and domesticated environments for one locus with two alleles (*A*, wild; and *a*, domesticated) of the domestication syndrome.

Wild environment $W_{AA} = W_{Aa} > W_{aa}$	Cultivated environment $W_{AA} = W_{Aa} < W_{aa}$
F_1 hybrids	
Possible heterosis	Farmer's selection (post-zygotic reproductive barrier) and possible heterosis (even if farmer's selection will eliminate F_1 plants whether they are heterotic or not)
Later generations	
Possible heterotic effect at linked loci for *Aa* genotypes	*A* alleles and linked loci easily eliminated
Slow purging of domesticated alleles and recombination with linked loci	

W_{AA}, W_{Aa} and W_{aa} are the relative fitness of *AA*, *Aa* and *aa*, respectively.

purging of domestication alleles in the wild environment because of dominance and possible action of heterosis. While heterosis may exist in both wild and cultivated environments, selection, for example for seed type, will eliminate F_1 plants whether they are heterotic or not. The result of this process will be recombination between the introgressed domestication alleles and neutral linked loci. In the domesticated environment, hybrids will be easily eliminated in the first generation by the combined effects of selection imposed by the agroecosystem and the farmer's seed selection. For more than one locus, in the absence of a possible role for epistasis, the effects of selection operating on the F_1 progeny will be the same as for one locus, while for segregating generations, selection will reduce the introgression in both environments (cultivated versus wild), but more intensively in the cultivated fields, increasing the asymmetry of introgression. In fact, with two loci (*AABB* wild and *aabb* domesticated genotype), Mendelian segregation predicts a phenotypic ratio of 9(*A_B_*):3(*A_bb*):3(*aaB_*):1(*aabb*) in the F_2 progeny, thus in the wild environment 56% of the hybrid progeny will show a phenotype identical to that of the wild parent (*A_B_*) compared with 75% when considering one locus (*A_*). In the cultivated environment, only 6% of the progeny will show a completely domesticated phenotype (*aabb*) as compared with 25% with one locus (*aa*). Overall, the ratio between domesticated and wild progeny presenting the 'right' phenotype in their respective environments will be 0.33 with one locus and 0.11 with two loci. Selection at multiple loci will also reduce the recombination between selected and linked neutral loci because the larger fraction of unfit genotypes leads to a reduction in effective population size. The effects of background selection (Charlesworth *et al.*, 1993, 1997) and selection at multiple loci will thus compound the effects of the autogamous breeding system and explain the large portion of the genome that is undergoing hitchhiking based on our observations.

Other Crops

Here, we will review the few studies that have analysed gene flow and the genetic diversity of wild and domesticated populations of both autogamous (barley) and allogamous (maize and lucerne) species that may be useful as a comparison with our data. Even if they did not formally address the directionality of gene flow and introgression, these studies can be discussed from this point of view. From this comparison, we can also obtain an indication of the relative importance of the different possible causes of asymmetry of gene flow and introgression.

Jarvis and Hodgkin (1999) reviewed several examples of farmer selection of genes introgressed from wild populations in order to improve upon local varieties of several crops. These examples suggest that gene flow is acting in both directions, but this does not necessarily mean that it is symmetric. Indeed, most examples were based on morphological traits, which may reflect the introgression of just one or a few genes, and may not be neutral, thus providing a biased estimate of introgression at target loci. Some useful traits may be retained by the farmer because of their agronomic importance or for other reasons, such as their novelty. After the initial introgression, subsequent backcrosses are likely to occur between the domesticated genotypes and the progeny of F_1 hybrids. The backcross generations will most probably be selected for the novel traits, in addition to the domesticated phenotype. To obtain an indication of directionality of gene flow, we need to analyse data obtained with neutral loci, such as molecular markers, and not, or not only, selected loci where both gene flow and selection influence their establishment in the populations.

Hordeum

Badr *et al.* (2000) studied the origin of barley domestication by analysing 367 genotypes of wild *H. vulgare* ssp. *spontaneum*. Using spike morphology traits, they identified 50 (14%) wild lines of putative hybrid origin with domesticated forms. Because the aim of the study was to identify a potential progenitor, these lines were excluded from the molecular analyses performed with AFLP markers and the remaining 317 were compared with 57 domesticated accessions. The domesticated accessions formed a single cluster, with none of them placed within the wild accessions. At the same time, Badr *et al.* (2000) indicated that early events of introgression from the wild during the expansion of the barley crop in the Far East had had a significant effect on the genetic structure of domesticated barley of the Himalayas and India, which indicated a possible role for gene flow from wild populations in the early stages of agricultural diffusion.

Medicago

Using random amplified polymorphic DNA (RAPD) and allozymes, Jenczewski *et al.* (1999) analysed the population structure of wild and domesticated lucerne (*Medicago sativa*), a strictly allogamous species, in Spain. They compared domesticated landraces, wild populations and morphologically intermediate wild populations. In contrast to the 'truly' wild populations, the intermediate populations did not show any significant genetic structure. There was no relationship between genetic and geographical distances because of the homogenizing effect of gene flow from the most widespread landrace. In addition, the domesticated populations were significantly differentiated from the wild populations and not from the intermediate forms.

Zea

A very low impact of gene flow from teosinte (*Zea mays* ssp. *mexicana*) was observed in maize in spite of the observation of frequent events of hybridization occurring in maize fields (Wilkes, 1977). The proportion of the maize gene pool that Matsuoka *et al.* (2002) attributed to gene flow from teosinte (ssp. *mexicana*) was 2.3% in areas of sympatry between wild and domesticated forms (high elevation), and only 0.4% in the allopatric situation (lowland), even if for some highland landraces the proportion was higher (~12%). These different levels of introgression may be explained by the possible action of the *teosinte crossing barrier 1* gene (*tcb1*) that prevents gene flow among some, but not all, populations of maize and its progenitor teosinte (Evans and Kermicle, 2001).

With respect to asymmetry of gene flow, the results of Badr *et al.* (2000) showed that a large portion of wild barley accessions derived from outcrossing events with domesticated forms. This may suggest that gene flow from domesticated barley is a frequent event. They used morphological traits, which may present a bias in gene flow assessment if we assume that selection favoured the introgression of domesticated traits even if it seems unlikely that domesticated spike morphology traits would be positively selected in the wild environment. However, a specific analysis using molecular markers would be needed to answer this question definitively.

More evidence comes from the data on gene flow from wild to domesticated populations, such as our results presented in Fig. 10.1, where the domesticated accessions form a single cluster separated from all the wild accessions. This suggests that gene flow from wild to domesticated accessions is not particularly significant on a genome-wide scale. Conversely, the presence of a few wild or weedy exceptions in the predominantly domesticated cluster further confirms the asymmetry of gene flow. These exceptions probably result from outcrossing from domesticated to wild types, followed by selection for a wild phenotype. Because most molecular markers are expected to be neutral, they can be dragged along after the initial hybridization event, but would not be affected by selection. More surprising are the results on maize, which also show a minor role for introgression from wild to domesticated types. An analysis of the distribution of introgression along the maize genome, particularly in relation to domestication loci, is lacking at this time, hence not allowing comparisons between these data and those presented for the common bean.

The results obtained in *M. sativa* by Jenczewski *et al.* (1999) showed that weedy populations have limited population structure and are more close to domesticated materials, as we showed in the common bean.

Conclusions

The information available on other crops besides the common bean is insufficient to determine whether asymmetric gene flow and introgression from crops into their wild relatives is a general phenomenon, and whether it may also displace the genetic diversity of the wild population in specific genomic regions, as seems to be the case in the common bean (Papa and Gepts, 2003). More data will be needed to provide a more definitive conclusion in this matter that is of significant practical importance from the standpoint of genetic conservation of genetic diversity among wild populations. On the other hand, the data available on genetic differentiation between wild and domesticated populations, the inheritance of the domestication syndrome and the demography of the two types of populations generally suggest that asymmetric gene flow and introgression is the most likely hypothesis to explain the observed patterns of genetic diversity in wild and domesticated populations not only in the common bean, but also in the other crops mentioned. Because both autogamous and allogamous species present a similar pattern of distribution of genetic diversity, selection is probably the major factor responsible for the 'resilience' against introgression in the domesticated populations. In allogamous species, gene flow is effective at a much greater geographical distance, which will reduce the effect of close-range gene flow that can be observed in autogamous species, such as the common bean.

In the first stages of agriculture, the role of gene flow from wild to domesticated populations was probably much more important, as also suggested by the results of Badr *et al.* (2000). They showed that early events of introgression from wild populations during the expansion of the barley crop into the Far East had had a significant effect on the genetic structure of domesticated barley of the Himalayas and India. A higher level of gene flow from wild to domesticated types during the early stages of agricultural diffusion can be explained in terms of both population size and selection. The amount of land destined to cropping was clearly more reduced, while the wild populations were colonizing much greater areas. Particularly during crop expansion and diffusion out of the centres of domestication, in the presence of populations of wild relatives, the relative population size of domesticated populations must have been very small, facilitating the introgression of wild alleles. In the same way, domesticated populations were much more similar to the wild populations, and the cultivated environment much more similar to the wild than at the present time, with a reduced relative selection pressure in both directions. In contrast, a much higher asymmetry would be expected with modern varieties, which are also based on a single genotype (F_1 hybrids, pure lines, clones). As seeds are replaced from the original source (plant breeders and seed companies) every year or after a few years, gene flow from modern cultivars to wild populations will present the major, if not unique, direction of gene flow, because wild to crop hybrids will only be used for subsistence consumption (if at all), and not as seed. For these reasons, conservation *in situ* of genetic diversity present in wild populations should take place in natural reserves (Frankel *et al.*, 1995), where cultivation activities can be closely managed. In parallel, particularly for autogamous species such as the common bean where gene flow appears to be much more important in close-range sympatry, a strict control of wild/weedy populations in field and field borders should be useful, because it will drastically reduce the amount of migration from domesticated to wild populations due to pollen flow.

References

Badr, A., Muller, K., Schafer-Pregl, R., El Rabey, H., Effgen, S., Ibrahim, H.H., Pozzi, C., Rohde, W. and Salamini, F. (2000) On the origin and domestication history of barley (*Hordeum vulgare*). *Molecular Biology and Evolution* 17, 499–510.

Bertorelle, B. and Excoffier, L. (1998) Inferring admixture proportion from molecular data. *Molecular Biology and Evolution* 15, 1298–1311.

Burke, J.M., Tang, S., Knapp, S.J. and Rieseberg, L. (2002) Genetic analysis of sunflower domestication. *Genetics* 161, 1257–1267.

Cavalli-Sforza, L.L. and Bodmer, W.F. (1971) *The Genetics of Human Populations.* W.H. Freeman, San Francisco, California.

Chakraborty, R., Kamboh, M.I., Nwankwo, M. and Ferrel, R.E. (1992) Caucasian genes in American Blacks: new data. *American Journal of Human Genetics* 50, 145–155.

Charlesworth, B., Morgan, M. and Charlesworth, D. (1993) The effects of deleterious mutations on neutral molecular variation. *Genetics* 134, 1289–1303.

Charlesworth, B., Nordborg, M. and Charlesworth, D. (1997) The effects of local selection, balanced polymorphism and background selection on equilibirum patterns of genetic diversity in subdivided populations. *Genetical Research* 70, 155–174.

Doebley, J., Stec, A., Wendel, J. and Edwards, M. (1990) Genetic and morphological analysis of a maize–teosinte F_2 population: implications for the origin of maize. *Proceedings of the National Academy of Sciences USA* 87, 9888–9892.

Ellstrand, N.C., Prentice, H.C. and Hancock, J. (1999) Gene flow and introgression from domesticated plants into their wild relatives. *Annual Review of Ecological Systems* 30, 539–563.

Evans, M.M.S. and Kermicle, J.L. (2001) Teosinte crossing barrier 1, a locus governing hybridization of teosinte with maize. *Theoretical and Applied Genetics* 103, 259–265.

Frankel, O.H., Brown, A.D.H. and Burdon, J.J. (1995) *The Conservation of Plant Biodiversity*. Cambridge University Press, Cambridge.

Gepts, P. (1988) Phaseolin as an evolutionary marker. In: Gepts, P. (ed.) *Genetic Resources of* Phaseolus *Beans*. Kluwer, Dordrecht, The Netherlands, pp. 215–241.

Gepts, P. (1993) The use of molecular and biochemical markers in crop evolution studies. *Evolutionary Biology* 27, 51–94.

Gepts, P. and Papa, R. (2003) Possible effects of (trans) gene flow from crops on the genetic diversity from landraces and wild relatives. *Environmental Biosafety Research* 2, 89–103.

Hails, R.S. (2000) Genetically modified plants – the debate continues. *Trends in Ecology and Evolution* 15, 14–18.

Hammer, K. (1984) Das Domestikationssyndrom. *Kulturpflanze* 32, 11–34.

Harrison, S. (1991) Local extinction in a metapopulation context: an empirical evaluation. *Biological Journal of the Linnean Society* 42, 73–88.

Ibarra-Pérez, F., Ehadaie, B. and Waines, G. (1997) Estimation of outcrossing rate in common bean. *Crop Science* 37, 60–65.

Jarvis, D.I. and Hodgkin, T. (1999) Wild relatives and crop cultivars: detecting natural introgression and farmer selection of new genetic combinations in agroecosystems. *Molecular Ecology* 8, 159–173.

Jenczewski, E., Prosperi, J.M. and Ronfort, J. (1999) Differentiation between natural and cultivated populations of *Medicago sativa* (Leguminosae) from Spain: analysis with random amplified polymorphic DNA (RAPD) markers and comparison to allozymes. *Molecular Ecology* 8, 1317–1330.

Jenczewski, E., Ronfort, J. and Chèvre, A.M. (2003) Crop-to-wild gene flow, introgression and possible fitness effects of transgenes. *Environmental Biosafety Research* 1, 9–24.

Koinange, E.M.K., Singh, S.P. and Gepts, P. (1996) Genetic control of the domestication syndrome in common-bean. *Crop Science* 36, 1037–1045.

Matsuoka, Y., Vigouroux, Y., Goodman, M.M., Jesus Sanchez, G., Buckler, E. and Doebley, J. (2002) A single domestication for maize shown by multilocus microsatellite genotyping. *Proceedings of the National Academy of Sciences USA* 30, 6080–6084.

Nei, M. (1978) Estimation of average heterozygosity and genetic distance from a small number of individuals. *Genetics* 89, 583–590.

Nei, M. and Li, W.H. (1979) Mathematical models for studying genetic variation in terms of restriction endonuclease. *Proceedings of the National Academy of Sciences USA* 76, 5269–5273.

Papa, R. and Gepts, P. (2003) Asymmetry of gene flow and differential geographical structure of molecular diversity in wild and domesticated common bean (*Phaseolus vulgaris* L.) from Mesoamerica. *Theoretical and Applied Genetics* 106, 239–250.

Poncet, V., Lamy, F., Devos, K., Gale, M., Sarr, A. and Robert, T. (2000) Genetic control of domestication traits in pearl millet (*Pennisetum glaucum* L., Poaceae). *Theoretical and Applied Genetics* 100, 147–159.

Pritchard, J.K., Stephens, M. and Donnelly, P. (2000) Inference of population structure from multilocus genotype data. *Genetics* 155, 945–959.

Roberts, D. and Hiorns, R. (1965) Methods of analysis of the genetic composition of hybrid populations. *Human Biology* 37, 38–43.

Servedio, M.R. and Kirkpatrick, M. (1997) The effect of gene flow on reinforcement. *Evolution* 51, 1764–1772.

Singh, S.P., Gepts, P. and Debouck, D.G. (1991) Races of common bean (*Phaseolus vulgaris* L., Fabaceae). *Economic Botany* 45, 379–396.

Taylor, P.D. (1995) Sex-ratio bias with asymmetric exchange of pollen between demes. *Evolution* 49, 1119–1124.

Wilkes, H.G. (1977) Hybridization of maize and teosinte, in Mexico and Guatemala and the improvement of maize. *Economy Botany* 31, 254–293.

Wright, S. (1931) Evolution in Mendelian populations. *Genetics* 16, 97–159.

Xiong, L., Liu, K., Dai, X., Xu, C. and Zhang, Q. (1999) Identification of genetic factors controlling domestication-related traits of rice using an F_2 population of a cross between *Oryza sativa* and *O. rufipogon*. *Theoretical and Applied Genetics* 98, 243–251.

11 Crop to Wild Gene Flow in Rice and its Ecological Consequences

BAO-RONG LU, ZHI-PING SONG AND JIA-KUAN CHEN

Ministry of Education Key Laboratory for Biodiversity and Ecological Engineering, Institute of Biodiversity Science, Fudan University, Shanghai, 200433, China; E-mail: brlu@fudan.edu.cn

Abstract

Gene flow is the major pathway for transgene escape from crops to their wild relatives (including weedy biotypes). Transgenes that spread to and persist in the environment will probably lead to ecological consequences. Those genes that resist biotic and abiotic stresses could considerably enhance ecological fitness of the wild relative species, causing unwanted environmental consequences. If transgenic crop varieties are released into the environment, transgene escape to wild relatives through outcrossing will probably occur. In the origin and diversity centres of crop species and their wild relatives, the possibility of transgene escape to the wild species will be high and, as a consequence, the ecological impact of transgene escape might also be great. It is generally understood that the possible crop to wild transgene escape must satisfy three conditions, these are: (i) spatially, transgenic crops and their wild relatives should have an overlapping distribution and be in close contact; (ii) temporally, the flowering time of transgenic crops and their wild relatives should coincide; and (iii) the transgenic crop and the target wild relative species should have sufficiently close biological relationships and insufficient reproductive barriers. This chapter presents studies of crop to wild gene flow in rice based on data of geographical distribution, flowering habit, interspecific hybridization and gene flow from cultivated rice (*Oryza sativa*) to its closely related wild relatives, to estimate the opportunity for crop to wild transgene escape. The general expectations of transgenic escape in rice are also discussed in terms of its potential ecological consequences.

Introduction

Rice (*Oryza sativa* L.) is one of the world's most important cereal crops that provides staple food for nearly one half of the global population (Khush, 1997; Lu, 1998). It serves as the number one crop in terms of food supply and cultural practices in many Asian countries. Rice is the primary food in China, where the largest rice-planting area is found and the most intensive hybrid rice cultivation was practised. In China, the rice-planting area is estimated to be over 13 Mha and rice yield to be over 180 Mt/year. However, as in many other countries in the world, the further enhancement of rice production in China has been significantly constrained by many factors including biotic and abiotic stresses, such as the diseases (blast and bacteria blight) and insect pests (stem borer, leaf folder and brown plant hopper) that have seriously reduced rice yields and qualities. Rice therefore became one of the earliest world's crop species to which transgenic biotechnology has been applied effectively and successfully for its genetic improvement (Ajisaka *et al.*, 1993; Yahiro *et al.*, 1993). Although no transgenic rice variety has yet been officially approved under the biosafety regulations for extensive commercial cultivation anywhere in the world, genes conferring traits such as high protein content, disease and insect resistance, virus resistance, herbicide resistance and salt tolerance have been transferred successfully into different rice varieties through transgenic techniques (Ajisaka *et al.*, 1993; Yahiro *et al.*, 1993; Matsuda, 1998; Messeguer *et al.*, 2001).

Like many other countries in Asia, China has developed its biotechnology, including transgenic technology, rapidly in the past decades, and gene transfer through biotechnology has been applied extensively in rice breeding programmes. Almost all provinces in China are involved in transgenic research or breeding. Based on our incomplete survey, there are more than 50 major institutions involved in transgenic rice production, including the Chinese Academy of Sciences (CAS), Chinese Academy of Agricultural Sciences, universities, Provincial Agricultural Academies, National and Ministry key laboratories and the National Rice Research Institute. A number of transgenic rice varieties that are resistant to major rice diseases and insects, tolerant to salinity and rich in proteins have been successfully developed, e.g. stem borer-resistant rice with *Bt* (*Bacillus thuringiensis*) and *CpTI* (cowpea trypsin inhibitor) genes, bacterial blight-resistant rice with the *Xa21* gene isolated from an African wild rice *Oryza longistaminata*, and blast resistance rice with the *agp* gene encoding ADP-glucose pyrophosphorylase. These varieties are either currently grown under laboratory/greenhouse conditions or have already been released into a confined environment for yield and other types of testing. In addition, some herbicide-resistant (with *bar* and *EPSPs* genes) transgenic rice varieties have also been produced (Tu *et al.*, 2000; Huang *et al.*, 2002). We are confident that, as an important world cereal crop, transgenic rice varieties will be released into the environment for commercial production in the near future, with official approval from competent authorities after their biosafety assessments are conducted and their standard satisfactorily meets the biosafety requirements.

Concerns about Transgene Escape in Rice

The biosafety issues raised in relation to transgenic biotechnology and its products have been extensively debated worldwide in recent years

(Bergelson *et al.*, 1998; Schiermeier, 1998; Crawley *et al.*, 2001; Ellstrand, 2001; Prakash, 2001). Biosafety has become the 'bottle-neck' to further development of transgenic biotechnology and wider application of transgenic products. It is obvious that many genetically modified rice varieties with different genes are available and waiting to be released into the environment. Biosafety issues of transgenic rice varieties have also become increasingly addressed. Among these, transgene escape from the genetically modified rice varieties is one of the concerns across the world (Lu *et al.*, 2003). The so-called crop to wild transgene escape refers generally to a gene or a group of genes introduced into a crop variety by genetic engineering moving to its wild relative species (including its weedy biotype) through gene flow (including pollen flow and seed dissemination). Cross-pollination between transgenic and wild relatives is the major pathway for transgene escape. The establishment and spread of transgenes into the environment may ultimately lead to unwanted ecological consequences, if those genes that resist biotic and abiotic stresses could significantly enhance the ecological fitness of the wild relative species. This might make the wild relative species become some kinds of weeds difficult to control, or may cause other unwanted environmental effects. If transgenic rice varieties are released into the environment, transgene spread to wild relatives through outcrossing will be likely to occur.

These aspects can be summarized as the following questions. Will transgenic products pose a safety problem to the environment? Can transgene escape occur through outcrossing and will the genes persist in the environment? Will transgene escape have significant ecological consequences? All these questions relating to the biosafety of transgenic crops need to be addressed scientifically. Effective strategies to minimize transgene escape and its ecological consequences can only be carried out when a better understanding of the pathways and consequences of transgene escape is achieved, which will in turn lead to a safer and more efficient use of transgenic crops. Our chapter discusses to what extent transgene escape in rice might have unwanted ecological effects, and what is the general consequence of crop to wild gene flow.

Normally, the successful occurrence of transgene escape needs to meet the following three prerequisites: (i) spatially, transgenic rice and its wild rice relatives should be sympatrically distributed, i.e. grow in the same vicinity, and also in close contact; (ii) temporally, the flowering time (including flowering duration within a year and flowering time within a day) of transgenic rice and its wild relatives should overlap considerably; and (iii) biologically, transgenic rice and its wild relative species should have a sufficiently close relationship, and the resulting interspecific hybrids should be able to reproduce normally. It is therefore necessary to be aware of geographical distribution patterns and flowering habits of cultivated rice and wild relative species, as well as to understand genetic relationships and actual gene flow frequencies between the cultivated and wild rice species concerned. This will facilitate the effective prediction of transgene escape and its potential ecological consequences, and the development of strategies to minimize the escape of transgenes.

When transgenes escape to and are expressed normally in wild relatives and weedy species of rice, the transgenes will persist and disseminate within the wild or weedy populations through sexual reproduction and/or vegetative propagation. If the transgenes are encoding traits, such as high protein content, special vitamins and better grain quality, which are not associated with the ecological fitness of the wild or weedy species and not linked to the survival of wild plant species by natural selection, the ecological consequences caused by escape of these genes will be minimal. However, if the transgenes mediate resistance to biotic and abiotic stresses (such as drought and salt tolerance and herbicide resistance), and the genes significantly enhance ecological fitness of wild and weedy species, escape of these genes will probably cause ecological effects. If several of such fitness-enhancing genes are stacked in the same individual wild or weedy species, the ecological consequences might become more significant through formation of aggressive weeds, causing unpredictable changes to local ecosystems. When these transgenes escape to populations of wild relatives through outcrossing, the persistence and rapid spread of the resulting hybrids and their transgene-carrying progeny will lead to introgression of transgenes in populations of the wild relatives, and in the worst case to extinction of endangered populations of the wild relatives in local ecosystems (Kiang et al., 1979; Ellstrand and Elam, 1993). The escape and persistence of transgenes in the environment will make effective in situ conservation of wild genetic resources more difficult. In addition, perennial hybrids of cultivated species and their wild relatives carrying transgenes may serve as a bridge to spread their transgenes through outcrossing to other wild related species, causing even more significant ecological effects.

The Close Wild Relatives of Rice in Asia

Cultivated rice is classified in the genus *Oryza* L. of the tribe *Oryzeae* in the grass family (Poaceae). The genus *Oryza* includes two cultivated species and over 20 wild species widely distributed in the pan-tropics and subtropics (Lu, 1996). The Asian cultivated rice *O. sativa* L. had its origin in South and South-east Asia, and is grown worldwide in the tropics, subtropics and some temperate regions, whereas the African cultivated rice *O. glaberrima* Steud. was domesticated in western Africa and is now cultivated only in local agricultural ecosystems in West Africa (Lu, 1996). The cultivated rice that we discuss here is only referred to as *O. sativa*.

Species in the genus *Oryza* included ten different genome types, i.e. the *AA, BB, CC, BBCC, CCDD, EE, FF, GG, JJHH* and *JJKK* genomes (Vaughan, 1994; Ge et al., 1999; Lu, 1999). Species containing different genomes have significant reproductive barriers. Therefore, genetically, they are distantly related, and spontaneous hybridization between species with different genomes is extremely rare. Asian cultivated rice contains the *AA* genome and is relatively easy to cross with its close relative species (including weedy rice) that also contain the *AA* genome. In theory, transgene escape from transgenic rice varieties will only occur to species with the *AA* genome. Therefore, this chapter only concerns the *AA* genome *Oryza* species.

There are eight diploid ($2n = 2x = 24$) *Oryza* species containing the *AA* genome. Apart from the two cultivated rice species, the perennial common wild rice *O. rufipogon* Griffith and the annual common wild rice *O. nivara* Sharma et Shastry from Asia, the perennial *O. longistaminata* A. Chev. et Roehr and annual *O. barthii* A. Chev. from Africa, the perennial *O. glumaepatula* Steud. from Latin America, and the annual *O. meridionalis* Ng from northern Australia and New Guinea all possess the *AA* genome (Lu, 1996, 1998; Lu and Silitonga, 1999). The geographical distribution of all the *AA* genome wild rice species overlaps significantly with the cultivation areas of Asian cultivated rice, *O. sativa* (Fig. 11.1). Weedy rice occurring in Asia usually has its origin from hybridization between cultivated and wild rice species or degenerated individuals of cultivated rice. It is mostly found in the rice field alongside cultivated rice, but also occurs in the vicinity of rice fields, in ditches or in sympatric regions of cultivated and wild rices (Vaughan, 1994). It is evident in Asia that these wild relatives of rice, i.e. *O. rufipogon*, *O. nivara* and *O. spontanea* (weedy rice), will be the target species for crop to wild transgene escape through gene flow. Our study clearly demonstrates that *O. rufipogon*, *O. nivara* and *O. spontanea* are distributed across a significantly wide geographical region, and cultivated rice is grown sympatrically with these wild rice species in many areas of South and South-east Asia (Lu *et al.*, 2003). It is concluded from the geographical distribution data that on a spatial scale, transgenes from cultivated rice are likely to escape to its wild relative species through gene flow.

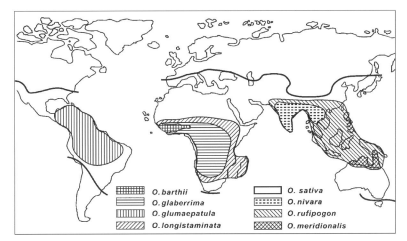

Fig. 11.1. The global *Oryza sativa* cultivation regions versus geographical distribution of the *AA* genome of the *Oryza* species that form a potential for the crop to wild transgene escape through gene flow. The map is drawn based on Lu *et al.* (2003).

Biosystematic Relationships of Rice and its Close Relatives

Biosystematic relationships of the *AA* genome *Oryza* species can be estimated from the following aspects: (i) crossability between cultivated rice and its wild relatives; (ii) meiotic chromosome pairing in the F_1 interspecific hybrids; and (iii) fertility of the F_1 hybrids. If the cultivated rice has relatively high crossability with its wild relatives, normal meiosis formation in the F_1 hybrids, where chromosomes from different parental species will pair and genetic recombination will take place, as observed in F_1 hybrids, and if the F_1 hybrids have comparatively high fertility, the transgenes will easily escape to wild relative species through cross-pollination. The transgenes are likely to spread by sexual as well as vegetative propagation if the hybrids and their progeny are perennial.

Research on crossability between cultivated and wild rice species has been reported extensively (Nezu *et al.*, 1960; Pental and Barnes, 1985; Langevin *et al.*, 1990). We also conducted experiments of interspecific hybridization between eight *AA* genome *Oryza* species under greenhouse conditions (Naredo *et al.*, 1997, 1998). The results from interspecific hybridization show that most of the *AA* genome wild rice species have a relatively high compatibility with the cultivated rice, although with comparatively large variations among species. Only the Australian *O. meridionalis* had low crossability with cultivated rice (< 5%). *Oryza rufipogon* had 6–10% and *O. nivara* had more than 10% crossability with cultivated rice.

The chromosome pairing ability at metaphase I in meiosis of the F_1 hybrids reflects the genetic relationships of the parental species. The high frequency of meiotic pairing indicates a close genetic relationship between their two parents. In this case, exchange of genetic materials between the parents will occur during meiosis through genetic recombination. Extensive studies showed that a very high frequency of meiotic pairing was found in F_1 hybrids between cultivated rice and its *AA* genome wild relatives (Nezu *et al.*, 1960; Pental and Barnes, 1985; Majumder *et al.*, 1997). Our cytogenetic studies also indicated that nearly all chromosomes from parental species formed 12 ring bivalents in meiosis of the hybrids with the *AA* genome (Lu *et al.*, 1997, 1998).

Spikelet fertility of the F_1 hybrids indicates whether the hybrids will continue to multiply through sexual reproduction. A large variation in spikelet fertility of F_1 hybrids between cultivated rice and its *AA* genome wild relatives was reported (Nezu *et al.*, 1960; Langevin *et al.*, 1990). Results from our spikelet fertility investigation in F_1 hybrids between cultivated and wild *AA* genome rice species show a similar rate of panicle fertility (Naredo *et al.*, 1997, 1998; Lu *et al.*, 2000). Our data indicated that spikelet fertility of the F_1 hybrids from all available combinations was relatively high under bagged self-pollination conditions. Hybrids of the cultivated rice with *O. rufipogon* and *O. nivara* produced highly fertile spikelets with fertility over 11%.

Flowering Habits of Cultivated Rice and its Wild Relatives

Flowering habits of cultivated rice grown in different parts of the world vary considerably depending on local cultivation time and seasons, and differences between varietal types (such as photoperiod and thermal sensitivity). The flowering and pollinating time of different wild rice species or different populations of the same species also varies significantly across different geographical regions. In general, the flowering habit of wild rice species is characterized by a protracted flowering period. In other words, different individuals within the same population, and different tillers and spikelets of the same individuals, will flower at considerably different times. For example, *O. rufipogon* usually starts its flowering at the beginning of September and terminates its flowering in December or towards the end of January in the next year; *O. nivara* has a relatively earlier and shorter flowering period from the middle of August to the end of October; and the weedy rice mimics the flowering time of the cultivated rice in the same field, but usually 1–2 weeks earlier. Obviously, the comparison of flowering habits becomes meaningful only when specific wild rice species and cultivated rice varieties from the same location are selected for flowering studies under the same conditions. Data from a selected *O. rufipogon* population, and two cultivated rice varieties (a late-maturing local variety and an improved variety, Minghui-63) at a field site in Chaling of Hunan Province, China, showed that both the flowering period in a year and flowering time in a day had considerable overlap for *O. rufipogon* and the two rice varieties (Table 11.1). Our additional experimental data further demonstrated that pollen grains of *O. rufipogon* and a cultivated rice variety (Minghui-63) are viable in the air for more than 60 min (Song *et al.*, 2001). In summary, it is essentially possible that cross-pollination between *O. rufipogon* and cultivated rice will occur, if the two species have sympatric distribution and are grown near to each other.

Gene Flow between Cultivated Rice and *O. rufipogon*

It is evident that cultivated rice and *O. rufipogon* have sympatric distribution and overlap in flowering time in many Asian countries and regions, and

Table 11.1. Comparison of flowering time of the common wild rice (*Oryza rufipogon*), a late maturing local rice variety and a modern rice variety, Minghui-63, in Chaling, Hunan Province of China.

Species/rice variety	Sowing time	Flowering period	Flowering time
O. rufipogon	–	Beginning of September to mid-November	9:30 a.m. to 4:30 p.m.
Late maturing local variety	Mid-June	Mid-September to beginning of October	8:30 a.m. to 2:00 p.m. 4:00 p.m. to 5:30 p.m.
Minghui-63	Mid-June	Mid-September to mid-October	9:00 a.m. to 3:00 p.m.

genetically the two species have close relationships and low reproductive isolation. As a consequence, introgression between the two species occurs frequently in nature (Oka and Morishima, 1967; Chu and Oka, 1970; Langevin *et al.*, 1990). Figure 11.2 shows the outcome of natural interspecific hybridization between *O. rufipogon* and *O. sativa* in Nepal. Although crossability between the two species obtained under artificial hybridization is considerably high, our knowledge on gene flow between the two species is limited. Information on gene flow frequency between the two species becomes essential for the assessment of transgene escape from transgenic rice varieties to wild relatives. In order to obtain actual data for the maximum gene flow frequency between cultivated rice and *O. rufipogon* under natural conditions, we designed an experiment where gene flow between a cultivated rice variety, Minghui-63, and *O. rufipogon* was examined under controlled conditions. Four different experimental designs were made with 12 treatments for gene flow detection, in which the cultivated Minghui-63 was planted to encompass the wild *O. rufipogon*, or be surrounded by the wild rice species, Minghui-63 and *O. rufipogon* were planted alternately in rows, and *O. rufipogon* was planted downwind of Minghui-63. Simple sequence repeats (SSRs) were used as molecular markers to determine cross-pollination rates between Minghui-63 and *O. rufipogon* under different designs. The results demonstrated that the maximum frequency of gene flow from Minghui-63 to *O. rufipogon* could reach as high as 3% in natural habitats of Chaling in Hunan Province, clearly indicating that gene flow

Fig. 11.2.   The common wild rice (*Oryza rufipogon*), Asian cultivated rice (*O. sativa*) and their spontaneous interspecific hybrids are commonly found as a mixture in the rice fields where the wild rice occurs in the vicinity. The picture in the top right corner shows the wild rice (left), hybrid progeny (middle) and cultivated rice (right).

between cultivated rice and the widely distributed *O. rufipogon* occurs at a considerable rate under natural conditions (Song *et al.*, 2003).

General Expectations of Gene Flow and Ecological Consequences

Transgene escape and its environmental impacts have become increasingly challenging biosafety issues worldwide, and should receive attention from the public, scientists and government agencies. It is difficult to monitor ecological problems caused by the transgene escape within a limited period. Most transgenes carried by genetically modified agricultural products are not from crops, but instead they are from other organisms or micro-organisms, even from an artificially synthesized origin. These genes may significantly alter the natural habit of crop species and significantly change wild relatives of the crop species when transgenes are picked up by the wild species. As a consequence, the environmental safety of these genes, particularly in agroecosystems, should be thoroughly addressed.

The crop to wild gene flow is a common phenomenon and is a part of plant evolution. Results from our own studies suggested that cultivated rice and its wild relative *O. rufipogon* have sympatric distribution and over-lapped flowering times, which meets the spatial and temporal conditions for transgene escape from cultivated rice to its wild relatives. A similar situation is found in many other crop species and their wild relatives (Ellstrand *et al.*, 1999). Our experimental data further show that most of the *AA* genomic wild *Oryza* species have a relatively close biosystematic relationship and high crossability with cultivated rice, particularly *O. rufipogon*, *O. nivara* and weedy rice (*O. spontanea*). They do not have a sufficient reproductive isolation with the cultivated rice, and spontaneous introgression with the cultivated rice occurs with considerable frequency in the field. Also, it was reported by Japanese scientists that outcrossing rates between *O. rufipogon* (including weedy rice) and cultivated rice could be as high as 50% in nature, although with considerable variation (Oka and Morishima, 1967). Data obtained from our experiment further showed that the frequency of gene flow from cultivated rice (Minghhui-63) to *O. rufipogon* in Chaling of Hunan Province could reach as much as 3%. All experimental data from previous reports and our studies support the fact that crop to wild transgene escape through gene flow will occur if transgenic rice varieties are grown in the vicinity of the wild relative species, and if no effective isolation measures are taken.

It is general knowledge that pollen flow is a major pathway for trans-gene escape because pollen can act as a vehicle to disseminate transferred genes in nature. Data from our studies on pollen flow of cultivated rice – a typical wind-pollinated species – showed that the dispersal range of rice pollen grains increased with the increase of wind speed and that the maximum distance of rice pollen flow could be as far as 110 m downwind when the wind speed reached 10 m/s (Song *et al.*, 2004). Therefore, to avoid or minimize transgene escape through pollen flow to the wild relatives of rice effectively, it is recommended to have a buffering isolation zone wider than 110 m or to use tall crops such as sugarcane as an effective buffer between transgenic rice and its wild relatives, given the fact that the spatial, temporal and biological conditions for rice transgene escape are present in many rice-producing countries or regions.

It is generally recognized that a better understanding of species bio-systematic relationships, pollen flow and gene flow will facilitate efficient prediction of transgene escape and its potential ecological consequences, as well as appropriate management of ecological consequences. Although an effective buffering isolation zone between transgenic crops and their closely related wild species is an important biosafety strategy to avoid or significantly minimize transgene escape temporarily, implementation of transgene containment at large scales is nearly impossible. Therefore, more scientific questions concerning the ecological impact of transgene escape need to be properly addressed and thoroughly studied. These questions include:

- whether transgenic hybrids and their progeny have higher ecological fitness than parental species if introgression between transgenic rice and its wild relative species does occur;
- whether transgenes can be normally expressed in the interspecific hybrids and their progeny;
- whether transgenes will change genetic structures and population dynamics of the natural wild rice populations;
- whether wild rice hybrids carrying transgenes will become a 'genetic bridge' that further passes transgenes to other wild plant species;
- what kind of ecological consequences will the individuals and populations carrying transgenes cause;
- what the adequate method is for assessing and managing the biosafety consequences related to environmental change.

The appropriate answers to all these scientific questions will assist us in effectively assessing and managing the potential ecological consequences resulting from rice transgene escape through gene flow, which in turn will also promote the safe use of transgenic rice varieties in the future.

Acknowledgements

This study was supported by the National Natural Science Foundation of China for Distinguished Young Scholars (Grant No. 30125029), Shanghai Commission of Science and Technology (Grant No. 02JC14022 and 03dz19309) and 211 Project (Project name: Biodiversity and Regional Ecosafety).

References

Ajisaka, H., Maruta, Y. and Kumashiro, T. (1993) Evaluation of transgenic rice carrying an antisense glutelin gene in an isolated field. In: Jones, D.D. (ed.)

Proceedings of the 3rd Internal Symposium on the Biosafety Results of Field Tests of Genetical Modified Plants and Microorganisms. University of California, Oakland, California, pp. 291–298.

Bergelson, J., Purrington, C.B. and Wichman, G. (1998) Promiscuity in transgenic plants. *Nature* 395, 25.

Chu, Y. and Oka, H.I. (1970) Introgression across isolating barriers in wild and cultivated *Oryza* species. *Evolution* 24, 344–355.

Crawley, M.J., Brown, S.L., Hails, R.S., Kohn, D.D. and Rees, M. (2001) Transgenic crops in natural habitats. *Nature* 409, 682–683.

Ellstrand, N.C. (2001) When transgenes wander, should we worry? *Plant Ecology* 125, 1543–1545.

Ellstrand, N.C. and Elam, D.R. (1993) Population genetic consequences of small population-size – implications for plant conservation. *Annual Review of Ecology and Systematics* 24, 217–242.

Ellstrand, N.C., Prentice, H.C. and Hancock, J.F. (1999) Gene flow and introgression from domesticated plants into their wild relatives. *Annual Review of Ecology and Systematics* 30, 539–563.

Ge, S., Sang, T., Lu, B.R. and Hong, D.Y. (1999) Phylogeny of rice genomes with emphasis on origins of allotetraploid species. *Proceedings of the National Academy of Sciences USA* 96, 14400–14405.

Huang, J.K., Rozelle, S., Pray, C. and Wang, Q.F. (2002) Plant biotechnology in China. *Science* 295, 674–677.

Khush, G.S. (1997) Origin, dispersal, cultivation and variation of rice. *Plant Molecular Biology* 35, 25–34.

Kiang, Y.T., Antonvics, J. and Wu, L. (1979) The extinction of wild rice (*Oryza perennis formosa*) in Taiwan. *Journal of Asian Ecology* 1, 1–9.

Langevin, S.A., Clay, K. and Grace, J.B. (1990) The incidence and effects of hybridisation between cultivated rice and its related weed red rice (*Oryza sativa*). *Evolution* 44, 1000–1008.

Lu, B.R. (1996) Diversity of the rice gene pool and its sustainable utilization. In: Zhang, A.L. and Wu, S.G. (eds) *Floristic Characteristics and Diversity of East Asian Plants*. China Higher Education Press, Berlin, pp. 454–460.

Lu, B.R. (1998) Diversity of rice genetic resources and its utilization and conservation. *Chinese Biodiversity* 6, 63–72.

Lu, B.R. (1999) Taxonomy of the genus *Oryza* (Poaceae): a historical perspective and current status. *International Rice Research Notes* 24, 4–8.

Lu, B.R. and Silitonga, T.S. (1999) Wild rice *Oryza meridionalis* was first found in Indonesia. *Internal Rice Research Notes* 24, 28.

Lu, B.R., Naredo, M.E., Juliano, A. and Jackson, M.T. (1997) Hybridization of AA genome rice species from Asia and Australia. II. Meiotic analysis of *Oryza meridionalis* and its hybrids. *Genetic Resources and Crop Evolution* 44, 25–31.

Lu, B.R., Naredo, M.E., Juliano, A. and Jackson, M.T. (1998) Taxonomic status of *Oryza glumaepatula* Steud., a diploid wild rice species from the New World. III. Assessment of genomic affinity among rice taxa from South America, Asia and Australia. *Genetic Resources and Crop Evolution* 45, 215–223.

Lu, B.R., Naredo, M.E., Juliano, A. and Jackson, M.T. (2000) Preliminary studies on taxonomy and biosystematics of the AA genome *Oryza* species (Poaceae). In: Jacobs, S.W.L. and Everett, J. (eds) *Grasses: Systematics and Evolution*. CSIRO Publishers, Australia, pp. 51–58.

Lu, B.R., Song, Z.P. and Chen, J.K. (2003) Can transgenic rice cause ecological risks through transgene escape? *Progress in Natural Science* 13, 17–24.

Majumder, N.D., Ram, T. and Sharma, A.C. (1997) Cytological and morphological variation in hybrid swarms and introgressed population of interspecific hybrids (*Oryza rufipogon* Griff. × *O. sativa* L.) and its impact on evolution of intermediate types. *Euphytica* 94, 295–302.

Matsuda, T. (1998) Application of transgenic techniques for hypo-allergenic rice. In: *Proceedings of the International Symposium on Novel Foods Regulation in The European Union – Integrity of The Process of Safety Evaluation*. Berlin, Germany, pp. 311–314.

Messeguer, J., Fogher, C., Guiderdoni, E., Marfa, V., Catala, M.M., Baldi, G. and Mele, E. (2001) Field assessment of gene flow from transgenic to cultivated rices (*Oryza sativa* L.) using a herbicide resistance gene as tracer marker. *Theoretical and Applied Genetics* 103, 1151–1159.

Naredo, M.E., Juliano, A., Lu, B.R. and Jackson, M.T. (1997) Hybridization of AA genome rice species from Asia and Australia. I. Crosses and development of hybrids. *Genetic Resources and Crop Evolution* 44, 17–24.

Naredo, M.E., Juliano, A., Lu, B.R. and Jackson, M.T. (1998) Taxonomic status of *Oryza glumaepatula* Steud., a diploid wild rice species from the New World. II. Hybridization among South American, Asian, and Australian AA genome species. *Genetic Resources and Crop Evolution* 45, 205–214.

Nezu, M., Katayama, T.C. and Kihara, H. (1960) Genetic study of the genus *Oryza*. I. Crossability and chromosomal affinity among 17 species. *Seiken Zihô* 11, 1–11.

Oka, H.I. and Morishima, H. (1967) Variations in the breeding systems of a wild rice, *Oryza perennis*. *Evolution* 21, 249–258.

Pental, D. and Barnes, S.R. (1985) Interrelationship of cultivated rices *Oryza sativa* and *O. glaberrima* with wild *O. perennis* complex. *Theoretical and Applied Genetics* 70, 185–191.

Prakash, C.S. (2001) The genetically modified crop debate in the context of agricultural evolution. *Plant Physiology* 126, 8–5.

Schiermeier, P. (1998) German transgenic crop trails face attack. *Nature* 394, 819.

Song, Z.P., Lu, B.R. and Chen, J.K. (2001) A study of pollen viability and longevity in *Oryza rufipogon*, *O. sativa* and their hybrid. *International Rice Research Notes* 26, 31–32.

Song, Z.P., Lu, B.R., Zhu, Y.G. and Chen, J.K. (2003) Gene flow from cultivated rice to the wild species *Oryza rufipogon* under experimental field conditions. *New Phytologist* 157, 657–665.

Song, Z.P., Lu, B.R. and Chen, J.K. (2004) Pollen flow of cultivated rice measured under experimental conditions. *Biodiversity and Conservation* 13, 579–590.

Tu, J.M., Zhang, G.A., Datta, K., Xu, C.G., He, Y.Q., Zhang, Q.F., Khush, S.G. and Datta, S.K. (2000) Field performance of transgenic elite commercial hybrid rice expressing *Bacillus thuringiensis* α-endotoxin. *Nature Biotechnology* 18, 1101–1104.

Vaughan, D.A. (1994) *The Wild Relatives of Rice: a Genetic Resources Handbook*. International Rice Research Institute, Los Baños, Philippines.

Yahiro, Y., Kimura, Y. and Hayakawa, T. (1993) Biosafety results of transgenic rice plants expressing rice stripe virus-coat protein gene. In: Jones, D.D. (ed.) *Proceedings of the 3rd Internal Symposium on the Biosafety Results of Field Tests of Genetically Modified Plants and Microorganisms*. University of California, Oakland, California, pp. 23–36.

12 Potential for Gene Flow from Herbicide-resistant GM Soybeans to Wild Soya in the Russian Far East

DMITRY DOROKHOV[1], ALEXANDER IGNATOV[1],
ELENA DEINEKO[2], ALEXANDER SERJAPIN[4], ALEXANDER ALA[3]
AND KONSTANTIN SKRYABIN[1]

[1]Centre 'Bioengineering' of the Russian Academy of Sciences, Pr. 60-letiya Oktyabrya, 7/1, Moscow, Russia; [2]Institute of Cytology and Genetics, Siberian Branch of the Russian Academy of Sciences; [3]Soybean Institute, Russian Academy of Agricultural Sciences, Blagoveschensk; [4]Monsanto CE Europe, Moscow Branch, Russia; E-mail: dorokhov@biengi.ac.ru

Abstract

We review the potential for related soya species to cross-pollinate in field and greenhouse conditions. Several local soybean cultivars were fertilized successfully by pollen from wild soya. However, when the GM soybean cv. Stine 2254 RR (GTS 40-3-2) was used as a pollen 'donor' and plants of *Glycine soja* were the pollen 'trap', no herbicide-resistant plants were obtained during two growing seasons. Thus, natural cross-pollination between plants of the soybean species would probably be extremely rare, with a frequency below the sensitivity of this experiment. More data are needed to evaluate fully the extent of transfer of herbicide tolerance genes from widespread cultivation of soybeans to wild soya that might occur in this region.

Introduction

The genus *Glycine* includes several wild and one cultivated species (Hermann, 1962). As a cultivated plant, *Glycine max* (L.) Merr. is one of the major sources of oil, high-protein human and animal food, medicinal and industrial plants, and fertilizer. Soybeans are grown primarily for the production of seed and represent one of the major sources of edible vegetable oil and proteins used for livestock feed. Until recent times, cultivated soybeans have been grown almost entirely in Asia (Komarov, 1958; Hymowitz, 1970); however, soybeans have become an important crop in North America (in

both the USA and Canada) and in South America. The major world producers of soybeans are the USA, China, North and South Korea, Argentina and
Brazil.

Soybeans are cultivated in two regions of the Russian Federation: the
far east along the Chinese border, and the area south of the European part
of Russia. The far east is also an area where wild relatives of soybeans
occur naturally (Komarov and Klubkova-Alisova, 1932; Komarov, 1958;
Kuientsova, 1962). Thus there is a potential risk in introducing genetically
modified (GM) cultivars to the region. Among GM cultivars, herbicide-
tolerant soybeans are the most prevalent. The acreage planted with
herbicide-tolerant soybeans increased from 17% of total farmed soybean
acreage in 1997 to 75% in 2002 (Carpenter *et al.*, 2002).

Taxonomy and wild relatives of cultivated soybeans

Cultivated soybeans (*G. max* (L.) Merr.) belong to the family Fabaceae, the
subfamily Papilionoideae, the tribe Phaseoleae, the genus *Glycine* Willd. It is
a diploidized tetraploid ($2n = 40$) (Hermann, 1962; Newell and Hymowitz,
1978). *G. max* belongs to the subgenus *Soja*, which also contains *G. soja* and
G. gracilis Skvortsov. *G. soja* Sieb. & Zucc., a wild species of soybean, can be
found in fields and hedgerows and on roadsides and riverbanks in Asian
countries, including China, Korea, Japan, Taiwan and in the Russian far east
(Skvortsov, 1927; Hood and Allen, 1980). Some other plants of the genus
Glycine are common in the Far East and Australasian regions (Tindale, 1984,
1986), but most of them cannot hybridize naturally (Palmer, 1965; Hood and
Allen, 1980; Grant, 1984; Singh and Hymowitz, 1985; Singh *et al.*, 1987;
Palmer *et al.*, 2000).

Cytological, morphological, and molecular data and data on cross-
pollination suggest that *G. soja* is the probable ancestor of *G. max* (Komarov,
1953; Ahmad *et al.*, 1977). *G. gracilis* is considered to be a weedy or semi-wild
form of *G. max*, with some phenotypic characteristics intermediate between
those of *G. max* and *G. soja*. *G. gracilis* may be an intermediate in the
speciation of *G. max* from *G. soja* (Fekuda, 1933) or it may be a hybrid
between *G. soja* and *G. max* (Karasawa, 1952; Hymowitz, 1970).

Geographical occurrence and genetic variability of wild soya in the far east of the Russian Federation

Genetic variability of wild plant species has been studied intensively
(Chemov, 1991). Intraspecies diversity, an indicator of evolutionary
stability, is a product of the interaction between genotype and environment
(Hymowitz and Newell, 1981; Broue *et al.*, 1982; Abdelnoor *et al.*, 1995).
Intraspecies genetic polymorphism of many important plants, including
cultivated and wild soybeans, has been investigated intensively (Fekuda,
1933; Shimamoto *et al.*, 2000). *G. soja* grows in the far east of Russia and is

endemic in the southeastern part of the Amurskaya oblast (region), in the south of the Khabarovsk region and in the Primorsky region. The densest population of the species was found near Lake Hanka on the Russian–Chinese border. Wild soybeans also occur in China, Korea and Japan (Skvortsov, 1927; USSR Academy of Science Publications, 1934; Melnikova, 1986). In each of these regions, soybean plants can be found only in particular locations (Ala, 1984; Melnikova, 1986; Dymina *et al.*, 2001).

Methods

Several field expeditions aimed at collecting wild soybean plants in their native areas were carried out during 1998–2000 by the Centre 'Bioengineering' (RAS, Moscow) and the Institute of Cytology and Genetics (Siberian Branch of RAS, Novosibirsk), with the participation of the Pacific Institute of Bio-organic Chemistry (Far Eastern Branch of RAS, Vladivostok) and the Soybean Institute (Far East Branch of RAS, Blagoveschensk). Plants were collected in the Amursk and Primorsk regions, especially in the Zeisko-Bureinskaya and Suifuno-Prihankaiskaya lowlands, in the Hasansky district and on the islands of the Gulf of Peter the Great (Table 12.1). Collections were made during the last half of September, when soybean seeds were maturing. Geobotanical descriptions of phytocenoses where wild soybeans occurred were made according to previously published procedures (Dymina *et al.*, 2001). Based on the analysis of 100 phytocenoses, four major phyto-cenosis groups for wild soybeans were found. The first group, the pioneer phytocenosis group, is characterized by plants from early stages of plant succession. The second group is composed of plants growing near water barriers and terraces. The third group is composed of plants inhabiting wet meadows. The fourth group is composed of plants inhabiting marshy meadows. The latter phytocenosis group was found in the Primorsk region only.

A total of 215 wild soybean accessions (seed samples from individual plants) representing 40 local populations (Table 12.1) from the Amursk and Primorsk regions of the Russian far east were collected during 1998–1999 by

Table 12.1. Origin of plant material of *G. soja*.

No. of subpopulation	Origin
1–8	Near Blagoveschensk, Amursk region
9–11, 15–20	Near Gomotaezhnoe, river Komarovka, Ussurian reserve
12–14	Near Slavyanka, coast of Japanese Sea, Primorsk area
21–27	Near lake Khanka, Primorsk region
28–30	Near Gulf Idol, coast of Japanese Sea
31–39	Near Kraskino, Russian–Chinese border, Primorsk area
40	P. Gamova, coast of Japanese Sea

scientists from the Centre 'Bioengineering' and the Institute of Cytology and Genetics. Five local and foreign soybean cultivars were obtained from genetic collections in Russia and from Monsanto Co., USA. Hybrids between cultivated and wild soybeans were obtained from the Institute of Cytology and Genetics. Plants representing 40 wild soybean populations and five cultivated soybean varieties (*G. max*) were used to estimate genetic polymorphisms between genotypes and inter-population diversity by means of random amplification of polymorphic DNA (RAPD) markers.

Five cultivars of *G. max* (Table 12.2) were crossed with plants of ten accessions of *G. soja* by spontaneous hybridization (Ala, 1984). Lines of *G. max* nos 1/83, 44/86, 712 and 28 were selected from cvs Portage (USA), Yantarnaya (Russia), Amurskaya 310 (Russia) and MK1 (Russia), respectively; line no. 686 was obtained from mass hybridization stock, and line no. 79/83 was selected from the interspecies hybrid *G. max* × *G. soja*. Wild accessions KZ571, KZOJ4, KZ6323, KZS350, KZ6352, KB1M, KB194, KPL155, KPL115 and KM136 were collected in the Amur region of the Russian far east. Wild soya possesses a number of dominant phenotypic marker traits and was used as a pollinator. Identification of hybrids was made on the violet colour of hypocotyls controlled by gene *Wi*.

Table 12.2. Frequency of spontaneous hybridization between *G. max* and *G. soja*.

	Cultivated type	Semi-cultivated type	Wild type	Total
***G. max* × *G. soja* (average)**				
Line 1/83	43	24	30	97
44/86	163	31	24	218
686	452	31	42	525
28	647	28	18	693
79/83	346	6	14	366
G. max* (average) x *G. soja				
KZ571	106	7	18	131
KZ634	83	2	8	93
KZ6350	317	36	19	362
KZ6323	148	8	10	166
KZ6352	218	3	11	232
KB154	177	4	10	191
KB194	109	6	12	127
KBL153	227	5	20	252
KBL115	207	12	10	249
KM136	59	26	15	102
Total number of plants	1651	109	133	1893
%	87.2	5.8	7.0	100
Seeds per cross	33.7	2.2	2.7	38.6

The average percentage is grouped for cultivated and wild plants.

Results

Dymina *et al.* (2001) found four types of phytocenosis groups typical of wild soybeans. These phytocenoses differed by landscape and by the composition and density of wild soybean plant communities (soybean coverage varied from 1 to 15%). Wild soybeans were not a predominant weed species in fields.

No variation was found in the number of chromosomes between the wild and cultivated soybeans. However, a high level of genetic polymorphism was found between plants, varying by sampling point and location where the plant was collected. Fifty-one RAPD bands, 46% of the total, were polymorphic between at least two plants. In total, there were 8.5 polymorphic bands per primer. Data on single plants were summarized, and the frequency of each allele and the genetic distance between sampling points were calculated. The genetic distances between samples of wild soybean varied from 0.27 to 0.69. Average genetic distance between samples of *G. soja* equalled 0.48. The genetic distance between wild and cultivated soybeans varied from 0.47 to 0.51. The genetic distance between cultivated soybean varieties Severnaya 4 and Stine 2254RR was 0.22. A cluster analysis indicated that most of the wild soybean accessions were grouped according to their geographical locations. It was clear from our data that wild soybean populations vary according to the regions in which they occur. The wild soybean population in the Amur region was the most uniform genetically. Our data and those of other researchers showed that genetic diversity among soybean populations is low (Bernard and Weiss, 1973; Thomson *et al.*, 1998; Seitova *et al.*, 2004). Higher variation was detected in the *G. soja* than in the *G. max* sample series.

Our results show that genetic distance between the most diverse wild soybean accession (0.69) was close to that between the species *G. soja* and *G. tabacina* + *G. falcata* (0.71) and between *G. tabacina* and *G. falcata* (0.68). Obviously there is a genetic trend associated with the ecological adaptation of soybean populations. For instance, cluster analysis of cultivated soybean ancestor lines that grouped the ancestors primarily according to the place of origin showed that these ancestors corresponded to three agricultural regions of China that are defined by climate (Bernard and Weiss, 1973; Hymowitz and Harlan, 1983).

Soybean subpopulations found in the geographically isolated Amursk region, near the settlement of Gamova, were the most genetically distant from the main population of the Russian far east. Interestingly, samples collected within the Gamova settlement were the most polymorphic among the studied soybean plants, even though the subpopulation from the Amursk region as a whole was the most uniform genetically.

Previous researchers have postulated that the diversity centre, or centre of origin, of cultivated soybeans was located in the area of the Yangtze River (Hymowitz and Harlan, 1983). Soybean plants in the Russian far east may have passed through one or several genetic 'bottle-necks', and this may explain the relative genetic uniformity of soybean plants there.

Spontaneous hybridization between *G. max* and *G. soja*

Two percent of F_1 hybrids exhibited a phenotype similar to the female parent, 7.0% exhibited a phenotype similar to the male parent, and 5.8% exhibited intermediate types (Table 12.2). The average number of putative hybrid plants with traits of wild soya in the hybrid population did not exceed 12.8%; this finding was similar to results of other researchers. The data also reveal much higher variation controlled by the female genotype than by the pollinator. The genetic diversity of the female *G. max* had significantly greater influence on success of hybridization (62.6% of variation) than did the polymorphism of the male parent *G. soja* (44.2%).

Spontaneous hybridization between *G. soja* and *G. max*

Spontaneous pollination of *G. soja* by pollen from *G. max* cv. Stine 2254RR in the field in the Krasnodar region (south of Russia) revealed that the theoretical frequency of hybridization was below 0.28% with 99% probability, and the frequency of artificial pollination (with emasculated female flowers) should be below 4.1% (Dymina *et al.*, 2001). Artificial pollination in the greenhouse produced 3.7% hybrid seeds.

Discussion

Outcrossing of soybeans

Soybeans have a very high percentage of self-fertilization; cross-pollination is usually below 1% (Caviness, 1966). The soybean flower stigma is receptive to pollen approximately 24 h before anthesis and remains receptive 48 h after anthesis. The anthers mature in the bud and pollination occurs inside the flower. Artificial hybridization is used for cultivar breeding. Under certain conditions, such as high temperature and low relative humidity, flowers may open before pollination and fertilization. Under such circumstances, insect pollinators may penetrate the flowers and cross-pollinate different cultivars or species.

Between species, fertile hybrids between *G. max* and *G. soja*, and between *G. max* and *G. gracilis* have been easily obtained. Spontaneous hybridization between soybean cultivars varies from less than 1% to over 2.5%, while the outcrossing rate of wild soya was between 2.4 and 3%, and 9 and 19% (Garner and Allard, 1920; Karasawa, 1952; Ahmad *et al.*, 1977). Frequency of hybridization in reciprocal combinations of *G. max* and *G. soja* varied from 1.8 to 11.1% (Singh and Hymowitz, 1985). Ala (1984) obtained an average of 34 fertile hybrid seeds per plant from artificial crosses between five cultivars of *G. max* and ten lines of *G. soja*. Hybridization in the opposite direction (*G. soja* × *G. max*) was less successful because of technical difficulties in the artificial pollination of the smaller flowers of *G. soja*.

Weediness of *G. soja*

It is not clear whether soybean is a weedy (invasive) plant or not. *G. max* is not found outside of cultivation in most countries. In managed ecosystems, soybeans do not compete effectively with cultivated plants or become primary colonizers. *G. max* is not listed as a noxious weed in the Weed Seed Order (1986). It is not reported as a pest or weed in managed ecosystems, nor is it recorded as being invasive in natural ecosystems. In summary, there is no evidence that *G. max* has weed or pest characteristics anywhere in the world. Skvortsov (1927) mentioned wild soya as a plant of abandoned fields. Komarov (1958) described wild soya as a semi-weed in wheat stands. *G. soja* is more a ruderal plant than a field weed in that it occupies disturbed soil left without further cultivation. Biological traits of wild soya do not satisfy minimal qualifications for being defined as a weed in that it needs a long growing season of around 4 months; it is late in flowering and late in producing mature seed, and has few seeds (100–250 per plant). If soya plants cover up to 2.5% of an area, they have no means to spread their seed over long distances (Dymina *et al.*, 2001). Spring flooding is probably the most important means by which soybean seeds are spread over long distances. Riverbanks and wet meadows are the most common niches for wild soya in the Russian far east region. Soya is also common in disturbed phytocenoses, but rare in farmed fields.

Gene flow from cultivated to wild soya: range of frequency

The probability of transfer of glyphosate resistance from glyphosate-tolerant soybeans to wild relatives of the soybean is very small. It is not considered a risk in the USA and Latin America, which account for about 83% of the total soybean acreage worldwide, because there are no sexually compatible relatives of soybeans growing wild in the Americas. Although wild soybean plants are grown in research plots, there are no reports of their escape to unmanaged habitats. In areas where wild relatives of the soybean naturally occur, such as the Philippines, China, Taiwan, Australia and the South Pacific, the risk of transfer of glyphosate resistance is minimal because soybeans are self-pollinated. The chance of the spread of glyphosate resistance in these regions is decreased further by other limiting factors such as lack of flowering synchrony between cultivated soybeans and wild relatives, the low extent of sexual compatibility, abundance, method and distance of pollen spread and environmental conditions pertinent to cross-pollination.

No reports on transfer of glyphosate resistance from glyphosate-tolerant soybeans to their wild relatives have yet appeared in the scientific literature. Usually, cross-pollination of soybean flowers produces less than 1% of seeds (Caviness, 1966). However, several studies have demonstrated significant potential for cross-pollination between the cultivated soybean *G. max* and either perennial or annual wild soya, including *G. soja* (Ahmad *et al.*, 1977; Ladizinsky *et al.*, 1979; Singh and Hymowitz, 1985; Singh *et al.*, 1987; Palmer *et al.*, 2000). Thus, introduction of GM soybean cultivars to regions containing wild relatives may lead to gene flow into wild populations. A model study of hybridization frequency between wild soya and GM cultivars is needed to estimate the risk.

To design an adequately sensitive experiment, we would need to know the range in frequencies of outcrossing and the number of plants that need to be evaluated in order to achieve statistical significance. From the previously reported data, we know the frequencies of outcrossing within and between *G. max* and *G. soja* (Table 12.3).

Thus, the expected chance of hybridization between *G. soja* and *G. max* ranges from 0.0336% (obtained by multiplying 1.4%, the hybridization rate for *G. soja* × *G. max*, by 2.4%, the outcrossing rate for *G. soja*), to 0.73%, obtained by experimental methods (Nakayama and Yamaguchi, 2002).

A simple equation shows the necessary number of plants to test in order to find at least one hybrid if the frequency of hybridization is known:

$$\alpha = 1 - (1 - p)^N$$

Where α = the probability of obtaining one or more hybrid plants from the population, p = the frequency of hybridization, and N = the number of plants.

Thus, to obtain a 99% probability of finding one or more hybrid plants with frequency $\geq 0.0336\%$, we must test almost 14,000 plants. To obtain a 99.9% probability, we must test over 20,000 plants.

Conclusion

Outcrossing (gene flow) refers to the transfer of genetic material from one crop to another or from a crop to a weed. The implications of gene flow from a crop to a weed depend on many factors. In the presence of selection pressure, the crop–weed hybrids may or may not have a greater adaptive advantage compared with their parents. If hybrids are more competitive, there can be an increase in weediness (Chèvre *et al.*, 1999). The chances of the transfer of glyphosate resistance to wild soya in regions of Asia and the Russian far east are limited by other factors such as flowering asynchrony between soybean and its relatives, extent of sexual compatibility, abundance,

Table 12.3. The ranges of outcrossing within and between *G. max* and *G. soja* at artificial and spontaneous pollination.

Male Female	Artificial pollination		Spontaneous pollination	
	G. max	*G. soja*	*G. max*	*G. soja*
G. max	ND	Up to 11%	< 1 to > 2.5%	12.8%
G. soja	From 1.4%	ND	0.73%	2.4–3% to 9–19%

method and distance of pollen spread, and environmental conditions pertinent to cross-pollination.

The Certified Seed Regulations do not specify a minimum distance between different soybean cultivars in the fields. Thus, they accept that the probability of transfer of genetic material from GM soybean to wild relatives or other soybeans is small. Our results completely support this point of view.

Acknowledgements

This study was supported by a special grant from Monsanto Co. for the field expeditions aimed at collecting wild soybean plants, and by the Russian Foundation for Basic Research (grant No. 03-04-48997) for the preparation of this chapter. We are grateful to Professors V.K. Shumny, P.G. Gorovoj and E.V. Korotkov, and to Drs G.D. Dymina, A.M. Seitova and T.P. Suprunova for their contributions to this investigation. Thanks are also extended to B.D. Dorokhov and I.V. Deineko, and many other investigators and students, for their assistance in the laboratory and field studies.

References

Abdelnoor, R.V., Barros, E.G. and Moreira, M.A. (1995) Determination of genetic diversity within Brazilian soybean germplasm using random amplified polymorphic DNA techniques and comparative analysis with the data of genetic tree. *Brazilian Journal of Genetics* 18, 265–273.

Ahmad, Q.N., Britten, E.J. and Byth, D.E. (1977) Inversion bridges and meiotic behaviour in species hybrids of soybeans. *Journal of Heredity* 68, 360–364.

Ala, A.I. (1984) Variability and inheritance of quantitative traits of wild Ussurian soy. In: *Germplasm of Wild Soy and its Symbiotic Behaviour*. Sci.-Techn. Bul. – VASKHNIL, Siberian Branch, Novosibirsk, pp. 5–25.

Bernard, R.L. and Weiss, M.G. (1973) Qualitative genetics. Soybeans, production and uses. In: Caldwell, B.E. (ed.) *Agronomy Series*. American Society of Agronomy, Madison, Wisconsin, pp. 117–154.

Broue, P., Douglass, J., Grace, J.P. and Marshall, D.R. (1982) Interspecific hybridisation of soybeans and perennial *Glycine* species indigenous to Australia via embryo culture. *Euphytica* 31, 715–724.

Carpenter, J., Felsot, A., Goode, T., Hammig, A., Onstad, D. and Sankula, S. (2002) *Comparative Environmental Impacts of Biotechnology-derived and Traditional Soybean, Corn, and Cotton Crops*. Council for Agricultural Science and Technology, Ames, Iowa. Available at: http://www.cast-science.org

Caviness, C.E. (1966) Estimates of natural cross-pollination in jackson soybeans in Arkansas. *Crop Science* 6, 211–219.

Chemov, Y.I. (1991) Biological diversity: origin and problems. *Advances in Modern Biology III* 4, 499–507.

Chèvre, A.M., Eber, F. and Renard, M. (1999) Gene flow from oilseed rape to weeds. In: Lutman, P.W. (ed.) *Gene Flow and Agriculture Relevance for Transgenic Crops*. Symposium Proceedings no. 72, British Crop Protection Council, Farnham, UK, pp. 125–130.

Dymina, G.D., Gorovoj, P.G., Deineko, E.V., Seitova, A.M., Ignatov, A.N., Suprunova, T.P., Serjapin, A.A., Ala, A.I., Dorokhov, D.B., Shumny, V.K. and Skryabin, K.G. (2001) Genetic polymorphism in wild soybean population of Russian Far East, collection and primary characterization of collection. *International Conference on Genetic Collections, Isogenic and Alloplasmic Lines – 2001, Institute of Cytology and Genetics, Siberian Branch of the Russian Academy of Sciences, Novosibirsk,* pp. 146–148.

Fekuda, Y. (1933) Cytological studies on the wild and cultivated Manchurian soybeans (*Glycine* L.). *Japanese Journal of Botany* 6, 489–506.

Garner, W.W. and Allard, H.A. (1920) Effect of the relative length of day and night and other factors of the environment on growth and reproduction of plants. *Journal of Agricultural Research* 18, 553–606.

Grant, J.E. (1984) Interspecific hybridization in *Glycine* Willd. Subgenus *Glycine* (Leguminosae). *Australian Journal of Botany* 32, 655–663.

Hermann, F.J. (1962) A revision of the genus *Glycine* and its immediate allies. *USDA Technical Bulletin* 1268, 1–79.

Hood, M.J. and Allen, F.L. (1980) Interspecific hybridization studies between cultivated soybean, *Glycine max* and a perennial wild relative, *G. falcata*. *Agronomy Abstracts, Proceedings of the American Society of Agronomy,* 58 pp.

Hymowitz, T. (1970) On the domestication of the soybean. *Economic Botany* 24, 408–421.

Hymowitz, T. and Harlan, J.R. (1983) Introduction of soybeans to North America by Samuel Bowen in 1765. *Economic Botany* 37, 371–379.

Hymowitz, T. and Newell, C.A. (1981) Taxonomy of the genus *Glycine*, domestication and uses of soybeans. *Economic Botany* 37, 371–379.

Karasawa, K. (1952) Crossing experiments with *Glycine soja* and *G. gracilis*. *Genetica* 26, 357–358.

Komarov, V.L. (1953) Types of flora at South-Ussuriisk region. *USSR Academy of Science Publications* 9, 545–742.

Komarov, V.L. (1958) Origin of cultivated plants. *M.-L.: USSR Academy of Science Publications* 12, 7–256.

Komarov, V.L. and Klubkova-Alisova, E.N. (1932) Plants of the Far East region. *L.: USSR Academy of Science Publications* 2, 623–1175.

Kuientsova, G.E. (1962) Flora of Prihankaiskaya valley and highland. *M.-L.: USSR Academy of Science Publications,* 137 pp.

Ladizinsky, G., Newell, C.A. and Hymowitz, T. (1979) Wild crosses in soybeans: prospects and limitations. *Euphytica* 28, 421–423.

Melnikova, A.B. (1986) Higher plants. Flora and plants of Bolshehehtsky reservation (Khabarovsk region) Vladivostok. *USSR Academy of Science Publications, Far-East Branch,* pp. 102–183.

Nakayama, Y. and Yamaguchi, H. (2002) Natural hybridisation in wild soybean (*Glycine max* ssp. *soja*) by pollen flow from cultivated soybean (*G. max* ssp. *max*) in a designed population. *Weed Biology and Management* 2, 25–30.

Newell, C.A. and Hymowitz, T. (1978) A reappraisal of the subgenus *Glycine*. *American Journal of Botany* 65, 168–179.

Palmer, R.G. (1965) Interspecific Hybridization in the Genus *Glycine*. MSc thesis. Universiy of Illinois, Urbana, Illinois.

Palmer, R.G., Sun, H. and Zhao, L.M. (2000) Genetics and cytology of chromosome inversions in soybean germplasm. *Crop Science* 40, 683–687.

Seitova, A.M., Ignatov, A.N., Suprunova, T.P., Tsvetkov, I.L., Deineka, E.V., Dorokhov, D.B. and Skryabin, K.G. (2004) Study of soybean (*Glycine soja* Sieb. &

Zucc.) in Far East region of Russia. Genetic diversity of wild soy. *Russian Journal of Genetics* 40, 1–8.

Shimamoto, Y., Fukushi, H., Abe, J., Kanazawa, A., Gai, J., Gao, Z. and Xu, D. (2000) Characterizing the cytoplasmic diversity and phyletic relationship of Chinese landraces of soybean, *Glycine max*, based on RFLPs of chloroplast and mitochondrial DNA II. *Genetic Resources and Crop Evolution* 47, 611–617.

Singh, R.J. and Hymowitz, T. (1985) An intersubgeneric hybrid between *Glycine tomentella* Hayata and the soybean, *G. max*. (L.) Merr. *Euphytica* 34, 187–192.

Singh, R.J., Kollipara, K.P. and Hymowitz, T. (1987) Intersubgeneric hybridization of soybeans with a wild perennial species, *Glycine clandestine* Wendl. *Theoretical and Applied Genetics* 74, 391–396.

Skvortsov, B.V. (1927) Wild and cultivated soybean of Eastern Asia. Manchuria Rep. *Kharbin* 9, 35–43.

Thomson, J.A., Nelson, R.L. and Vodkin, L.O. (1998) Identification of diverse soybean germplasm using RAPD markers. *Crop Science* 38, 1348–1355.

Tindale, M.D. (1984) Two eastern Australian species of *Glycine* Willd. (Fabaceae). *Brunonia* 7, 207–213.

Tindale, M.D. (1986) Taxonomic notes in three Australian and Norfolk island species of *Glycine* Willd. (Fabaceae: Phaseoleae) including the choice of a neotype of *G. clandestine* Wendl. *Brunonia* 9, 179–191.

USSR Academy of Science Publications (1934) Weeds of USSR. *USSR Academy Science Publications* 3, 447 pp.

13 Analysis of Gene Flow in the Lettuce Crop–Weed Complex

CLEMENS VAN DE WIEL[1], ANDREW FLAVELL[2], NAEEM SYED[2], RUDIE ANTONISE[3], JEROEN ROUPPE VAN DER VOORT[3] AND GERARD VAN DER LINDEN[1]

[1]Plant Research International, PO Box 16, 6700 AA Wageningen, The Netherlands; [2]University of Dundee, School of Life Sciences, Division of Biochemistry, MSI/WTB Complex, Dundee DD1 4HN, UK; [3]Keygene NV, Molecular Marker Services, PO Box 216, 6700 AE Wageningen, The Netherlands; E-mail: clemens.vandewiel@wur.nl

Abstract

Cultivated lettuce, *Lactuca sativa*, and wild prickly lettuce, *L. serriola*, have been shown to be closely related, if not conspecific. Even though both species are regarded as basically self-pollinating, outcrossing does occur, however to an unknown extent. In the context of an EU-funded project (acronym 'ANGEL'), an attempt is made to assess the level of gene flow between cultivated and wild forms by comparing the two, using the molecular marker systems of amplified fragment length polymorphism (AFLP) and the retrotransposon-based sequence-specific amplified polymorphism (SSAP). In addition, a marker system targeting disease resistance genes and gene analogues, called NBS (nucleotide-binding site)-directed profiling, is implemented in order to screen variation in genomic regions expected to be relevant for plant fitness and to play an important role in plant breeding.

Introduction

Cultivated lettuce's (*Lactuca sativa* L.) immediate wild relative, prickly lettuce (*L. serriola* L.), has extended its distribution considerably in recent decades, in particular into north-western Europe (Bowra, 1991; Frietema de Vries *et al.*, 1994). Being of Mediterranean origin, *L. serriola*'s expansion could simply be related to recent gradual warming of the climate in this part of Europe; a recent inventory of floral changes in The Netherlands showed an overall increase of plant species indicative for a warmer climate (Van Oene *et al.*, 2001). An alternative hypothesis recently put forward is the possible contribution of traits introgressed from the cultivated species that confer an advantage to the wild species, e.g. disease resistances (e.g. de Vries *et al.*,

1994). If this is the case, it makes this species pair an excellent baseline study model in the ongoing debate on the possible consequences of gene flow from transgenic crops to wild forms.

The species pair *L. sativa/L. serriola* has been designated a crop–weed complex by van Raamsdonk and van der Maesen (1996), implying significant gene flow between the two members. Nevertheless, their precise relationship has not been fully explored yet, as was pointed out in the compilation of the Botanical Files for The Netherlands by de Vries *et al.* (1992). These botanical files contain basic information on dispersal for risk assessment of GM plants, which was gathered from data in herbaria and the floristic literature. Since the data for lettuce were inconclusive, de Vries *et al.* (1994) performed an additional field study showing that *L. sativa* and *L. serriola* overlap morphologically. In combination with the strong cross-compatibility between the two species (de Vries, 1990), they could be regarded as conspecific. This view was challenged by de Vries and van Raamsdonk (1994) on the basis of their numerical morphological analysis on *L. sativa* and its allies.

Both lettuce species are considered to be basically self-pollinating, but heterozygotes are known to occur in natural populations. The only report in the literature on lettuce outcrossing frequency known to us (Thompson *et al.*, 1958) showed an outcrossing frequency on average of about 1% and a maximum of 2.87% in multiplications of cultivars in close proximity. In practice, frequencies will vary, depending on parameters such as the presence of pollinators and wind direction. This means that exact estimations of outcrossing frequency in particular cases will not suffice to predict introgression from crop to wild under natural conditions. Klinger (2002) showed that even for an obligatory outcrossing species such as *Raphanus sativus*, frequencies and distances over which outcrossing occurs may vary to such an extent that predicting rates of hybridization, a prerequisite for risk assessment, becomes rather problematic.

An alternative strategy to study the actual occurrence of outcrossing and subsequent introgression in practice is comparing wild and cultivated populations in the same area using neutral population genetic marker systems. Several studies have appeared (e.g. Whitton *et al.*, 1997 on sunflower; Bartsch *et al.*, 1999 on beet; Hansen *et al.*, 2001 on rape; Westman *et al.*, 2001 on strawberry), most of them using the well known systems of isozymes, restriction fragment length polymorphisms (RFLPs), amplified fragment length polymorphisms (AFLPs) and microsatellites (simple sequence repeats; SSRs).

When tracing introgression in this way, several problems need to be taken into account. Even though markers can be selected that are specific for a panel of varieties of the cultivated form, rare occurrences of these seemingly cultivar-specific markers in the wild form may still be attributable to common ancestry, and not to recent introgression, since the cultivated form will most probably be domesticated from historical populations of this wild relative. On the other hand, wild relatives are widely used in breeding, e.g. for the introduction of disease resistance genes, and, thus, specific genomic segments from particular wild accessions recently have been introgressed into the cultivated form, instead of the other way around. In their turn, these genomic segments may end up in wild forms growing in the neighbourhood of such cultivars, and may even confer an advantage by way of these disease resistance traits.

As part of a recently started EU project, acronym 'ANGEL' (QLK3-CT-2001-01657, Plant Research International, 2001), three molecular methods are tested to address the problems of tracing introgression: in order to obtain maximum informativeness, two efficient multi-locus marker systems from which the markers will be selected by linkage mapping, to achieve a good coverage of the genome. In this way, it should be possible to identify specific chromosomal segments shared by wild and cultivated populations, which could be taken as an indication for recent introgression. The multi-locus marker systems used are AFLP, and the more recently developed system, sequence-specific amplified polymorphism (SSAP), which is based on retro-transposons occurring ubiquitously in eukaryotic genomes. The latter have the advantage of providing markers for which the direction of change can be inferred from the way they move through the genome. Retrotransposons are copied into new parts of the genome through RNA intermediates; each new insertion will most probably be at a unique place in the genome, whilst progenitor insertions will stay in place. Such new insertions will be good evidence for a common ancestry of the chromosomal parts of the plants/populations in which they are found (cf. Waugh *et al.*, 1997; Tatout *et al.*, 1999).

In order to be able to trace genomic regions most likely to be involved in breeding efforts, a third multi-locus marker system is tested, called nucleotide-binding site (NBS) profiling. NBS profiling is screening for variation in and around disease resistance genes containing the conserved NBS region, which make up the great majority of resistance genes known so far (van Tienderen *et al.*, 2002; van der Linden *et al.*, unpublished results). In lettuce, a lot of effort has been put into introducing resistance against pathotypes of the downy mildew, *Bremia lactucae*. At least one of these resistance factors, Dm3, has recently been shown to be encoded by a gene belonging to the NBS-leucine-rich repeat (LRR) type of resistance genes (Meyers *et al.*, 1998). In the following, preliminary results on the implementation of the three marker methods are reviewed.

AFLP

AFLP was used to compare three wild populations of *L. serriola* sampled in 2001 in southern Italy (Taranto), southern Germany (Baden-Württemberg) and northern Germany (Mecklenburg-Vorpommern), respectively, and a broad test set of cultivated lettuce containing representatives of the main crop types butterhead, latin, crisp, cos, cutting, stalk and oilseed. More than 700 polymorphic bands were scored using 11 primer combinations. As expected, cluster analysis separated the plants into a cultivated and a wild group, with the oilseeds as an intermediate group. From the polymorphic bands, approximately 300 could be traced back on an integrated linkage map at Keygene based on about 1000 loci and four separate crosses. MACP (marker-assisted chromosome painter, Keygene) software was used to visualize closely linked markers (haplotypes) in both cultivated and wild lettuce individuals. Preliminary analysis identified a chromosomal segment in common between the cultivated types and one of the wild populations. Nevertheless, the level of detail achieved by the density of genome coverage of the markers used was judged as yet too low for ascertaining introgression (van de Wiel *et al.*, 2003).

SSAP

Variation in retrotransposable elements can be accessed by various methods (cf. Kumar and Hirochika, 2001), from which SSAP is relatively straight-forward to develop for a given species. Insertion sites of retro-elements in the genome are visualized by polymerase chain reaction (PCR) using one primer derived from the transposable element in combination with an adaptor primer similar to those used in AFLP (Waugh *et al.*, 1997). Like AFLP, a multi-locus pattern of markers basically distributed over the whole genome is obtained. In order to be able to design the primer from the transposable element, its terminal sequences (long terminal repeats, LTRs) need to be cloned and sequenced. LTR sequences were isolated in two ways: (i) chromosome walking from the conserved RNase H motif close to the LTRs of Ty1-*copia* type retrotransposons using degenerate primers; and (ii) chromosome walking from retrotransposon sequences mined from lettuce expressed sequence tag (EST) libraries (sequences obtained from R. Michelmore, UC Davis, California). Both methods produced LTR sequences from which primers were designed to be tested in SSAP. In these tests, genomic DNA was digested with the restriction enzymes *Pst*I and *Mse*I. A combination of two selective bases attached to the *Mse*I adaptor primer and one base added to the LTR primer produced clear banding profiles in both cultivated lettuce and wild *L. serriola*. This is considerably lower than the usual number of selective bases of three, each implying a relatively low copy number of the retro-elements targeted (*gypsy* and *copia*) as compared with other crops such as wheat and maize (van de Wiel *et al.*, 2003).

NBS Profiling

The NBS profiling method targets variation in and around the most prominent family of disease resistance genes (*R*-genes) described so far, the NBS-LRR family (Young, 2000). Primers are designed from conserved sequences of the NBS domain. Genomic DNA is digested with a single restriction enzyme, and adaptors are ligated to the fragments. Genetic variation is assessed by *R*-gene-specific PCR with the selective primer and an adaptor primer. The resulting multi-locus banding pattern will be highly enriched in RGA (resistance gene analogue)-like sequences (van Tienderen *et al.*, 2002; van der Linden *et al.*, unpublished results). Based on known *Lactuca* RGA sequences, several of the universal primers already working successfully in other crops, such as potato, tomato, apple and barley, were applied on a test set consisting of lettuce cultivars and accessions of three wild *Lactuca* species (*L. serriola*, *L. saligna* and *L. virosa*) using *Rsa*I as restriction enzyme. This resulted in marker profiles of between 60 and 90 bands, half of which showed polymorphisms among lettuce cultivars. The degree of variation was higher among different species. The *L. serriola* profiles resembled most the cultivated lettuces, whereas the other two wild species were clearly different.

A number of bands from each profile were sequenced and compared with database sequences using XBLAST and NBLAST. Several bands could be considered as derived from RGAs based on homology to GenBank sequences, examples of which are shown in Fig. 13.1. A band common to both *L. sativa* and *L. serriola* (5.06) showed high similarity to the RCG1 family (RCG = resistance candidate gene), another (5.45) to the RCG4 family of lettuce in Shen *et al.* (1998). Moreover, bands 5.05 and 5.04 were related to the RCG2 family, from which one member has been identified as a gene conferring resistance to a *Bremia lactucae* pathotype, namely *Dm3* (Meyers *et al.*, 1998).

For only half the bands, significant similarity to known RGAs or any other sequence could be established. This is less than previously found for other crops, such as potato, in which about 80% were found to be homologous to published RGA sequences. However, far fewer RGA sequences have been published for Asteraceae, such as lettuce, than for the Solanaceae to which the much studied potato and tomato belong. Therefore, part of the lettuce sequences may still belong to as yet unknown types of RGAs for which there are no sequence homologues known in the nucleotide databases.

Concluding Remarks

In principle, it proved possible to compare haplotypes of the cultivated lettuce species with those of the wild *L. serriola* using 'neutral' multi-locus AFLP markers previously mapped in an integrated genomic linkage map. However, insufficient detail in genome coverage was attained up until now to trace recent introgression from cultivated to wild. Therefore, additional mapping of polymorphic markers will be performed. This will also be done for the novel marker systems successfully developed for introgression analysis in lettuce, the other 'neutral' system SSAP and the resistance gene-targeted NBS profiling. The NBS profiling is promising in that it appears to be capable of screening for variation in RGAs related to *R*-genes known to confer resistance to the downy mildew *B. lactucae*. A preliminary comparison of NBS profiling band patterns and *Bremia* pathotype scores in the *Bremia* differentials test set of lettuce accessions described by van Ettekoven and

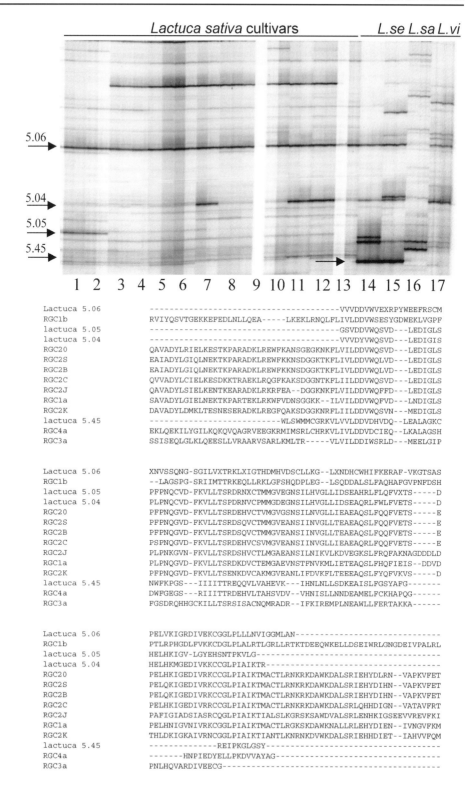

```
Lactuca 5.06   --------------------------------------VVVDDVWVEXRPYWEEFRSCM
RGC1b          RVIYQSVTGEKKEFEDLNLLQEA-----LKEKLRNQLFLIVLDDVWSESYGDWEKLVGPF
lactuca 5.05   ----------------------------------GSVDDVWQSVD---LEDIGLS
lactuca 5.04   ---------------------------------VVVDYVWQSVD---LEDIGIS
RGC20          QAVADYLRIELKESTKPARADKLREWFKANSGEGKNKFLVILDDVWQSVD---LEDIGLS
RGC2S          EAIADYLGIQLNEKTKPARADKLREWFKKNSDGGKTKFLIVLDDVWQLVD---LEDIGLS
RGC2B          EAIADYLGIQLNEKTKPARADKLREWFKKNSDGGKTKFLIVLDDVWQLVD---LEDIGLS
RGC2C          QVVADYLCIELKESDKKTRAEKLRQGFKASDGGNTKFLIILDDVWQSVD---LEDIGLS
RGC2J          QAVADYLSIELKENTKEARADKLRKRFEA--DGGKNKFLVILDDVWQFFD---LEDIGLS
RGC1a          SAVADYLGIELNEKTKPARTEKLRKWFVDNSGGKK--ILVILDDVWQFVD---LNDIGLS
RGC2K          DAVADYLDMKLTESNESERADKLREGFQAKSDGGKNRFLIILDDVWQSVN---MEDIGLS
lactuca 5.45   --------------------------WLSWMMCGRKVLVVLDDVDHVDQ--LEALAGKC
RGC4a          EKLQEKILYGILKQKQVQAGRVEEGKRMIMSRLCHRKVLIVLDDVDCIEQ--LKALAGSH
RGC3a          SSISEQLGLKLQEESLLVRAARVSARLKMLTR-----VLVILDDIWSRLD---MEELGIP


Lactuca 5.06   XNVSSQNG-SGILVXTRKLXIGTHDMHVDSCLLKG--LXNDHCWHIFKERAF-VKGTSAS
RGC1b          --LAGSPG-SRIIMTTRKEQLLRKLGFSHQDPLEG--LSQDDALSLFAQFGVPNFDSH
lactuca 5.05   PFPNQCVD-FKVLLTSRDRNXCTMMGVEGNSILHVGLLIDSEAHRLFLQFVXTS-----D
lactuca 5.04   PLPNQCVD-FKVLLTSPDRNVCPMMGDEGNSILHVGLLIDSEAQRLFWLFVETS-----D
RGC20          PFPNQGVD-FKVLLTSRDEHVCTVMGVGSNSILNVGLLIEAEAQSLFQQFVETS-----E
RGC2S          PFPNQGVD-FKVLLTSRDSQVCTMMGVEANSIINVGLLTEAEAQSLFQQFVETS-----E
RGC2B          PFPNQGVD-FKVLLTSRDSQVCTMMGVEANSIINVGLLTEAEAQSLFQQFVETS-----E
RGC2C          PSPNQGVD-FKVLLTSRDEHVCSVMGVEANSIINVGLLIEAEAQRLFQQFVETS-----E
RGC2J          PLPNKGVN-FKVLLTSRDSHVCTLMGAEANSILNIKVLKDVEGKSLFRQFAKNAGDDDLD
RGC1a          PLPNQGVD-FKVLLTSRDKDVCTEMGAEVNSTPNVKMLIETEAQSLFHQFIEIS--DDVD
RGC2K          PFPNQGVD-FKVLLTSENKDVCAKMGVEANLIFDVKFLTEEEAQSLFYQFVKVS-----D
lactuca 5.45   NWFKPGS---IIIITTREQQVLVAHEVK---IHNLNLLSDKEAISLFGSYAFG-------
RGC4a          DWFGEGS---RIIITTRDEHVLTAHSVDV--VHNISLLNNDEAMELFCKHAPQG------
RGC3a          FGSDRQHHGCKILLTSRSISACNQMRADR--IFKIREMPLNEAWLLFERTAKKA------


Lactuca 5.06   PELVKIGRDIVEKCGGLPLLLNVIGGMLAN----------------------------
RGC1b          PTLRPHGDLFVKKCDGLPLALRTLGRLLRTKTDEEQWKELLDSEIWRLGNGDEIVPALRL
lactuca 5.05   HELHKIGV-LGYEHSNTPKVLG---------------------------------
lactuca 5.04   HELHKMGEDIVKKCCGLPIAIKTR--------------------------------
RGC20          PELHKIGEDIVRKCCGLPIAIKTMACTLRNKRKDAWKDALSRIEHYDLRN--VAPKVFET
RGC2S          PELQKIGEDIVRKCCGLPIAIKTMACTLRNKRKDAWKDALSRIEHYDIHN--VAPKVFET
RGC2B          PELQKIGEDIVRKCCGLPIAIKTMACTLRNKRKDAWKDALSRIEHYDIHN--VAPKVFET
RGC2C          PELHKIGEDIVRRCCGLPIAIKTMACTLRNKRKDAWKDALSRLQHHDIGN--VATAVFRT
RGC2J          PAFIGIADSIASRCQGLPIAIKTIALSLKGRSKSAWDVALSRLENHKIGSEEVVREVFKI
RGC1a          PELHNIGVNIVRKCCGLPIAIKTMACTLRGKSKDAWKNALLRLEHYDIEN--IVNGVFKM
RGC2K          THLDKIGKAIVRNCGGLPIAIKTIANTLKNRNKDVWKDALSRIEHHDIET--IAHVVFQM
lactuca 5.45   --------------REIPKGLGSY-------------------------------
RGC4a          -------HNPIEDYELLPKDVVAYAG---------------------------------
RGC3a          PNLHQVARDIVEECG-------------------------------------------
```

van der Arend (1999) indicated, in a few instances, correlation between bands and a specific pathotype resistance. Breeding efforts are focused on resistance against downy mildew. *Bremia* resistance is therefore an important type of trait to address in studying introgression from the crop because of its potential to influence fitness of wild populations. Future work will address whether these three types of markers in combination with mapping data can reach a degree of detail sufficient for tracing significant introgression in the lettuce crop–weed complex.

Acknowledgements

The authors thank Dr R. Michelmore (UC Davis, California) for providing transposable element sequences from lettuce, and Dr René Smulders for critically reading an earlier draft of the manuscript.

References

Bartsch, D., Lehnen, M., Clegg, J., Pohl Orf, M., Schuphan, I. and Ellstrand, N.C. (1999) Impact of gene flow from cultivated beet on genetic diversity of wild sea beet populations. *Molecular Ecology* 8, 1733–1741.

Bowra, J.C. (1991) Prickly lettuce (*Lactuca serriola*) – a population explosion in Warwickshire. *Botanical Society of the British Isles News* 60, 12–16.

de Vries, F.T., Van der Meijden, R. and Brandenburg, W.A. (1992) Botanical Files. A study of the real chances for spontaneous gene flow from cultivated plants to the wild flora of The Netherlands. *Gorteria Supplement* 1, 1–100.

de Vries, F.T., Van der Meijden, R. and Brandenburg, W.A. (1994) Botanical files on lettuce (*Lactuca sativa*) – on the chance for gene flow between wild and cultivated lettuce (*Lactuca sativa* L. including *L. serriola* L., Compositae) and the generalized implications for risk-assessments on genetically modified organisms. *Gorteria Supplement* 2, 1–44.

de Vries, I.M. (1990) Crossing experiments of lettuce cultivars and species (*Lactuca* sect. *Lactuca*, Compositae). *Plant Systematics and Evolution* 171, 233–248.

de Vries, I.M. and van Raamsdonk, L.W.D. (1994) Numerical morphological analysis of lettuce cultivars and species (*Lactuca* sect. *Lactuca*, Asteraceae). *Plant Systematics and Evolution* 193, 125–141.

Fig. 13.1. *(Opposite)* Alignments of deduced protein sequences from bands in an NBS profiling gel. Bands indicated by arrows in the NBS profiling gel detail are aligned below to lettuce RGA sequences from sequence databases. All lanes in the gel are in duplicate to test the reproducibility of individual bands: lanes 1 and 2, two plants from cv. Madrilene; lanes 3, 4 and 5, three plants from cv. Hector; lane 6, Grise Maraîchère; lane 7, cv. Milly; lane 8, cv. Jaune d'Or; lane 9, cv. Balisto; lane 10, cv. Monet; lane 11, cv. Great Lakes 659; lane 12, cv. Karif; lane 13, Tianjin Big Stem; lane 14, *L. serriola* CGN04667; lane 15, *L. serriola* CGN15684; lane 16, *L. saligna* CGN15697; lane 17, *L. virosa* CGN13339.

Hansen, L.B., Siegismund, H.R. and Jørgensen, R.B. (2001) Introgression between oilseed rape (*Brassica napus* L.) and its weedy relative *B. rapa* L. in a natural population. *Genetic Resources and Crop Evolution* 48, 621–627.

Klinger, T. (2002) Variability and uncertainty in crop-to-wild hybridization. In: Letourneau, D.K. and Burrows, B.E. (eds) *Genetically Engineered Organisms: Assessing Environmental and Human Health Effects*. CRC Press, Boca Raton, Florida, pp. 1–15.

Kumar, A. and Hirochika, H. (2001) Applications of retrotransposons as genetic tools in plant biology. *Trends in Plant Science* 6, 127–134.

Meyers, B.C., Chin, D.B., Shen, K.A., Sivaramakrishnan, S., Lavelle, D.O., Zhang, Z. and Michelmore, R.W. (1998) The major resistance gene cluster in lettuce is highly duplicated and spans several megabases. *Plant Cell* 10, 1817–1832.

Plant Research International (2001) ANGEL. Available at: http://www.plant.wageningen-ur.nl/projects/angel

Shen, K.A., Meyers, B.C., Islam Faridi, M.N., Chin, D.B., Stelly, D.M. and Michelmore, R.W. (1998) Resistance gene candidates identified by PCR with degenerate oligonucleotide primers map to clusters of resistance genes in lettuce. *Molecular Plant–Microbe Interactions* 11, 815–823.

Tatout, C., Warwick, S., Lenoir, A. and Deragon, J.M. (1999) SINE insertions as clade markers for wild crucifer species. *Molecular Biology and Evolution* 16, 1614–1621.

Thompson, R.C., Whitaker, T.W., Bohn, G.W. and Van Horn, C.W. (1958) Natural cross-pollination in lettuce. *Proceedings of the American Society for Horticultural Science* 72, 403–409.

van de Wiel, C., van der Linden, G., Den Nijs, J.C.M., Flavell, A., Syed, N., Jørgensen, R., Felber, F., Scotti, I., Rouppe van der Voort, J. and Peleman, J. (2003) An EU project on gene flow analysis between crop and wild forms of lettuce and chicory in the context of GMO biosafety: first results in lettuce. In: van Hintum, T., Lebeda, A., Pink, D. and Schut, J. (eds) *Eucarpia Leafy Vegetables 2003. Proceedings of the Eucarpia Meeting on Leafy Vegetables Genetics and Breeding*, Noordwijkerhout, The Netherlands 19–21 March 2003. Centre for Genetic Resources (CGN), Wageningen, The Netherlands, pp. 111–116.

van Ettekoven, K. and van der Arend, A.J.M. (1999) Identification and denomination of 'new' races of *Bremia lactucae*. In: Lebeda, A. and Kristkova, E. (eds) *Eucarpia Leafy Vegetables '99*. Palacký University, Olomouc, pp. 171–175.

Van Oene, H.H., Ellis, W.N., Heijmans, M.M.P.D., Mauquoy, D., Tamis, W.L.M., Van Vliet, A.J.H., Berendse, F., Van Geel, B., Van der Meijden, R. and Ulenberg, S.A. (2001) *Long-term Effects of Climate Change on Biodiversity and Ecosystems Processes*. Dutch National Research Programme on Global Air Pollution and Climate Change NOP report no. 410 200 089, Bilthoven, 306 pp.

van Raamsdonk, L.W.D. and van der Maesen, L.J.G. (1996) Crop–weed complexes: the complex relationship between crop plants and their wild relatives. *Acta Botanica Neerlandica* 45, 135–155.

van Tienderen, P.H., De Haan, A.A., van der Linden, C.G. and Vosman, B. (2002) Biodiversity assessment using markers for ecologically important traits. *Trends in Ecology and Evolution* 17, 577–582.

Waugh, R., McLean, K., Flavell, A.J., Pearce, S.R., Kumar, A., Thomas, B.B.T. and Powell, W. (1997) Genetic distribution of Bare-1-like retrotransposable elements in the barley genome revealed by sequence-specific amplification polymorphisms (S-SAP). *Molecular and General Genetics* 253, 687–694.

Westman, A.L., Levy, B.M., Miller, M.B., Gilles, G.J., Spira, T.P., Rajapakse, S., Tonkyn, D.W. and Abbott, A.G. (2001) The potential for gene flow from transgenic crops to related wild species: a case study in strawberry. *Acta Horticulturae* 560, 527–530.

Whitton, J., Wolf, D.E., Arias, D.M., Snow, A.A. and Rieseberg, L.H. (1997) The persistence of cultivar alleles in wild populations of sunflowers five generations after hybridization. *Theoretical and Applied Genetics* 95, 1–2.

Young, N.D. (2000) The genetic architecture of resistance. *Current Opinion in Plant Biology* 3, 285–290.

14 Introgression of Cultivar Beet Genes to Wild Beet in the Ukraine

OLEH SLYVCHENKO[1] AND DETLEF BARTSCH[2,3]

[1]Institute for Sugar Beet UAAS, Klinichna, 25, 03110 Kiev, Ukraine, E-mail: slyvchenko@hotmail.com; [2]Aachen University of Technology, Biology V, Worringerweg 1, D-52056 Aachen, Germany; [3]Present address: Robert Koch Institute, Center for Gene Technology, Wollankstraße 15–17, D-13187 Berlin, Germany, E-mail: BartschD@rki.de

Abstract

One concern about the use of genetically modified plants is the unintentional spread of new genes from cultivated plants to their wild relatives causing unwanted effects. The ecological and genetic consequences of gene flow depend on the amount and direction of gene flow as well as on the fitness of hybrids. Since wild relatives of cultivated plants are important plant genetic resources, the conservation of wild plants and their biodiversity has become an important task in providing biosafety of transgenic plants. In general, gene flow is hard to control in wind-pollinated plants. A well-studied subject in this respect is the genus *Beta*. We have shown recently that a century of gene flow from cultivated *Beta vulgaris* ssp. *vulgaris* has not altered the genetic diversity of wild *B. vulgaris* ssp. *maritima* in the Italian sugarbeet production area. Here, we present evidence of gene introgression from cultivated beet into wild *B. vulgaris* populations found on the Crimean Peninsula of the Ukraine. Crimean wild beets have two different origins, from European seabeet (*B. vulgaris* ssp. *maritima*) or from *Beta trigyna*. Population level patterns of isozyme variation for wild Crimean seabeets revealed a significantly higher genetic diversity than other European or US populations only in terms of Nei's heterozygosity and Shannon's I. Estimations on gene flow based on a single quantitative marker locus specific for sugarbeet and *F*-statistics suggest that the level of hybridization and introgression is high for seabeet on the Crimean Peninsula in comparison with other areas in Europe and the USA. Since sugarbeet seed production in the Ukraine offers opportunities for recombinant genes to escape to the wild relative, future consequences of such gene flow must be assessed and should be monitored for the protection of plant genetic resources.

Introduction

One concern about the use of genetically modified plants is the unintentional spread of the new genes from cultivated plants to their wild relatives causing unwanted effects. Most cultivated plants can hybridize naturally with wild relatives. The consequence of hybridization can lead to gene flow, i.e. the introgression of genes into the gene pool of one population from one or more populations (Futuyma, 1998). When the amount of gene flow between groups is high enough, gene flow has the effect of homogenizing genetic variation over the groups. When gene flow is low, genetic drift, selection and even mutation in the separate groups may lead to genetic differentiation (Hedrick, 2000). A particular concern of homogenizing wild plant genetic resources is loss of rare alleles that could be beneficial for use in plant breeding programmes.

The importance of gene flow in evolution is not fully agreed upon. A variety of different views are held: one view considers gene flow as common and that a small amount of gene flow among different parts of a species' range effectively unifies the species and significantly affects the genetic distinctiveness in each part of the range. A different view is that gene flow is uncommon and that natural selection acts more or less independently in each part of the species' range. Both of these views are probably correct for some species (Slatkin, 1981).

Most cultivated plants hybridize at least somewhere in their cultivation range with one or another wild relative taxon. On the basis of these data, Ellstrand *et al.* (1999) argue that many wild relatives are not evolutionarily independent of the crops, and that the outcome of the genetic contacts could be twofold. First, a sort of hybrid vigour effect may lead to formation of new weeds, or increasing invasiveness in existing ones. The weed beet type within the species *Beta vulgaris* is an example of this phenomenon in Europe (Van Dijk, Chapter 5, this volume; Cuguen *et al.*, Chapter 15, this volume; Soukup and Holec, Chapter 16, this volume). Secondly, hybridization and introgression may lead to genetic swamping of the recipient (wild relative, eventually rare) taxon, although the wild beet–cultivar beet example demonstrates that gene flow and introgression do not necessarily have adverse effects on genetic diversity. This effect was recently reported for western European sugarbeet seed production areas in south-western France and north-eastern Italy (Bartsch *et al.*, 1999; Desplanque *et al.*, 1999). However, the south-eastern Mediterranean area including the Black Sea region has, as far as we know, never been investigated for cultivar–wild plant introgression, although this region is regarded as the centre of origin and diversity for the genus *Beta* (Letschert, 1993).

Estimating Gene Flow and Introgression in Ukrainian Beet

Beets (*B. vulgaris* ssp. *vulgaris*) have been cultivated for more than 2000 years in the eastern Mediterranean region, but the best known cultivar of this species – sugarbeet – is a relatively young crop introduced for planting less than 200 years ago. From a geographical point of view, the Ukraine is the biggest sugarbeet market in the world. Approximately 1 Mha of sugarbeets are grown by farmers from year to year for sugar production. In addition, about 9000 ha are seed production areas that are mainly concentrated in the Black Sea region using direct plantation methods, or in the central part using an indirect method of steckling cultivation (Fig. 14.1). Domesticated beet seed production in the Black Sea region has been going on for more than 100 years, with an intensification of sugarbeet seed producing since the 1970s. In Crimea, commercial sugarbeet seeds are produced on 4500 ha, each hectare containing approximately 100,000 flowering plants. Furthermore, small farmers in the region grow red beet and Swiss chard for private seed production, which may be an additional source of gene flow from cultivated to wild populations, since all subspecies of *B. vulgaris* are wind pollinated.

We examined wild beet accessions of Ukrainian origin according to the methods previously described by Bartsch and Ellstrand (1999). We have used isozymes to characterize the genetic diversity of wild beet populations on the Crimean Peninsula and to compare it with the diversity of the cultivated *Beta* and putative wild progenitors. Seed material was obtained from seed companies, from international plant genetic resource collections or from collecting directly from wild populations (Bartsch *et al.*, 1999). Accessions were selected so that most of the geographical range of wild and cultivated beets in Europe and USA was represented. The use of *F*-statistics and calculation of gene flow within population groups by N_m values leads to false-positive values especially if the genetic similarity between populations

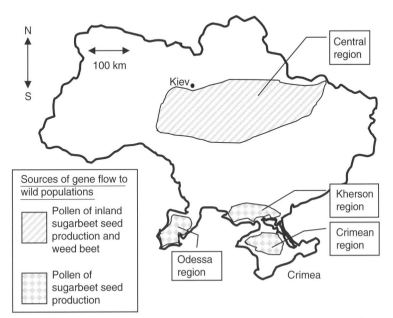

Fig. 14.1. Regions of potential gene flow from sugarbeet to wild beet populations in the Ukraine.

is relatively high. In the case of the Ukrainian wild–cultivar beet complex, we compared the relative difference in N_m values between distinct groups at a northern hemisphere scale. This comparison provides only indirect estimations of whether gene flow between cultivated and wild beet is (relatively) higher or lower in the Ukraine than in other areas of gene flow.

Plants and seed material of Crimean wild beet were sampled in 2001 at several locations of the Crimean Peninsula (Fig. 14.2). Based on morphology, we found two clearly distinct groups of wild beet: *B. vulgaris* ssp. *maritima* and *B. trigyna*. To evaluate the genetic origin and relatedness of Crimean wild beets, we first constructed an UPGMA dendrogram to elucidate the genetic relationships among 11 major groups of wild and cultivated beet populations based on Nei's (1978) genetic distances (Fig. 14.3). The dendrogram generally separates the accessions according to their taxonomic designations. Individual seabeet populations cluster in one group according to their geographical origin, and all sugarbeets group together independently of their breeding country origin (data not shown). Measured genetic divergence suggests that Crimean seabeet is intermediate between two different taxa, i.e. *B. vulgaris* ssp. *maritima* and *B. trigyna*, although they are morphologically not different from other European seabeet populations. As a first conclusion of our isozyme analysis, Ukrainian seabeets are clearly

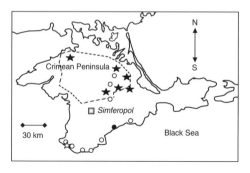

Fig. 14.2. Geographical distribution of wild *Beta* accessions on the Crimean Peninsula: stars, *Beta vulgaris* ssp. *maritima*; circles, *Beta trigyna*. Filled symbols represent accessions that were used for isozyme analyses. The local sugarbeet seed production area is surrounded by a dotted line. Simferopol is the district capital.

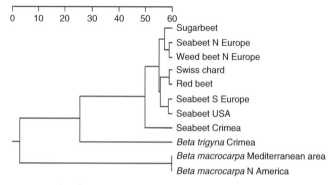

Fig. 14.3. UPGMA dendrogram of systematic relationships among 11 major groups of wild and cultivated beet based on Nei's (1978) genetic distances (relative scale) derived from allele frequencies at 13 polymorphic isozyme loci (unpublished data for Crimean beet; for other accession groups, see Bartsch *et al.*, 1999).

separated from European seabeets. However, we found only one unique allele that cannot be found in other *B. vulgaris* accession groups.

Sugarbeet as a crop has a relatively low variability, as its genetic base is very narrow. All modern cultivars were derived from the relatively young 'White Silesian' origin at the end of the 18th century (Fischer, 1989). The genetic variability in cultivated beet is also a result of the regular use of seabeet germplasm for breeding improvement. This practice has 'contaminated', for example, the sugarbeet gene pool with wild beet genes, which cannot be eliminated completely by backcrossing (Van Geyt *et al.*, 1990). This breeding history may also explain why diagnostic genetic markers, which can only be found in sugarbeet and not in seabeet, have not yet been described for isozymes. Only the quantitative frequency at two isozyme loci ('quantitative marker') provides evidence for introgression of cultivar alleles into the gene pool of local seabeet populations. We based our analyses on a recent publication by Bartsch *et al.* (1999) who found one allele, Mdh^{2-1}, that is specific to sugarbeet, relative to other cultivated types, and a second, Aco^{1-2}, that is present in high frequency in Swiss chard and red beet. Both alleles typically are rare in seabeet populations that are distant from seed production areas, but both are common in those that are near the Italian and Crimean cultivated beet seed production region, supporting the contention that gene flow from the crop to the wild species can be substantial when both grow in proximity (Table 14.1). We found surprisingly high levels of the Swiss chard/red beet-specific Aco^{1-2} allele within the seabeet accessions, although no particular commercial seed production of these cultivars is reported in this area. An explanation might be that local citizens produce seeds or allow flowering cultivar bolters on their own private land. We observed this phenomenon at least in private gardens at the Crimean south coast.

Our analysis was too limited to quantify gene flow or genetic diversity of *B. trigyna*. In the case of *B. vulgaris*, the results revealed significantly higher genetic diversity of Ukrainian seabeet populations in comparison with other

Table 14.1. Indication for gene flow from cultivated beet to wild beets. Two enzymes can be used as quantitative cultivar markers (Bartsch *et al.*, 1999): malate dehydrogenase (*Mdh*) isozyme[2-1] for sugarbeet and aconitase (*Aco*) isozyme[1-2] for Swiss chard and red beet.

	Isozyme frequency/group (%)	
	Mdh^{2-1}	Aco^{1-2}
Sugarbeet (14 varieties)	100	57
Swiss chard (5 varieties)	0	100
Red beet (7 varieties)	14	100
Seabeet (Europe, 19 accessions outside sugarbeet seed production areas)	21	53
Seabeet (NE Italy, 20 accessions near sugarbeet and red beet/Swiss chard seed production)	65	100
Seabeet (Crimea/Ukraine, 6 accessions)	100	100

beet accessions using either the unbiased estimate of gene diversity H (expected heterozygosity; Nei, 1978) (Table 14.2) or Shannon's I (Fig. 14.4), which were higher than that of the accessions growing far from seed production regions in northern Europe. The expected heterozygosity in Crimean seabeet is even higher than in sugarbeet, a cultivar known to express high levels of heterozygosity due to hybridization of high inbred lines (Tables 14.2). Independently of the supspecies comparison, the number of species-unique *B. vulgaris* alleles is 29, which is lower than the 37 of the southern European group and similar to the number of northern European and US groups.

Predictions that the crop to wild gene flow should result in different genetic variation (Ellstrand *et al.*, 1999) can be confirmed for Crimean seabeet. For Nei's gene diversity H, we found an increase in per-accession genetic variation in our Crimean wild beet populations compared with their counterparts from elsewhere in the range of the wild subspecies. There is obviously less diversity with respect to the number of unique alleles in comparison with the southern Europe group which consists mainly of Mediterranean accessions. However, it is unusual for a small portion of a wild taxon's range (in the order of a few hundred square kilometres) to hold as much as, or more diversity than the vast majority of its range

Table 14.2. Genetic diversity statistics for seven major groups of genus *Beta*.

	N^a	A	A_p	P	H	U	F_{ST}	N_m
B. vulgaris (all)	23.2	2.92	3.27	0.846	0.294	38	0.169	1.22
Sugarbeet (16 varieties)	29.8	2.08	2.36	0.846	0.300	27	0.171	1.21
Swiss chard (4 varieties)	30.0	1.76	2.22	0.692	0.200	23	0.140	1.54
Red beet (5 varieties)	24.0	2.00	2.56	0.692	0.195	26	0.189	1.07
Weed beet (1 accession)	30.0	2.00	2.40	0.769	0.260	26	–	–
Seabeet Europe North (11 accessions)	22.8	2.31	2.63	0.846	0.196	30	0.371	0.42
Seabeet Europe South (27 accessions)	21.7	**2.85**	**3.18**	0.846	0.294	37	0.309	0.56
Seabeet Ukraine (6 accessions)	20.0	2.23	2.45	0.846	**0.320**	29	0.091	**2.51**
Seabeet USA (13 accessions)	15.5	2.23	2.60	0.769	0.247	29	0.142	1.52
B. trigyna Crimea (1 accession)	7.0	1.92	2.33	0.692	0.335	19	–	–
B. marcocarpa N. America (10 accessions)	30.4	2.46	3.22	0.682	0.057	25	0.107	2.10
B. macrocarpa Mediterranean (4 accessions)	23.6	1.62	2.50	0.462	0.063	16	0.199	1.01

Abbreviations for gene diversity statistics: N^a, average number of plants sampled per accession; A, average number of alleles per locus; A_p, average number of alleles per polymorphic locus; P, proportion of loci which are polymorphic; H, gene diversity or expected heterozygosity *sensu* Nei; and U, number of unique alleles per group (38 found with the *B. vulgaris* alleles, 19 with *B. trigyna*, and 26 with *B. macrocarpa*). F_{ST}, Nei's (1978) summary F-statistics for population differentiation between populations within *B. vulgaris* and *B. macrocarpa*; N_m, estimated genetic migration from $1/4 (1-F_{ST})/F_{ST}$ (Slatkin and Barton, 1989). The highest values within *B. vulgaris* subgroups are in bold.

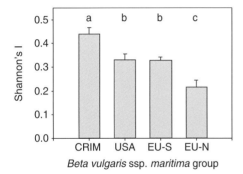

Fig. 14.4. Shannon's diversity index based on isozyme allele frequency at 13 loci (Bartsch *et al.*, 1999) within four population groups. Mean values of single populations within a geographical group ± SEM. Different letters over the bars reveal significant differences.

(encompassing tens of thousands of square kilometres). We do not know whether rare alleles have already been lost in Crimean populations due to gene flow and homogenization in the past. Gene flow, estimated as N_m, within Ukrainian accessions is relatively higher than in other accession groups (Table 14.2).

Our genetic information is insufficient to determine the extent to which hybridization of cultivated beet with seabeet and/or direct introduction of *B. vulgaris* ssp. *maritima* from outside Crimea may have contributed to the local wild beet evolution. Despite the postulated relatively higher amount of gene flow in the Crimean Peninsula area, local seabeets are genetically distinct from cultivated beets as well as from other seabeet populations. Hybridization and introgression has not led to genetic homogenization or gene swamping from whatever cultivated beet or interspecific wild beet gene pool. However, the highest genetic identity of Crimean seabeet was found with the southern European seabeet group and the sugarbeet group (Nei's genetic identity $I = 0.8535$) in comparison with the northern European seabeet group and the sugarbeet group ($I = 0.7834$ and $I = 0.8167$).

The Crimean Peninsula is a 'centre of diversity' for *B. vulgaris* ssp. *maritima*. It is not clear whether the diversity we observed may have been influenced by crop to wild gene flow in the past. Given the fact that seabeets on the Crimean Peninsula grow so close to overwhelming numbers of cross-compatible, wind-pollinated crop plants, gene flow seems likely to be the most parsimonious explanation of the current genetic structure of wild seabeet populations. Indeed, we can suggest several reasons why gene flow from the crop may have led to homogenization of genetic diversity in this particular system.

1. Although most crops examined have low genetic diversity compared with their wild relatives (Ladizinsky, 1985; Doebley, 1989), beet cultivars typically have a level of genetic diversity roughly equivalent to that of their wild progenitor (Table 14.2). Thus, if gene flow from sugarbeet were to proceed to equilibrium, we would expect similar levels of diversity in cultivar and wild beets.
2. Seabeets in Crimea may have received gene flow from many cultivars over the last century as new varieties have emerged (Van Geyt *et al.*, 1990). If each new cultivar contained less variation than the previous one, we might

expect an erosion in diversity in the wild populations receiving gene flow. However, if the new cultivars were well differentiated from the previous ones, we might see at least a temporary accumulation of different alleles in the wild populations.

3.   In this particular system, seabeets have received gene flow from both sugarbeet and red beet/Swiss chard, as our data have demonstrated (Table 14.1). These different groups of cultivars are genetically distinct (Table 14.2). A population receiving gene flow from two well-differentiated sources would be expected to evolve more diversity than one receiving gene flow from a single source.

4.   The system is very likely not to be in equilibrium. Under high levels of unidirectional gene flow from a source fixed for a novel allele, recipient populations would experience a transient increase in allelic richness as that allele accumulates over generations. Fixation of that allele would eventually occur if not opposed by natural selection, mutation or gene flow from an alternative source.

Whether these populations are in equilibrium or not, it is worth noting that a century of crop to wild gene flow had an evolutionary effect on the wild populations. Although natural selection undoubtedly plays a role in limiting the establishment of certain domesticated alleles, we acknowledge that gene flow also may be somewhat limited. Many descriptive genetic studies on other crop species have provided evidence for crop to wild introgression (reviewed by Doebley, 1989).

Consequences for GM Beet

Sugarbeet is an important crop of Europe, North America, the Near East, Chile and Japan. Consequently, it has become subjected to transformation by recombinant DNA technology. There has been concern that transgenic traits may cause unwanted effects after they escape via hybridization into seabeet populations (Bartsch et al., 2003). Transgenes may be more likely to alter the fitness of hybrid or introgressed individuals than supposedly neutral alleles such as isozymes. Therefore, the introgression of transgenes into wild populations may change their niche relationships (Ellstrand, 1992). Given that neutral crop alleles apparently move with ease into populations of the seabeets of north-eastern Italy, south-western France and the Crimean Peninsula, we suggest that these populations should be monitored after seed production of transgenic beets starts in this region. Our study provides baseline data on the current gene flow and genetic diversity prior to the introductions of GMOs.

It is hard to judge the ecological genetic impact of a century of gene flow from traditionally bred cultivated beets into the seabeet populations of the Crimean Peninsula. We have not quantified the crop to wild gene flow rates in this system, nor have we quantified the fitness consequences of crop alleles in the wild. Therefore, we do not have the parameters to assess the long-term evolutionary impact of unilateral crop gene flow into the wild beet populations (Ellstrand *et al.*, 1999).

A comprehensive assessment of potential effects of genetically modified plants has to be performed taking into account that conventional crops also cross with wild plants (Saeglitz and Bartsch, 2002). We know that containment strategies for the introduction of transgenic cultivars do not work properly (Sukopp and Sukopp, 1993; Saeglitz *et al.*, 2000). The ecological and genetic consequences of gene flow depend on the amount and direction of gene flow as well as on the fitness of transgenic hybrids (Bartsch and Schuphan, 2002). An assessment of environmental effects can realistically only be based on end-points and consequences of gene introgression, which include socioeconomic values of biodiversity in littoral and other ecosystems containing wild beet. In general, there is still a great need to harmonize worldwide monitoring and assessment systems by the development of appropriate methods to evaluate the environmental impact of introgressed transgenes. We need more detailed studies on the impacts of cultivar–wild gene flow on genetic diversity and genetic resources.

Acknowledgement

We thank the European Science Foundation AIGM programme for financial support of our studies in the Ukraine.

References

Bartsch, D. and Ellstrand, N.C. (1999) Genetic evidence for the origin of Californian wild beets (genus *Beta*). *Theoretical and Applied Genetics* 99, 1129–1130.

Bartsch, D. and Schuphan, I. (2002) Lessons we can learn from ecological biosafety research. *Journal of Biotechnology* 98, 71–77.

Bartsch, D., Lehnen, M., Clegg, J., Pohl-Orf, M., Schuphan, I. and Ellstrand, N.C. (1999) Impact of gene flow from cultivated beet on genetic diversity of wild sea beet populations. *Molecular Ecology* 8, 1733–1741.

Bartsch, D., Cuguen, J., Biancardi, E. and Sweet, J. (2003) Environmental implications of gene flow from sugar beet to wild beet – current status and future research needs. *Environmental Biosafety Research* 2, 105–115.

Desplanque, B., Boudry, P., Broomberg, K., Saumitou-Laprade, P., Cuguen, J. and Van Dijk, H. (1999) Genetic diversity and gene flow between wild, cultivated and weedy forms of *Beta vulgaris* L. (*Chenopodiaceae*), assessed by RFLP and microsatellite markers. *Theoretical and Applied Genetics* 98, 1194–1201.

Doebley, J. (1989) Isozymic evidence and the evolution of crop plants. In: Soltis, D.E. and Soltis, P.S. (eds) *Isozymes in Plant Biology*. Dioscorides Press, Portland, Oregon, pp. 87–105.

Ellstrand, N.C. (1992) Gene flow by pollen: implications for plant conservation genetics. *Oikos* 63, 77–86.

Ellstrand, N.C., Prentice, H.C. and Hancock, J.F. (1999) Gene flow and introgression from domesticated plants into their wild relatives. *Annual Review of Ecology and Systematics* 30, 539–563.

Fischer, H.E. (1989) Origin of the 'Weisse Schlesische Rübe' (white Silesian beet) and resynthesis of sugar beet. *Euphytica* 41, 75–80.

Futuyma, D.J. (1998) *Evolutionary Biology*, 3rd edn. Sinauer Press, Sunderland, Massachusetts.

Hedrick, P.W. (2000) *Genetics of Populations*, 2nd edn. Jones and Bartlett Publishers, Sudbury.

Ladizinsky, G. (1985) Founder effect in crop-plant evolution. *Economic Botany* 39, 191–199.

Letschert, J.P.W. (1993) *Beta* section *Beta*: biogeographical patterns of variation and taxonomy. *Wageningen Agricultural University Papers* 93, 1–137.

Nei, M. (1978) Estimation of average heterozygosity and genetic distance from small number of individuals. *Genetics* 89, 583–590.

Saeglitz, C. and Bartsch, D. (2002) Plant gene flow consequences. *AgBiotechNet* 4, ABN 084.

Saeglitz, C., Pohl, M. and Bartsch, D. (2000) Monitoring gene escape from transgenic sugar beet using cytoplasmic male sterile bait plants. *Molecular Ecology* 9, 2035–2040.

Slatkin, M. (1981) Estimating levels of gene flow in natural populations. *Genetics* 99, 323–335.

Slatkin, M. and Barton, N.H. (1989) A comparison of three indirect methods for estimating average levels of gene flow. *Evolution* 43, 1349–1368.

Sukopp, H. and Sukopp, U. (1993) Ecological long-term effects of cultigens becoming feral and of naturalization of non-native species. *Experientia* 49, 210–218.

Van Geyt, J.P.C., Lange, W., Oleo, M., De Bock, T.S.M. (1990) Natural variation within the genus *Beta* and its possible use for breeding sugar beet: a review. *Euphytica* 49, 57–76.

15 Crop–Wild Interaction Within the *Beta vulgaris* Complex: a Comparative Analysis of Genetic Diversity Between Seabeet and Weed Beet Populations Within the French Sugarbeet Production Area

JOËL CUGUEN[1], JEAN-FRANÇOIS ARNAUD[1],
MAXIME DELESCLUSE[1] AND FRÉDÉRIQUE VIARD[1,2]

[1]UMR CNRS 8016, Laboratoire de Génétique et Evolution des Populations
Végétales, Bâtiment SN2, Université de Lille 1, 59655 Villeneuve d'Ascq
cedex, France; E-mail: joel.cuguen@univ-lille1.fr; [2]Present address: UMR
7127, EGPM, Station Biologique, Place Georges-Teissier, BP 74, 29682
Roscoff cedex, France

Abstract

Among the commonly listed risks associated with environmental release of genetically engineered cultivars is the hybridization of transgenic plants with wild relatives. The beet complex is of particular interest, as crop (*Beta vulgaris* ssp. *vulgaris*), wild (*B. vulgaris* ssp. *maritima*) and weedy forms are all interfertile and can be found in sympatry in various places in Europe, hence heightening the likelihood of accidental hybridization events. In northern France, sugarbeet fields can be found close to the coastline, together with the common wild seabeet populations. Nine wild populations and 12 weedy populations were sampled and examined for diversity of chloroplastic and nuclear DNA using polymerase chain reaction–restriction fragment length polymorphism (PCR–RFLP) and microsatellite length variation. Most of the weed populations were characterized by a unique haplotype, named Owen CMS and characteristic of cultivated lineages. Although no diagnostic allele was depicted using microsatellite markers, a highly significant genetic differentiation was found between weed and wild forms. Furthermore, weed populations displayed a significantly lower allelic diversity in addition to a systematic heterozygote deficit compared with Hardy–Weinberg expectations. This suggests either an introgression between genetically differentiated gene pools within the region of seed production (i.e. a spatial Wahlund effect, as well as a temporal one), or the presence of self-compatibility alleles commonly used in breeding programmes. Taken together, our results seem to indicate a low overall level of gene flow from weed to seabeet populations in the French sugarbeet production area. Nevertheless, gene exchanges from crop to wild relatives were depicted for some seabeet populations, indicating the possibility of local introgression and highlighting the importance of case-by-case and long-term monitoring surveys.

Introduction

Investigating the genetic structure of plant populations is of major concern to elucidate the relative effects of fundamental evolutionary consequences of genetic drift, gene flow or selective pressures on genetic diversity (Hamrick and Nason, 1996; Raybould *et al.*, 2002). The genetic features of plant populations also have important implications for the management and conservation of genetic resources (Frankham *et al.*, 2002; Pullin, 2002) as well as for biosafety studies such as the assessment of crop–wild interactions and, consequently, the likelihood of transgene escape from crop to wild relatives (Jarvis and Hodgkin, 1999; Bartsch and Schuphan, 2002). Indeed, over the last decade, the development of transgenic plants has been accompanied by increasing concerns about the potential effects of the commercial release of genetically modified crops (Raybould and Gray, 1994; Gray and Raybould, 1998; Hails, 2000; Ellstrand, 2001). The escape of a transgenic trait from cultivated to wild relatives is a three-step process involving hybridization, a subsequent spread, and establishment, that require pre-mating and post-mating barriers to be overcome. An increasing number of studies have documented gene exchanges from crops into their wild relatives (reviewed in Ellstrand *et al.*, 1999). Furthermore, many plant species can be found both as crop and wild types, but also as weedy types. If crops, their wild and weedy relatives co-occur, overlap in flowering period, share pollinators and are cross-compatible, then hybridization events have a strong potential to introduce transgenes into weedy and/or wild populations (Raybould and Gray, 1994; van Raamsdonk and van der Maesen, 1996; Gray and Raybould, 1998; Treu and Emberlin, 2000).

Concerning the potential effects of gene flow and introgression between domesticated plants and their wild relatives, the *Beta vulgaris* complex is of immediate concern as wild, crop and weedy forms are all present in Europe (Letschert, 1993; Gray and Raybould, 1998; Bartsch *et al.*, 1999; Desplanque *et al.*, 1999). Four major groups can be described within the *B. vulgaris* complex: (i) the common wild seabeet *B. vulgaris* ssp. *maritima* that occurs naturally throughout the eastern Mediterranean region and Asia Minor and has spread along Europe's Atlantic coast to the North and Baltic Seas; (ii) the cultivated beet *B. vulgaris* ssp. *vulgaris*, including Swiss chard, red garden beet, fodder and sugarbeet; (iii) inland or ruderal beets typically encountered within markedly man-disturbed area and present within the seed production areas located in northern Italy and southern France (similar populations occur in other sugarbeet-growing regions) (Hornsey and Arnold, 1979); and (iv) weed beets infesting sugarbeet fields and mostly resulting from recurrent accidental hybridization events between cultivated seed bearers and nearby ruderal or wild beets during the seed production (Hornsey and Arnold, 1979; Santoni and Berville, 1992; Boudry *et al.*, 1993; Longden, 1993; Desplanque *et al.*, 1999; van Dijk and Desplanque, 1999). Weed beets can be a serious problem in sugarbeet fields, with significant effects on sugarbeet yield and quality at high density (Longden, 1989).

Weedy beets are also of particular interest because they can act as an indirect route for gene transfer from crops into natural populations of seabeets (Treu and Emberlin, 2000; Desplanque *et al.*, 2002; Eastham and Sweet, 2002; Viard *et al.*, 2002). Indeed, commercial varieties of sugarbeets are typically biennial, producing in the first year a conical shaped root used for sugar production and being harvested before the flowering and pollination period the next year. The biennial nature of cultivated beets is under the genetic control of the non-bolting allele *b* (Boudry *et al.*, 1994; Hansen *et al.*, 2001; El-Mezawy *et al.*, 2002). In contrast, the weed beets do not need any vernalization requirement because they inherited from ruderal forms the dominant bolting *B* allele, enabling the possibility of bolting and flowering in their first year (Boudry *et al.*, 1993). These contaminant weed beet hybrids have given rise to weedy lineages, whose crucial novelty of first-year flowering pre-adapts them for invasive success in disturbed agricultural environments (van Dijk and Desplanque, 1999; Ellstrand and Schierenbeck, 2000). Together with the 'variety bolters' (cultivar bolters arising due to early sowing and/or cold weather conditions), weed beets can flower, reproduce and shed large amounts of seeds (Mücher *et al.*, 2000; Treu and Emberlin, 2000; Bartsch and Schuphan, 2002; Viard *et al.*, 2002). As cultivated beets, wild seabeets as well as weedy beets are all highly interfertile, wind-pollinated and obligatory outcrossing species, gene flow from crop to wild may be expected with a high probability (Eastham and Sweet, 2002).

It is now well established that the main source of weediness in beet is caused by pollen transfer from wild or ruderal forms of *B. vulgaris* into seed production fields of cultivated forms (Boudry *et al.*, 1993; Bartsch *et al.*, 1999; Desplanque *et al.*, 1999). However, up to now, there were no published studies investigating the genetic features of wild and weedy populations within the sugarbeet production areas. The French sugarbeet production area is located in northern regions, and many fields can be found along the coastlines of northern France. Consequently, numerous crops bordering the coast can be found close to wild seabeet populations encountered along the sea shore (high water margins) as well as in estuarial zones. Therefore, the present study was devoted to carry out a comparative analysis of the genetic diversity of: (i) typical wild seabeet populations collected along the coastline; and (ii) nearby weed populations sampled within sugarbeet fields of the French production area. To address this issue, we used maternally (chloroplastic DNA) and biparentally (nuclear microsatellite loci) inherited genetic markers that hold the potential to gain further insights into the genetic inter-relationships within the complex *Beta*.

Materials and Methods

Sampling collection

Nine wild seabeet populations were collected together with 12 populations
of weed beets sampled as close as possible to the French coast (Fig. 15.1). The
distance between wild and weed populations ranges between 1 and 10 km.
A total of 803 individuals were collected with a mean sample size of 38
individuals per population (309 seabeet and 494 weed beet individuals,
respectively). Detailed sample sizes as well as geographical coordinates of
studied populations can be found in Table 15.1. As notified by Treu and
Emberlin (2000) and Viard *et al.* (2002), it is important to distinguish the
weed beets, that result from the seed bank and bolt due to the presence of the
dominant *B* allele, from sugarbeet bolters that arise due to low vernalization
requirement and/or cold weather conditions. Consequently, only weed
individuals bolting outside the sowing line were sampled. These 'out-row'
(following the terminology of Viard *et al.* (2002)) weed beets originate from
the seed bank and are the result of several generations of intercrossing.

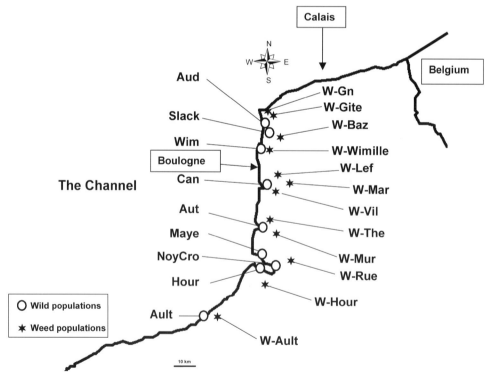

Fig. 15.1. Map of sampled populations located along the French coast of the Channel in
the sugarbeet production area. The sampling collection consists of: (i) nine wild seabeet
populations collected along the seashore; and (ii) 12 weed beet populations collected as
close as possible within sugarbeet fields.

Table 15.1. Populations and their acronyms, geographical coordinates of sampling locations, and some genetic diversity parameters calculated over all loci.

Population	Acronym	Latitude	Longitude	N	A_r	Ho	He	F_{is}
Weed populations								
Cap Gris-nez	W-Gite	N 50°51.0′	E 1°35.2′	39	5.12	0.44	0.56	0.209
Falaise Gris-nez	W-Gn	N 50°50.4′	E 1°35.2′	40	3.84	0.47	0.53	0.126
Bazinghem	W-Baz	N 50°47.3′	E 1°38.6′	40	4.94	0.53	0.58	0.084
Wimille	W-Wim	N 50°45.9′	E 1°38.3′	40	5.49	0.39	0.62	0.371
Lefaux	W-Lef	N 50°34.3′	E 1°38.8′	39	5.96	0.52	0.60	0.131
Maresville	W-Mar	N 50°31.3′	E 1°42.6′	42	5.63	0.42	0.61	0.320
Villiers	W-Vil	N 50°29.7′	E 1°39.0′	39	6.12	0.47	0.62	0.234
Tigny	W-The	N 50°21.0′	E 1°41.7′	42	4.51	0.42	0.54	0.216
Le Muret	W-Mur	N 50°20.0′	E 1°39.0′	40	6.22	0.48	0.65	0.276
Le Champ Neuf	W-Rue	N 50°15.8′	E 1°37.0′	41	4.62	0.41	0.52	0.225
Le Hourdel	W-Hour	N 50°11.2′	E 1°33.2′	40	6.09	0.49	0.58	0.170
Hable d'Ault	W-Ault	N 50°09.6′	E 1°29.6′	52	4.89	0.41	0.55	0.263
Mean					5.28	0.45	0.58	0.222
Wild seabeet populations								
Audresselles	Aud	N 50°49.4′	E 1°35.4′	30	7.27	0.54	0.61	0.104
Ambleteuse	Slack	N 50°48.5′	E 1°36.4′	37	6.65	0.57	0.63	0.095
Wimereux	Wim	N 50°46.1′	E 1°36.6′	38	7.18	0.61	0.61	−0.006
Etaples	Can	N 50°30.9′	E 1°38.4′	39	6.99	0.56	0.64	0.124
La Madelon	Aut	N 50°22.3′	E 1°37.6′	33	5.12	0.53	0.56	0.064
Maye	Maye	N 50°15.7′	E 1°37.2′	36	5.79	0.52	0.56	0.074
Noyelles Crotoy	NoyCro	N 50°12.8′	E 1°40.1′	34	5.79	0.50	0.55	0.088
Le Hourdel	Hour	N 50°12.9′	E 1°33.9′	27	4.86	0.50	0.48	−0.052
Ault	Ault	N 50°06.4′	E 1°27.2′	35	5.93	0.54	0.64	0.164
Mean					6.18	0.54	0.59	0.073

N, sample size; A_r, allelic richness estimated following the rarefaction method developed by El Mousadik and Petit (1996); *Ho*, mean observed heterozygosity; *He*, mean expected heterozygosity (gene diversity); F_{is}, intrapopulation fixation index according to Weir and Cockerham (1984).

Genetic data collection

DNA was extracted and purified from dried leaves by using a DNeasy® 96 Plant Kit and following the standard protocol for isolation of DNA from plant leaf tissue outlined in the DNeasy 96 Plant protocol handbook (QIAGEN Inc.).

The maternal cultivated origin was assessed by means of a chloroplastic polymerase chain reaction–restriction fragment length polymorphism (PCR–RFLP) marker defined by Ran and Michaelis (1995). Weedy lineages generally carry a chloroplastic haplotype which is characteristic of cultivated lines, i.e. the Owen cytoplasmic male sterility (CMS) universally used in sugarbeet breeding programmes (Owen, 1945; Boudry *et al.*, 1993). *Hind*III digests of cpDNA from non-Owen CMS and Owen CMS lines are

characterized by one *Hind*III fragment of 563 bp and two *Hind*III fragments
of 454 and 109 bp, respectively (Ran and Michaelis, 1995). Note that the
non-Owen CMS group comprises different haplotypes, not distinguishable
with this technique (Desplanque *et al.*, 2000). Primers used, PCR conditions
and DNA digestion were applied as described by Ran and Michaelis (1995).
Digested DNA products were separated by 0.8% agarose gel electrophoresis
and visualized after ethidium bromide staining under UV light.

The nuclear polymorphism was assessed using highly polymorphic
microsatellite DNA markers, which have become the markers of choice for
many applications including population genetic structure (Beaumont and
Bruford, 1999). Microsatellites are short stretches of tandemly repeated DNA
sequence (no more than six bases long) that are ubiquitously interspersed in
eukaryotic genomes, and generally show an exceptional variability in most
species. In this study, seven loci (CT4, GTT1, GCC1, GAA1, BVM3, CAA1
and CA2) were used among those previously described in Mörchen *et al.*
(1996), Viard *et al.* (2002) and Arnaud *et al.* (2003). They were selected
for their high polymorphism and their unambiguous amplification pattern.
Electrophoresis and genotyping were carried out on a LI-COR automated
sequencer (LI-COR Inc., Nebraska).

Statistical analyses

For each population, genetic polymorphism was examined by calculating
allelic frequencies for each locus, allelic richness (A_r) (El Mousadik and Petit,
1996), the observed heterozygosity (*Ho*), the genetic diversity *sensu* Nei
(1978) (expected heterozygosity under Hardy–Weinberg expectations, *He*)
and the unbiased intrapopulation fixation index (F_{is}, a measure of departure
from panmixia within populations) over all loci using either GENEPOP
version 3.3 (Raymond and Rousset, 1995, available at http://www.
cefe.cnrs-mop.fr/) or FSTAT version 2.9.3.2. (Goudet, 1995, available at
http://www.unil.ch/izea/softwares/fstat.html). Differences in mean *He*, F_{is}
and allelic richness were tested among weed and wild groups by using
permutation procedures implemented in FSTAT. Genotypic linkage disequilibrium was estimated prior to other analyses using GENEPOP. Bonferroni
adjustments for simultaneous statistical tests were applied (Rice, 1989).

There is currently no clear consensus concerning alternative methods for
quantifying genetic differentiation and genetic distances using microsatellite
data. Theoretical studies suggest that methods taking into account the mutational differences among alleles are most appropriate for estimating genetic
substructuring or phylogenetic relationships, using the stepwise model of
mutation thought to fit microsatellite evolution (reviewed in Estoup and
Cornuet, 1999; Rousset, 2001; Raybould *et al.*, 2002). However, the general
applicability of these models to data derived from natural populations
remains unclear, especially when some particular microsatellite loci may not
follow a strict stepwise pattern. All of the loci used in this study had pairs
of alleles that differed by less than one repeat unit, and thus cannot be

considered as 'perfect' loci. As such, we chose to assess population differentiation using classical Wright's *F*-statistics. *F*-statistics were computed according to the ANOVA procedure of Weir and Cockerham (1984) and tested for significance using permutation procedures implemented in FSTAT. Genetic relationships among wild and weed populations were also depicted using an unrooted neighbour-joining (NJ) tree constructed with the genetic distance introduced by Reynolds *et al.* (1983) based on allele frequencies. This distance makes no assumption regarding mutation rates among loci, assumes a pure genetic drift model as the evolutionary process and appears, therefore, most appropriate for closely related populations. Bootstrapped confidence values on depicted branches were determined using random resampling replications over loci (1000 replicates).

Results

Chloroplastic diversity

In the present study, a tremendous difference was found between weed and seabeet populations for the presence of the Owen CMS cytoplasm: most of the weed individuals carried the Owen CMS type (89.5%), whereas seabeet individuals were only exceptionally found to have this cytoplasm (0.9%; Fig. 15.2). None the less, three out of the 12 weed populations were found to be polymorphic, with a high frequency of non-Owen CMS haplotypes (*W-Gite*, 87%; *W-Mar*, 19%; and *W-Ault*, 15%). Among the nine seabeet populations sampled, only two contained a few individuals carrying the Owen CMS cytoplasm (3 and 4% in *Aut* and *Ault* populations, respectively), evidencing the trace of gene flow through seeds from crop to wild populations.

Microsatellite polymorphism

Microsatellite loci displayed high levels of polymorphism, with a total number of alleles per locus varying from four (GTT1 and GCC1), five (GAA1), seven (CA2), 15 (CT4) and 30 (CAA1) to 31 (BVM3) for an overall total of 96 alleles. Exact tests for genotypic linkage disequilibrium within each population depicted five (1.13%) significant *P*-values out of 441 comparisons, i.e. a proportion comparable with that expected by chance alone. By testing each pair of loci across all population samples (Fisher's method), multiple probability tests only gave two significant *P*-adjusted values out of the 21 possible pairs of loci, involving GTT1/GCC1 and GTT1/BVM3 (χ^2_{42} = 74.06 and χ^2_{42} = 94.03, respectively, $P < 0.05$ after Bonferroni correction).

Summary statistics describing the genetic diversity within the 21 sampled populations are given in Table 15.1. At the intrapopulation level, mean observed heterozygosity (*Ho*) per population ranged from 0.41 to 0.61. No significant difference (*P* = 0.56) was depicted between wild and weedy lineages concerning the unbiased estimates of genetic diversity (*He*).

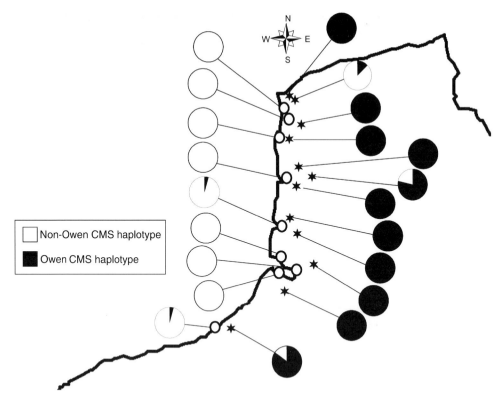

Fig. 15.2. Distribution of chloroplast diversity within wild and weed beet populations. Black refers to the percentage of Owen CMS haplotypes (specific for cultivated lines), whereas white indicates the percentage of non-Owen CMS haplotypes.

However, mean allelic richness (A_r) as well as mean *Ho* were significantly higher for wild beet populations as compared with weedy beet populations ($P < 0.001$, after 10,000 permutations). Conversely, weed beet populations exhibited a mean F_{is} significantly higher ($P < 0.001$, after 10,000 permutations) than that observed for the wild group. These high values of F_{is} within weed populations suggest consistent heterozygote deficiencies, whereas wild seabeet populations generally fit the Hardy–Weinberg expectations (see Table 15.1). Although we did not find any obvious diagnostic alleles to discriminate cultivated/weed lineages from wild types of *B. vulgaris*, some loci displayed contrasting allele frequencies. Striking examples of microsatellite genetic diversity as well as differences in allelic frequency distributions between wild/weed populations are shown in Fig. 15.3.

A significant amount of genetic differentiation ($F_{st} = 0.115$, $P < 10^{-4}$) occurred among all the 21 populations. Overall genetic substructuring did not differ significantly between wild and weed populations when these two groups were considered separately (mean F_{st} was equal, respectively, to 0.063 and 0.062, $P < 10^{-4}$, within each group of populations). None the less, by partitioning the genetic variance between wild and weed lineages, 11.3%

$(P < 10^{-4})$ of the total genetic variation is apportioned among these two lineages. This interlineage divergence in allele frequencies differed between loci, and was most evident at loci GTT1 and CA2 that exhibited highly significant individual F_{st} values of 0.278 and 0.129, respectively (see also Fig. 15.3). In contrast, locus GCC1 was not informative enough to reveal a clear genetic differentiation between wild and weedy lineages ($F_{st} = 0.003$). Overall, the substantial interlineage differentiation is mirrored by the NJ tree of population relationships that depicted a non-ambiguous genetic cleavage between wild seabeet populations and their weedy relatives (Fig. 15.4). The topology of the tree further did not indicate any obvious concordance between the genetic clustering and the spatial location of populations within both groups, i.e. no clear geographical pattern emerged such as isolation by distance effects. The *Ault* seabeet population is of special interest as it occupies an intermediate position on the NJ tree. The examination of its genetic variation reveals the existence of some individuals exhibiting the Owen CMS cytoplasm and the presence at relatively high frequency of microsatellite alleles generally characteristic of weed beet populations. This is particularly spectacular for the GTT1 locus (see Fig. 15.3A). This intermixing of different alleles should indicate the trace of gene exchanges from weed to wild through pollen and seed flow.

Discussion

Genetic diversity of weed/seabeet populations and crop to wild gene flow

The usual view is that invasive species such as weeds should be relatively genetically depauperate owing to the bottle-necks associated with their recurrent colonization dynamics. As a result, subsequent hybridization with wild relatives may lead to genetic erosion and/or genetic pollution of the natural gene pool of the recipient population (Raybould and Gray, 1994; Bartsch *et al.*, 1999; Frankham *et al.*, 2002). Consequently, there are concerns that crop to wild gene exchanges would cause a decrease in genetic variation in wild populations. Sugarbeet is a relatively recently cultivated crop, originating from Silesian fodder beet dating back only 200 years (Treu and Emberlin, 2000). All modern sugarbeets are thought to originate from a few cultivars and, as such, are expected to be characterized by relatively low levels of genetic diversity. However, extensive use of seabeet gene resources in conventional breeding programmes (Ford-Lloyd and Williams, 1975; Lange *et al.*, 1999), and spontaneous hybridization as well as contamination of sugarbeet with the wild beet types may have contributed to the high genetic diversity found within sugarbeet lines (see Bartsch *et al.*, 1999). Furthermore, weed beets originally result from recurrent hybridization events between cultivated and inland ruderal beets during the sugarbeet seed multiplication (Boudry *et al.*, 1993; Treu and Emberlin, 2000). As a result, hybridizations between genetically well-differentiated populations from different geographical locations (i.e. crop–ruderal hybrid seed coming

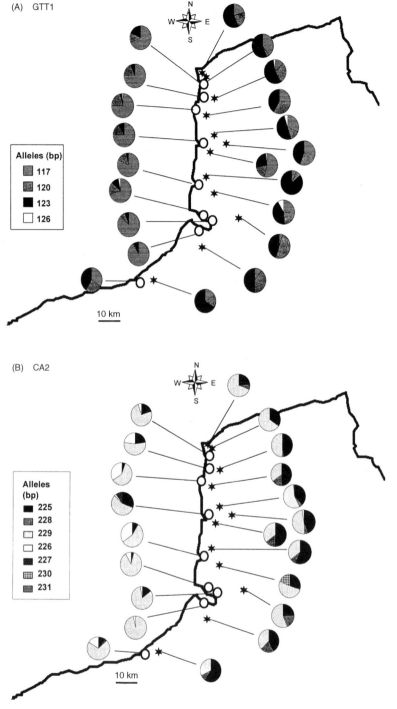

Fig. 15.3. Allelic frequency distributions for loci (A) GTT1, (B) CA2, (C) CT4 and (D) CAA1 for wild and weed beet populations. Alleles are denoted by their total size in base pairs (bp).

(C) CT4

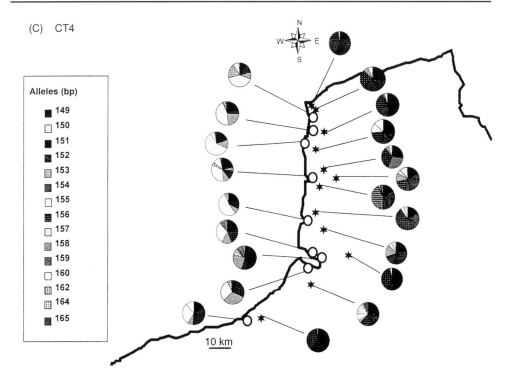

Alleles (bp)

- ■ 149
- □ 150
- ■ 151
- ▦ 152
- ▤ 153
- ▨ 154
- □ 155
- ▦ 156
- □ 157
- ▨ 158
- ▨ 159
- □ 160
- ▦ 162
- ▦ 164
- ▨ 165

10 km

(D) CAA1

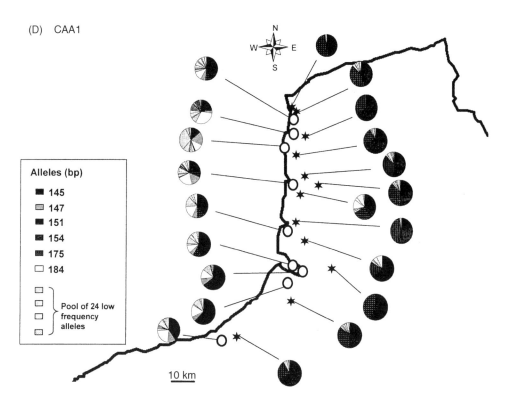

Alleles (bp)

- ■ 145
- ▨ 147
- ■ 151
- ▦ 154
- ▦ 175
- □ 184

- □
- □   Pool of 24 low
- □   frequency
- □   alleles

10 km

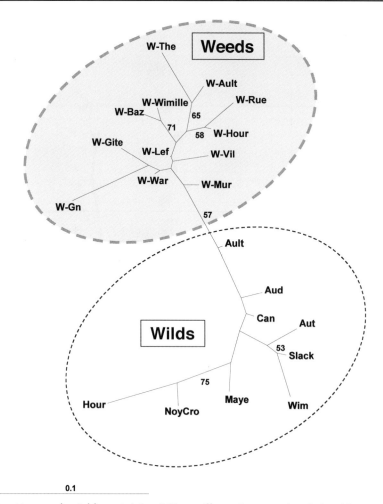

Fig. 15.4. Unrooted neighbour-joining (NJ) tree illustrating genetic relationships between the 21 sampled populations of wild and weedy relatives within the complex of *B. vulgaris*. The NJ tree was based on genetic distance from Reynolds *et al.* (1983) using the information from seven microsatellite loci. Numbers on the right of the nodes refer to the percentage bootstrap values (values < 50% are not shown).

from the seed production area, and secondary hybridization events within sugarbeet fields involving backcrossed crop–wild hybrids) may lead to high levels of within-population polymorphism (Viard *et al.*, 2002). This 'signature' of multiple hybridization events is well demonstrated by the present study: all weed beet populations displayed a relatively high genetic diversity as measured by estimates of allelic richness or gene diversity (5.28 alleles on average, ranging from 3.84 to 6.22; compared with seabeet, 6.18 alleles on average, ranging from 4.86 to 7.27; see Table 15.1 for details). The variance in allelic richness and genetic diversity among populations is similar for weed and seabeet populations. Using isozyme markers, Bartsch *et al.* (1999) studied the genetic structure of populations that were exposed to gene flow from traditionally bred cultivated beets into the wild seabeet populations of north-eastern Italy. They demonstrated that gene flow from crop to wild relatives does not necessarily result in a decrease in the genetic diversity of the wild plant, despite the fact that cultivated beets are less diverse and outnumbered the wild relatives by the factor 10,000 to 1. In our study, a similar result was found, with a limited genetic impact of neighbouring weed populations on the genetic diversity of seabeet populations.

Furthermore, the present work also shows that mean allelic richness is significantly higher in wild seabeet populations compared with weedy lineages (6.18 versus 5.28). In the study of Bartsch *et al.* (1999), beet cultivars hold a substantially higher level of genetic diversity than that of wild beet accessions coming from regions outside the Italian seed production area. Such a discrepancy with our data presumably may be due to the lower polymorphism of isozyme markers compared with microsatellite variation that enables improved comparison between cultivars and natural beet lines. Raybould *et al.* (1998) also found different results between isozymes, RFLPs and microsatellites when searching for isolation by distance patterns in *B. vulgaris*, and argued that such differences may be caused by distinct mutational processes (Raybould *et al.*, 1996; see also Chambers and MacAvoy, 2000).

Genotypic structure and heterozygote deficits within weed beet populations

In the present study, a significant heterozygote deficit (as shown by significant F_{is} values) was consistently found within weed beet populations, whereas seabeet populations appear to fit more closely to Hardy–Weinberg expectations (Table 15.1). These results are rather unexpected as *B. vulgaris* is a self-incompatible species (Larsen, 1977; Bruun *et al.*, 1995). Two main hypotheses can be put forward to explain these findings.

Besides the intrinsic self-incompatibility of *B. vulgaris* taxa, a Mendelian self-fertility factor has been identified by Owen (1942). It was then widely introduced in the cultivated germplasm to produce inbred lines (Mackay *et al.*, 1999). This factor behaves as dominant, and allows circumvention of the self-incompatibility system, causing an increasing selfing and consequently a reduction in heterozygosity. Some weed beets could have inherited this self-fertility factor from their cultivated maternal parent within the seed production area and, therefore, introduced it into the weed populations present in sugarbeet fields, with, in an associated way, significant heterozygote deficits.

Additionally, multiple contamination events within the seed production areas, as well as subsequent crop–weed backcrossing within sugarbeet fields, may explain the strong significant heterozygote deficiencies observed within populations of weed beets. These departures from Hardy–Weinberg expectations may result from the mixture of spatially and/or temporally genetically distinct individuals coming from several sources within the seed production area (see also Viard *et al.*, 2002). Consequently, within a particular field, recurrent immigration events from different contamination origins are likely to increase the genetic heterogeneity of weedy lineages as well as the F_{is} estimates, through a classical 'Wahlund' effect, owing to the genetic differentiation of weed beet patches within sugarbeet fields. Such a within-population structure could be expected given the life history of weedy lineages and could be associated with a migrant pool model of colonization where individual colonizers are drawn at random from a metapopulation involving recurrent introductions by different contaminated seed cultivar sources (for a review of genetic models of population structure, see McCauley, 1993; Thrall *et al.*, 2000).

Genetic differentiation between seabeet and weed beet populations and implication for gene flow

One of the most striking results emerging from the present study is the obvious genetic differentiation observed between weed and seabeet populations, despite the fact that they are separated by a few kilometres at most (Figs 15.1 and 15.4). Seabeets, weed beets and cultivated beets all belong to the same outcrossing species, are fully cross-compatible and mostly wind pollinated. Additionally, they share a recent common ancestry, reinforced by the recurrent use of wild genetic resources in sugarbeet breeding programmes (Ford-Lloyd and Williams, 1975; Lange *et al.*, 1999), and by the recent paternal ruderal origin of weed beets (Boudry *et al.*, 1993; Desplanque *et al.*, 1999). Recent estimates of pollen-mediated gene flow from crop to crop, and from crop to wild relatives in the seed production area suggest that gene dispersal far exceeds 1000 m (Saeglitz *et al.*, 2000; Lavigne *et al.*, 2002). Furthermore, the study of Bartsch *et al.* (1999) revealed the presence of two alleles, commonly found in cultivars, at unusually high frequencies in natural populations from north-east Italy exposed to gene flow from cultivated beets. Consequently, in this high-risk species for crop to wild pollen-mediated gene flow, a high level of gene exchange is expected between weed and seabeets in regions of close contact. Contrary to these expectations, our results show a low overall level of gene flow from weed to seabeet populations in the French sugarbeet production area, and a limited impact on the genetic diversity of natural populations. The maintenance of such an overall genetic distinctiveness between weed and seabeets can in some cases simply be due to geographical isolation between weed and wild populations. An alternative explanation is the time lag between their respective flowering periods: seabeets mainly flower during the first 15 days of June, whereas, due to their late germination in sugarbeet fields (April), weed beets begin to flower at the end of July. The flowering period overlap is then very limited, and pollen-mediated gene flow paradoxically may concern only a few individuals in this wind-pollinated outcrossing complex (see Arnaud *et al.*, 2003).

Nevertheless, some natural populations of seabeets seem to have been exposed to gene flow through seed. Indeed, the presence of rare individuals carrying the Owen CMS cytoplasm (characteristic of cultivated lineages) in two out of nine seabeet populations demonstrates that gene flow through seed occurs from weed to crop. This is particularly clear for the *Ault* population, which also displays microsatellite allele frequencies more suggestive of weedy lineages and appears clearly intermediate on the NJ tree (see Figs 15.3A and 15.4). Reciprocally, the presence of individuals carrying the non-Owen CMS cytoplasm in three weed populations could be the result of seed migration from natural populations to the fields. These findings clearly emphasize the importance of the often underestimated seed-mediated gene flow (see Cain *et al.*, 2000). It can take place through dispersal in space but also through dispersal in time via the soil seed bank (Arnaud *et al.*, 2003). Whereas the lifetime of pollen is limited from some hours to a few days, seeds can stay dormant within the soil and germinate several years after being shed.

Altogether, our results have revealed an unexpected overall genetic isolation of seabeet and weed beet populations in the sugarbeet production area of the French coast of the Channel. None the less, the few individuals showing the Owen CMS found in natural populations highlight the likelihood of seed-mediated gene flow. These findings underline the necessity for careful and fine-scaled monitoring surveys to identify the hot spots where transgenes could escape from the crop, be relayed in weed and finally be introduced into natural populations.

Acknowledgements

We thank Jacqueline Bernard for her expert technical assistance, Nicolas Czerwinski for dressing some figures, and Henk van Dijk and Pascal Touzet for helpful discussions and comments. This work was funded by a grant from the 'Bureau des Ressources Génétiques', the ACI 'Impact des OGM' from the French Ministry of Research and the 'Contrat de Plan Etat/Région Nord-Pas-de-Calais'. M.D. was supported by an 'INRA/Région Nord-Pas-de-Calais' fellowship.

References

Arnaud, J.-F., Viard, F., Delescluse, M. and Cuguen, J. (2003) Evidence for gene flow via seed dispersal from crop to wild relatives in *Beta vulgaris* (Chenopodiaceae): consequences for the release of genetically modified crop species with weedy lineages. *Proceedings of the Royal Society of London B* 270, 1565–1571.

Bartsch, D. and Schuphan, I. (2002) Lessons we can learn from ecological biosafety research. *Journal of Biotechnology* 98, 71–77.

Bartsch, D., Lehnen, M., Clegg, J., Pohl-Orf, M., Schuphan, I. and Ellstrand, N.C. (1999) Impact of gene flow from cultivated beet on genetic diversity of wild sea beet populations. *Molecular Ecology* 8, 1733–1741.

Beaumont, M.A. and Bruford, M.W. (1999) Microsatellites in conservation genetics. In: Goldstein, D.B. and Schlötterer, C. (eds) *Microsatellites: Evolution and Applications*. Oxford University Press, Oxford, pp. 165–182.

Boudry, P., Mörchen, M., Saumitou-Laprade, P., Vernet, P. and Van Dijk, H. (1993) The origin and evolution of weed beets: consequences for the breeding and release of herbicide resistant transgenic sugar beets. *Theoretical and Applied Genetics* 87, 471–478.

Boudry, P., Wieber, R., Saumitou-Laprade, P., Pillen, K., Van Dijk, H. and Jung, C. (1994) Identification of RFLP markers closely linked to the bolting gene B and their significance for the study of the annual habit in beets (*Beta vulgaris* L.). *Theoretical and Applied Genetics* 88, 852–858.

Bruun, L., Haldrup, A., Petersen, S.G., Frese, L., De Bock, T.S.M. and Lange, W. (1995) Self-incompatibility reactions in wild species of the genus *Beta* and their relation to taxonomical classification and geographical origin. *Genetic Resources and Crop Evolution* 42, 293–301.

Cain, M.L., Milligan, B.G. and Strand, A.E. (2000) Long-distance seed dispersal in plant populations. *American Journal of Botany* 87, 1217–1227.

Chambers, G.K. and MacAvoy, E.S. (2000) Microsatellites: consensus and controversy. *Comparative Biochemistry and Physiology, Part B* 126, 455–476.

Desplanque, B., Boudry, P., Broomberg, K., Saumitou-Laprade, P., Cuguen, J. and Van Dijk, H. (1999) Genetic diversity and gene flow between wild, cultivated and weedy forms of *Beta vulgaris* L. (Chenopodiaceae), assessed by RFLP and microsatellite markers. *Theoretical and Applied Genetics* 98, 1194–1201.

Desplanque, B., Viard, F., Bernard, J., Forcioli, D., Saumitou-Laprade, P., Cuguen, J. and Van Dijk, H. (2000) The linkage disequilibrium between chloroplast DNA and mitochondrial DNA haplotypes in *Beta vulgaris* ssp. *maritima* (L.): the usefulness of both genomes for population genetic studies. *Molecular Ecology* 9, 141–154.

Desplanque, B., Hautekèete, N.-C. and Van Dijk, H. (2002) Transgenic weed beets: possible, probable, avoidable? *Journal of Applied Ecology* 39, 561–571.

Eastham, K. and Sweet, J. (2002) *Genetically Modified Organisms (GMOs): the Significance of Gene Flow Through Pollen Transfer*. Environmental issue report. European Environment Agency, Copenhagen, Denmark.

El Mousadik, A. and Petit, R.J. (1996) High level of genetic differentiation for allelic richness among population of the argan tree [*Argania spinosa* (L.) Skeels] endemic to Morocco. *Theoretical and Applied Genetics* 92, 832–839.

Ellstrand, N.C. (2001) When transgenes wander, should we worry? *Plant Physiology* 125, 1543–1545.

Ellstrand, N.C. and Schierenbeck, K.A. (2000) Hybridization as a stimulus for the evolution of invasiveness in plants? *Proceedings of the National Academy of Sciences USA* 97, 7043–7050.

Ellstrand, N.C., Prentice, H.C. and Hancock, J.F. (1999) Gene flow and introgression from domesticated plants into their wild relatives. *Annual Review of Ecology and Systematics* 30, 539–563.

El-Mezawy, A., Dreyer, F., Jacobs, G. and Jung, C. (2002) High-resolution mapping of the bolting gene B of sugar beet. *Theoretical and Applied Genetics* 105, 100–105.

Estoup, A. and Cornuet, J.-M. (1999) Microsatellite evolution: inferences from population data. In: Goldstein, D.B. and Schlötterer, C. (eds) *Microsatellites: Evolution and Applications*. Oxford University Press, Oxford, pp. 49–65.

Ford-Lloyd, B.V. and Williams, J.T. (1975) A revision of *Beta* section *Vulgares* (Chenopodiaceae) with new light on the origin of cultivated beets. *Botanical Journal of the Linnaean Society* 71, 86–102.

Frankham, R., Ballou, J.D. and Briscoe, D.A. (2002) *Introduction to Conservation Genetics.* Cambridge University Press, Cambridge.

Goudet, J. (1995) FSTAT (Version 1.2). A computer program to calculate *F*-statistics. *Journal of Heredity* 86, 485–486.

Gray, A.J. and Raybould, A.F. (1998) Reducing transgene escape routes. *Nature* 392, 653–654.

Hails, R.S. (2000) Genetically modified plants – the debate continues. *Trends in Ecology and Evolution* 15, 14–18.

Hamrick, J.L. and Nason, J.D. (1996) Consequences of dispersal in plants. In: Rhodes, O.E., Chesser, R.K. and Smith, M.H. (eds) *Population Dynamics in Ecological Space and Time.* University of Chicago Press, Chicago, Illinois, pp. 203–236.

Hansen, M., Kraft, T., Ganestam, S., Sall, T. and Nilsson, N.O. (2001) Linkage disequilibrium mapping of the bolting gene in sea beet using AFLP markers. *Genetical Research* 77, 61–66.

Hornsey, K.G. and Arnold, M.M. (1979) The origin of weed beet. *Annals of Applied Biology* 92, 279.

Jarvis, D.I. and Hodgkin, T. (1999) Wild relatives and crop cultivars: detecting natural introgression and farmer selection of new genetic combinations in agrosystems. *Molecular Ecology* 8, S159–S173.

Lange, W., Brandenburg, W.A. and De Bock, T.M.S. (1999) Taxonomy and cultonomy of beet (*Beta vulgaris* L.). *Botanical Journal of the Linnean Society* 130, 81–96.

Larsen, K. (1977) Self incompatibility in *Beta vulgaris* L. 1. Four gametophytic, complementary S-loci in sugar beet. *Hereditas* 85, 227–248.

Lavigne, C., Klein, E.K. and Couvet, D. (2002) Using seed purity data to estimate an average pollen mediated gene flow from crops to wild relatives. *Theoretical and Applied Genetics* 104, 139–145.

Letschert, J.P.W. (1993) *Beta* section *Beta*: biogeographical patterns of variation and taxonomy. *Wageningen Agricultural University Papers* 93, 1–137.

Longden, P.C. (1989) Effects of increasing weed-beet density on sugar-beet yield and quality. *Annals of Applied Biology* 114, 527–532.

Longden, P.C. (1993) Weed beet: a review. *Aspects of Applied Biology* 35, 185–194.

Mackay, I.J., Gibson, J.P. and Caligari, P.D.S. (1999) The genetics of selfing with concurrent backcrossing in breeding hybrid sugar beet (*Beta vulgaris altissima* L.). *Theoretical and Applied Genetics* 98, 1156–1162.

McCauley, D.E. (1993) Genetic consequences of extinction and recolonization in fragmented habitats. In: Kareiva, P.M., Kingsolver, J.G. and Huey, R.B. (eds) *Biotic Interactions and Global Change.* Sinauer Associates Inc., Sunderland, Massachusetts, pp. 217–233.

Mörchen, M., Cuguen, J., Michaelis, G., Hanni, C. and Saumitou-Laprade, P. (1996) Abundance and length polymorphism of microsatellite repeats in *Beta vulgaris* L. *Theoretical and Applied Genetics* 92, 326–333.

Mücher, T., Hesse, P., Pohl-Orf, M., Ellstrand, N.C. and Bartsch, D. (2000) Characterization of weed beet in Germany and Italy. *Journal of Sugar Beet Research* 37, 19–38.

Nei, M. (1978) Estimation of average heterozygosity and genetic distance from a small number of individuals. *Genetics* 89, 583–590.

Owen, F.V. (1942) Inheritance of cross- and self-sterility and self-fertility in *Beta vulgaris*. *Journal of Agricultural Research* 64, 679–698.

Owen, F.V. (1945) Cytoplasmically inherited male-sterility in sugar beets. *Journal of Agricultural Research* 71, 423–440.

Pullin, A.S. (2002) *Conservation Biology*. Cambridge University Press, Cambridge.

Ran, Z. and Michaelis, G. (1995) Mapping of a chloroplast RFLP marker associated with the CMS cytoplasm of sugar beet (*Beta vulgaris*). *Theoretical and Applied Genetics* 91, 836–840.

Raybould, A.F. and Gray, A.J. (1994) Will hybrids of genetically modified crops invade natural communities? *Trends in Ecology and Evolution* 9, 85–89.

Raybould, A.F., Mogg, R.J. and Clarke, R.T. (1996) The genetic structure of *Beta vulgaris* ssp. *maritima* (sea beet) populations: RFLPs and isozymes show different patterns of gene flow. *Heredity* 77, 245–250.

Raybould, A.F., Mogg, R.J., Aldam, C., Gliddon, C.J., Thorpe, R.S. and Clarke, R.T. (1998) The genetic structure of sea beet (*Beta vulgaris* ssp. *maritima*) populations. III. Detection of isolation by distance at microsatellite loci. *Heredity* 80, 127–132.

Raybould, A.F., Clarke, R.T., Bond, J.M., Welters, R.E. and Gliddon, C.J. (2002) Inferring patterns of dispersal from allele frequency data. In: Bullock, J.M., Kenward, R.E. and Hails, R.S. (eds) *Dispersal Ecology*. Blackwell Science, Oxford, pp. 89–110.

Raymond, M. and Rousset, F. (1995) GENEPOP (Version 1.2): a population genetics software for exact tests and ecumenicism. *Journal of Heredity* 86, 248–249.

Reynolds, J., Weir, B.S. and Cockerham, C.C. (1983) Estimation of the coancestry coefficient: basis for a short-term genetic distance. *Genetics* 105, 767–779.

Rice, W.R. (1989) Analyzing tables of statistical tests. *Evolution* 43, 223–225.

Rousset, F. (2001) Genetic approaches to the estimation of dispersal rates. In: Clobert, J., Danchin, E., Dhondt, A.A. and Nichols, J.D. (eds) *Dispersal*. Oxford University Press, Oxford, pp. 18–28.

Saeglitz, C., Pohl, M. and Bartsch, D. (2000) Monitoring gene flow from transgenic sugar beet using cytoplasmic male-sterile bait plants. *Molecular Ecology* 9, 2035–2040.

Santoni, S. and Berville, A. (1992) Extramarital sex amongst the beets – evidence for gene exchanges between sugar beet (*Beta vulgaris* L.) and wild beets: consequences for transgenic sugar beets. *Plant Molecular Biology* 20, 578–580.

Thrall, P.H., Burdon, J.J. and Murray, B.R. (2000) The metapopulation paradigm: a fragmented view of conservation biology. In: Young, A.G. and Clarke, G.M. (eds) *Genetics, Demography and Viability of Fragmented Populations*. Cambridge University Press, Cambridge, pp. 75–95.

Treu, R. and Emberlin, J. (2000) *Pollen Dispersal in the Crops Maize* (Zea mays*), Oil Seed Rape* (Brassica napus *ssp.* oleifera*), Potatoes* (Solanum tuberosum*), Sugar Beet* (Beta vulgaris *ssp.* vulgaris*) and Wheat* (Triticum aestivum*)*. Report from the National Pollen Research Unit. Soil Association, University College, Worcester, UK.

van Dijk, H. and Desplanque, B. (1999) European *Beta*: crops and their wild and weedy relatives. In: van Raamsdonk, L.W.D. and Den Nijs, J.C.M. (eds) *Plant Evolution in Man-made Habitats*. Hugo de Vries Laboratory, Amsterdam, pp. 257–270.

van Raamsdonk, L.W.D. and van der Maesen, L.J.G. (1996) Crop–weed complexes: the complex relationship between crop plants and their wild relatives. *Acta Botanica Neerlandica* 45, 135–155.

Viard, F., Bernard, J. and Desplanque, B. (2002) Crop–weed interactions in the *Beta vulgaris* complex at a local scale: allelic diversity and gene flow within sugar beet fields. *Theoretical and Applied Genetics* 104, 688–697.

Weir, B.S. and Cockerham, C.C. (1984) Estimating *F*-statistics for the analysis of population structure. *Evolution* 38, 1358–1370.

16 Crop–Wild Interaction Within the *Beta vulgaris* Complex: Agronomic Aspects of Weed Beet in the Czech Republic

JOSEF SOUKUP AND JOSEF HOLEC

Department of Agroecology and Biometeorology, Faculty of Agonomy and Natural Resources, Czech University of Agriculture in Prague, Kamýcká 957, 165 21 Praha 6–Suchdol, Czech Republic; E-mail: soukup@af.czu.cz

Abstract

This review is focused on weed beet in the Czech Republic (Central Europe). In our conditions, beet is quite a new weed. Weed beet has occurred more frequently since the end of the 1980s when beet seed started to be imported from southeastern European countries. By the beginning of the 1990s, about 50% of fields used for sugarbeet production had already been contaminated by weed beet. Nowadays, the infestation of many fields used for intensive sugarbeet production is already so high that it is impossible to grow sugarbeet in them.

The growth habit of weed beet varies significantly from plagiotropic to orthotropic forms and from large to small root bodies. Our observations have shown a high degree of variance of morphological traits between localities as well as within populations. Genetic variability was high and unstable in all populations. This was due to different hybridizations between weedy and cultivated beet in different geographical localities where the seed was produced. Seed testing is an important method for the prevention of weed beet spread, and every year some seed lots with weed beet contamination higher than the permitted 0.05% can be found. In these cases, seed production companies have to pay the farmers for the additional costs of weed beet control. As a direct control method, hand pulling can still be used, especially on fields with low weed beet infestation. When higher levels of infestation occur, mechanical hoeing can be used, but weed beet plants survive within the crop rows. Chemical control using non-selective herbicides can be highly effective. Current control systems do not provide sufficient weed beet suppression in sugarbeet fields as the conditions of intensive sugarbeet crop production seem to favour the introduction and spread of weed beet. New management tools for effective control of weed beet populations are expected with the introduction of GM herbicide-tolerant sugarbeet cultivars. However, it will also be very important to study the influence of the agronomy and the cropping systems on the dynamics of the weed beet complex in order to quantify and limit gene flow from transgenic beet to weed beets.

Introduction

Weed beets are a serious economic problem for sugarbeet farmers, seed producers and sugarbeet processing factories. In situations with high infestations, the strongly branching weed beet plants compete with the crop and cause harvest complications. Any bolted stems cause problems at harvest due to their rotting roots, and affect the alignment of the row finders. After mid-August, any direct field treatment of weeds (stem cutting or herbicide spray) is too late. In addition, harvested weed beet roots cause technological difficulties in sugar factories, because they are woody, which leads to problems in the beet slicing and sugar extraction. Yield loss caused by weed beet can be relatively high. Longden (1989) observed 11.7% sugarbeet yield loss for each weed beet plant per square metre, tested at a weed beet abundance from 0.0 to 13.6 plants/m^2.

Another agronomic problem arises from the fact that these two related forms of *Beta vulgaris* are hosts of the same pathogens. This problem is particularly serious when weed beet has been present for some time, allowing the build-up of certain diseases. The most important diseases are rhizomania caused by beet necrotic yellow vein virus (BNYVV) and root and crown rot caused by *Rhizoctonia solani*. When the weed beet occurs in other crops in the rotation, there is no disruption of the pathogen life cycle and the likelihood of pathogen transfer to the next sugarbeet crop is considerably increased.

Particular attention is being paid to the potential gene flow from GM sugarbeet cultivars to weed beet and other wild beet relatives. The ecosystem biodiversity, conservation and ecological consequences of the introduction of GM sugarbeet have been studied at length (Desplanque *et al.*, 1999; Bartsch *et al.*, 2001; Bartsch and Schmitz, 2002; Cuguen *et al.*, Chapter 15, this volume). The transfer of pest or disease resistance genes into natural ecosystems could theoretically lead to a competitive fitness advantage for the recipient wild or weed form. Gene flow between wild and cultivated beet has been observed in the past (Bartsch *et al.*, 1999). Biosafety research has been carried out simulating the consequences of transgene flow from GM beet resistant to BNYVV (Pohl-Orf *et al.*, 2000; Bartsch *et al.*, 2001). In this particular case, transgenic wild hybrids were not more fit in terms of seed set even under conditions of virus infestation. Bartsch and co-authors also tested beet plants containing a transgenic herbicide tolerance marker (*bar* gene) in the absence of the herbicide phosphinothricin (glufosinate ammonium), and observed no selective fitness advantage in the GM beets. However, transfer of the herbicide tolerance gene to weeds will cause agronomic problems in agrosystems where the complementary herbicides will be applied.

Agronomic problems due to the introduction of GM sugarbeet into farming systems might arise because of the specific biology and food processing of sugarbeet in comparison with other crops.

- Beet plants are strictly cross-pollinated and hybridize easily with related wild species and subspecies of the genus *Beta* in areas of their co-occurrence.
- Domesticated forms of beet are biennial and do not flower during the first year. The likelihood of gene escape is therefore very low in the first year of growth.
- The seed is produced in areas different from those of root (leaf) production. In seed production areas, outcrossing is very likely since wild annual beets often co-occur in these areas. Native genes in wild populations of beet mediate the genetically dominant annual habit (Van Dijk, Chapter 5, this volume) in beet seed, and might result in combinations with GM traits in GM seed crops. Different risk assessments are required for seed and root production areas.
- Only a relatively small number of breeding companies breed sugarbeet and multiply beet seed. This provides a good opportunity to manage the seed quality and to direct liability for products.
- Chemical control of common sugarbeet weeds using glyphosate and glufosinate ammonium will probably decrease the importance and usage of preventive and mechanical control methods (rotational use of herbicides, seed bank management, soil tillage), which are necessary for efficient weed beet management.
- The evolution of populations of weed beet resistant to non-selective herbicides would complicate chemical control and management.
- The final product (sugar) does not contain either the DNA or specific protein products of transgenes. Any adventitious presence of GMO in the sugarbeet crop harvest cannot be detected analytically in the food chain.

The aim of this review is to provide technical and applied insight into the weed beet problem from an agronomic point of view. First, we will give a short overview of the development of weed beet in the Czech Republic. Secondly, we present data from recent studies on the morphology and genetic diversity of weed beet with special regard to identification characteristics. Thirdly, future agricultural management tools are discussed depending on the Czech agricultural situation. Finally, we discuss the introduction of GM sugarbeet as a weed beet management tool as well as the implication that gene flow to weed beet may have.

Origin and Spread of Weed Beet in the Czech Republic

The first reports of the weed beet evolution are from 1973–1977, when annual forms of beet with dormant seeds were observed in the UK (Longden in Maughan, 1984). In the years 1978–1981, 18–27% of fields were infested by

Beta seedlings (Maughan, 1984), which were accidentally introduced into the fields as a consequence of seed contamination and called 'weed beet'. Later, the weed beet was also observed in other European countries (Boudry *et al.*, 1993; Danovská, 1996; Jassem *et al.*, 1997; Mücher *et al.*, 2000; Van Dijk, Chapter 5, this volume).

In the Czech Republic, weed beet has occurred more frequently since the end of the 1980s when the beet seed started to be imported from southeastern European countries. In 1992, weed beet was classified as quarantine weed (Vášová, 1995). By the beginning of the 1990s, about 50% fields used for sugarbeet production had been contaminated by weed beet. Later imports were mainly from northern Italy and the South of France. Nowadays, sugarbeet is grown in the Czech Republic on approximately 75,000 ha, and all seeds needed are imported. The shares of the most important seed producers in the Czech import are given in Table 16.1. Although all imported seed lots are tested, weed beet infestation of many fields used for intensive sugarbeet production is already so high that it is impossible to grow sugarbeet there. Two factors supported the introduction and spread of weed beets in Europe: first, management of European sugarbeet cropping systems changed significantly in the last 30–40 years from labour-intensive low input systems to highly technical high input systems. Secondly, plant breeding introduced highly effective hybrid seed production techniques with cytoplasmic male-sterile (CMS) mother plants that favoured the evolution of weed beet types via introgression of wild genes causing annuality.

This latter factor needs special attention. Although sugarbeet varieties have always given a few bolters (< 1%), long rotations and hoeing generally controlled the problem until the mid-1960s. At about that time, breeders started to use monogerm hybrid varieties using CMS males and tetraploid pollinators, which are less able to avoid introgression of annuality traits by pollen from diploid wild beets. Introduction of monogerm seeds and effective herbicides changed the management during the vegetative growth. Hand weeding and hoeing, which can be sufficient for weed beet

Table 16.1. Area of sugarbeet crop production in the Czech Republic in relation to the import origin of the seed material from different seed companies.

Company	Year			
	2000 (ha)	2001 (ha)	2002 (ha)	2002 (%)
Danisco Seed	13,071	19,807	23,011	22.54
Delitzsch Pflanzenzucht GmbH	3,214	10,587	15,924	16.98
S.E.S. Europe N.V./S.A.	4,902	8,900	10,144	10.82
D.J. VAN DER HAVE BV	135	727	746	0.80
Kuhn & Co., International BV	–	48	3,918	4.17
Syngenta Seeds AB	8,140	8,927	8,109	8.65
Fr. Strube Saatzucht KG Söllingen	8,561	11,631	14,432	15.39
KWS SAAT AG	10,770	15,751	17,500	18.65

Source: Srba (2002).

eradication, are no longer used and there is a reliance on herbicides for weed control both in and between the sugarbeet rows. However, the herbicides do not differentiate between beet crops and weed beet.

Many farmers abandoned animal production and specialized in cash crops such as sugarbeet, which worsened the weed beet problem. Farmers started to narrow crop rotation time, leading to fewer plantation periods of other crops in which weed beet could be controlled indirectly. At the same time, these farmers started using herbicides instead of hoeing to control weeds in their beet crops. A rotational interval of 4–5 years was recommended for sugarbeet on the same field, but rotations were shortened to 2–3 years, favouring weed beet build-up in the soil seed bank. Maughan (1984) showed that 4 year or 5–8 year rotations decreased weed beet infestation by 30 and 80%, respectively, as compared with 3 year rotations.

Many people left agriculture at this time so that hand weeding of bolters and weed beet is now impossible on bigger farms. Beet crops are not used for feeding animals at present. Leaves are cut and spread over the soil surface after the harvest together with weed beet seeds. After the mechanized harvest, plant parts remain in the soil and, in the absence of ploughing, can regenerate and produce seeds in the following year.

Characteristics of Weed Beet in the Czech Republic

Morphological variability of weed beet

Weed beet has some characteristic physiological and morphological markers, which are not always expressed at the same time. One of the most important markers is annuality, which is genetically dominant. Associated characteristics are root branching, pleiotropic growth and anthocyan coloration of the hypocotyl and stem (Kohout, 1996).

Monitoring of weed beet morphological and reproductive variability in sugarbeet fields in the Czech Republic was conducted by Soukup *et al.* (2002) and Soukup and Holec (2002). Localities in areas most suitable for sugarbeet production were chosen in these studies (Fig. 16.1). Weed beets showed a high degree of phenotypic variability in the studied traits; coloration of roots, hypocotyl and stem, and root branching (Table 16.2).

The growth habit of weed beets varies significantly from plagiotropic to orthotropic forms and from large to small root bodies (Fig. 16.2). Plants with strongly branching roots are found in each locality, while the proportion of non-branching individuals varied from 10 to 70%. As far as the root colour is concerned, many plants were without any root anthocyan coloration and, in five localities, colorized plants rarely occurred or were absent. About 50% of plants exhibited hypocotyl and stem coloration. The growth habit of weed beets is given in Fig. 16.3. Observations performed at the end of the growing season (Table 16.3) have shown high variance between localities as well as within populations, especially in the plant weight and number of fruits. On average, each plant produced 1700 viable fruits, many of which remain

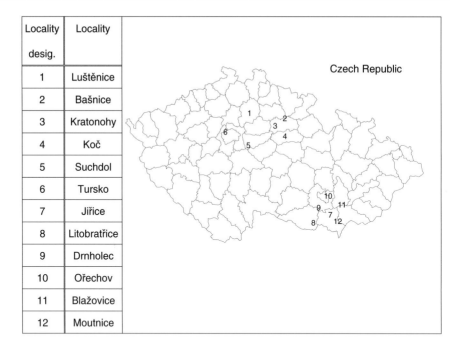

Locality desig.	Locality
1	Luštěnice
2	Bašnice
3	Kratonohy
4	Koč
5	Suchdol
6	Tursko
7	Jiřice
8	Litobratřice
9	Drnholec
10	Ořechov
11	Blažovice
12	Moutnice

Czech Republic

Fig. 16.1. Localities in which the occurrence, morphology and genetic differentiation of weed beet were measured. The locations 1–12 are representative of important sugarbeet crop plantation areas within the administrative districts of the Czech Republic (Soukup et al., 2002).

Table 16.2. Morphological variability of weed beet characteristics (shown by percentage of plants with a particular character) derived from locations given in Fig. 16.1.

Characteristics		1	2	3	4	5	6	7	8	9	10	11	12	Average
Root branching	None	30	70	30	10	10	20	60	20	50	30	40	10	32
	Low	30	10	30	30	50	40	30	10	40	10	20	10	26
	Middle	10	0	30	20	10	20	0	40	0	20	10	30	16
	High	30	20	10	40	30	20	10	30	10	40	30	50	27
Root coloration	White	90	70	100	80	50	80	50	90	70	50	100	100	78
	Pink	10	30	0	10	40	10	50	10	30	50	0	0	20
	Red	0	0	0	10	10	10	0	0	0	0	0	0	2
Hypocotyl colour	White	70	50	40	30	30	80	60	80	60	50	40	80	56
	Pink	30	50	40	50	50	10	40	20	40	50	60	20	38
	Red	0	0	20	20	20	10	0	0	0	0	0	0	6
Stem anthocyan colour	0%	70	90	30	0	10	70	70	70	40	60	10	80	50
	At nodes	20	10	60	20	10	20	20	20	10	0	60	20	23
	< 50%	10	0	10	60	40	10	0	10	10	0	20	0	14
	> 50%	0	0	0	20	40	0	10	0	40	40	10	0	13

Source: Soukup et al. (2002).

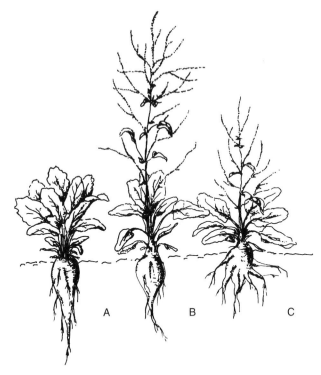

Fig. 16.2.   Sugarbeet morphology. (A) First year vegetative growth of biennial sugarbeet. (B) Annual physiological bolter (vernalized sugarbeet due to low spring temperatures). (C) Primary type of weed beet (F₁ cross-bred with wild seabeet) (illustration by Věra Kožnarová).

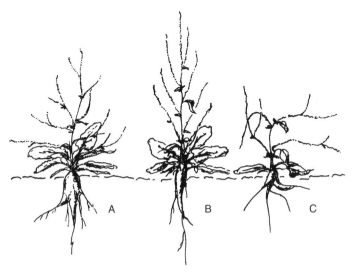

Fig. 16.3.   Various growth habits of weed beets: (A) intermediate, (B) orthotropic, (C) plagiotropic (illustration by Věra Kožnarová).

Table 16.3. Reproductive and morphological characteristics of fertile weed beet plants under field conditions.

Characteristics		Locality (according to Fig. 16.1)			
		3	4a	4b	5
No. of branches	Average	19.2	13.6	20.0	23.0
	SD	5.83	6.65	7.09	4.36
Plant weight	Average	159.3	239.2	419.5	419.3
(g fresh weight)	SD	94.88	110.28	188.39	172.35
Weight of 1000 fruits	Average	16.48	24.49	15.44	23.06
(g dry weight)	SD	3.23	7.88	2.95	7.23
No. of fruits per plant	Average	1103	1261	1894	2724
	SD	781	760	768	996
No. of seeds per seed	Average	2.3	2.2	2.2	2.5
cluster (germicity)	SD	0.80	1.15	0.75	0.86

Source: Soukup and Holec (2002).

viable for long periods so that the weed beet problem becomes progressively worse with every rotation.

Genetic variability of weed beet in the Czech Republic

Genetic variability of local populations was studied by Soukup *et al.* (2002) in the Czech Republic using randomly amplified polymorphic DNA–polymerase chain reaction (RAPD–PCR) analysis – a standard method for study of genetic variability, similarity, and describing genetic distances. For RAPD reactions, 60 decameric primers were tested from which four decamers with a high degree of band polymorphism were selected. The sequences of the primers used and their designations were as follows: F 42 (5′ CTC CTC CCC C 3′), OPB 11 (5′ TGC CCG TCG T 3′), OPL 14 (5′ GTG ACA GGC T 3′) and RAPD 1 (5′ ACG CAG GCA C 3′). Similarity between analysed genotypes was described using Dice's coefficient and UPGMA (unweighted pair group method with arithmetic mean) dendrograms. Statistical characteristics are presented in Table 16.4. Large genetic distances were observed between individuals in all populations studied. The maximal average value of Dice's coefficient of similarity (30.5) was detected in population 2 and the minimal average value (22.2) in population 11. In all populations except populations 2, 7 and 11, similarity to the co-cultivated variety was higher than 50. Maximal genetic similarity (71) to the co-cultivated variety was found in population 4, and minimal similarity (30) in population 2. Figure 16.4 shows the UPGMA dendrogram of genetic similarity within individuals of a weed beet in population 8 as an example: genetic similarity above 50% to co-cultivated sugarbeet was found in four of ten tested individuals originating from a single location.

Table 16.4. Statistical characteristics of obtained Dice's coefficients in individual weed beet populations.

Population	Maximum	Minimum	Average	Mode	Variation coefficient (%)	Frequency of value zero (%)	Maximal similarity with co-cultivated variety
1	60	0	23.4	0	68.3	16.4	50
2	67	0	30.5	50	51.3	5.5	30
3	62	0	29.8	33	50.1	6.7	57
4	92	0	23.3	38	84.8	7.3	71
5	67	0	26.1	25	59.4	8.9	50
6	63	0	24.6	0	60.9	10.6	59
7	67	0	29.9	0	63.4	14.6	47
8	63	0	22.3	0	72.3	14.6	53
9	60	0	27.3	13	57.6	8.9	53
10	76	0	30.3	22	52.7	3.6	56
11	56	0	22.2	0	69.3	19.7	44
12	78	0	27.1	0	61.8	13.7	63

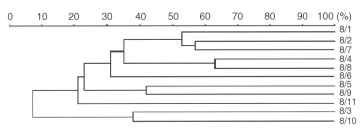

Fig. 16.4. UPGMA dendrogram of genetic similarity within individuals of a weed beet population (Location 8, see Fig. 16.1) (Soukup *et al.*, 2002); (sugarbeet, 8/1; weed beet individuals, 8/2–8/10).

The results showed that the genetic variability was high in all populations, and the populations were very unbalanced. The reason for this high genetic variability is that weed beets in individual populations have different geographical origins and there was hybridization between them and cultivated beet. As different cultivars have been planted over time, there were different genotypes hybridizing with the weed beet populations. It is important to note that these are only results of a first initial RAPD study with limited sample sizes. Our genetic data are principally consistent with those reported from France (Desplanque *et al.*, 1999), Germany and Italy (Mücher *et al.*, 2000). In conclusion, wild, domesticated and weedy beet forms worldwide create a genetically coherent group with both ecological and economic impacts.

Aspects of Future Weed Beet Management

Seed control system in the Czech Republic

At present, sugarbeet seeds in the Czech Republic are imported from areas where seeds can be produced better due to more suitable climatic conditions. The national seed control system is therefore the first control systems to detect any adventitious presence of weed beet in sugarbeet seeds lots. By law, the Czech Republic allows a maximum of 0.05% weed beet contamination in sugarbeet seed. Imported seed lots are tested for the presence of weed beet by the Central Control and Testing Agency every year based on performance tests under field conditions (ÚKZÚZ, 1990). Seed samples from every lot with more than 100 seed units (100 ha) are sown in testing stations, and the numbers of annual beet are counted. This is followed by an evaluation of whether they are physiological bolters (compared with weather conditions during the critical period) or weed beet. There are always additional decision criteria such as earliness (according to date of occurrence and growth stage), coloration (presence of anthocyan), root properties (thickness, colour, branching, rigidity, number of vascular bundles), multigermicity or monogermicity. When the weed beet content in the seeds is higher than 0.05%, the seed company has to compensate farmers for the additional costs for the eradication of weed beets from the field. High quality of imported seeds has been achieved in this way. However, one can still find some seed lots containing more than 0.05% of weed beet every year. Unfortunately, the results of these tests become available too late and at the time when early plants of weed beet are already producing seeds in the field and, as a result, introduction of the weed beets and their spread via seed importation has already taken place (see Table 16.5).

Direct control of weed beet

The direct control of the weed beet in sugarbeet fields is mostly done mechanically, using hand weeding or hoeing, and chemically, using non-selective herbicides.

Table 16.5. Adventitious presence of weed beets in sugarbeet seed lots according to the national Czech seed control system 1999–2002.

Year	No. of seed lots tested	No. of seed lots without contamination	No. of seed lots with permitted weed beet content	No. of seed lots above the permitted limit
1999	102	84	17	1
2000	91	83	7	1
2001	123	94	23	6
2002	131	106	22	3

Source: ÚKZÚZ (unpublished results).

Hoeing

Hoeing is the most suitable technology for the fields where weed beet occurrence from the soil seed bank (in inter-rows) is expected. Rotary or knife weeders are typically used for these purposes. The effectiveness of this method depends on the inter-row space, the size of the weed beet plants, the number of replications and the physical properties of the soil. It is advantageous to use the hoeing immediately after a crop emergence when sugarbeet rows become visible (two-leaf growth stage), since one can effectively go quite close to crop rows without destroying the very sensitive sugarbeet root system. Rotary hoeing of two- to four-leaf beets will generally reduce the stand by about 5–10%. The smallest beets should be at the two-leaf stage or greater, otherwise the risk of stand loss increases considerably. Emerging weed beet plants are controlled effectively by covering them with the tilled soil. At later sugarbeet growth stages, hoeing should be more distant from the crop row and can be deeper (3–4 cm) in order to damage weed beet plants via undercutting.

However, hoeing has several disadvantages: (i) it is expensive and time consuming (1 ha/h); (ii) it can be used at optimum soil moisture only; (iii) it can elevate weed seeds from the seed bank to the soil surface; and (iv) it is only effective between the rows. Hoeing is no longer a common part of the sugarbeet growing technology. It is used mainly for improving soil properties and for weed beet control. Its use is limited to the early growing season until canopy closure.

Hand selection

Hand weeding is very effective but too laborious. It is used mainly for selecting bolting individuals from contaminated seed and physiological bolters that occur in the row and cannot be removed by hoeing. The hand selection period is typically at the beginning of the generative phase of bolting beets. Early flowering weed beets emerging from the soil seed bank are regulated at the same time. Hand-selected plants should be removed from the field to prevent seed ripening and enriching the soil seed bank. Hand selection is used as an additional tool after hoeing or herbicide treatment. It takes about 2–10 h/ha and its overall effectiveness depends on the amount of weed in the field. Hand treatment of fields with a large amount of weed beet infestation is practically impossible.

Cutting

Stems of bolters and weed beets, which grow above the sugarbeet, can be reduced by cutting in the growth stages of stem elongation or flowering. For this purpose, machines built on the principle of a rotary mower are used, such as the Weed Beet Surfer (CTM, 2003) machine, which is towed on a standard tractor pickup hitch or other suitable drawbar arrangement. A typical machine mesh is about 6.4 m to cover 12 rows, and 10 × 4 blade rotors accomplish the weed cutting. The machine has variable height adjustment provided by two wheels. By positioning in the appropriate machine brackets, the wheels provide both road transport and cut height control. The main disadvantage of the cutting method is that mechanically injured plants can easily regenerate so that the action must be repeated several times. Also, plagiotropic weed beets remain untouched. From this point of view, it is only appropriate to use the cutter for the control of physiological bolters with erect stems and for control of other tall weeds. The active mesh is quite narrow (6.4 m) and the cutter can be used only until crop canopy closure.

Chemical control

Weed beet chemical control is one of the most important tools, not only in sugarbeet stands but also in other crops. Chemical control of common weed species in the sugarbeet fields is based on herbicides, which are applied in the early growth stages (cotyledons, first true leaves). Pre-emergent herbicides (metamitron and chloridazon) and a sequence of post-emergence herbicides (phenmedipham, desmedipham, ethofumesate, clopyralid, lenacil and triflusulfuron-methyl) are usually used for these purposes. However, these herbicides do not affect the weed beet, and remove weed competition so that they have even more space to develop within the crop. Due to the close relatedness of the sugarbeets and weed beets, a herbicide specific to the weed beet has not yet been developed. Non-selective herbicides such as glyphosate can be used against the weed beet either in inter-row application or in weed wipers that operate above the level of the crop touching only the flowering beet (Longden, 1989).

The most advanced contact herbicide applicator and weed wiping system for sugarbeet and root crops is ROTOWIPER™ (Rotowiper GmbH, 2003). A special roller material releases the herbicide to the underside of leaves and stems when the plant touches the roller. This rotating roller does not damage plants, so that the systemic herbicide active ingredient can be transported to the roots. Unlike the earlier wick applicators, the roller rotation prevents dripping, herbicide usage is related to the density of weed beets, and weed beet size determines the dose of herbicide automatically. The herbicide is applied on the undersides of leaves, where it is well absorbed and gives the most effective control. Weed beets (or bolters) must be ≥ 10 cm taller, and the wiper ≥ 5 cm higher than the desired sugarbeet stand. Dense populations must be treated twice, in opposite directions. Bolting beet requires three applications, 2 weeks apart, from early July to early August. Roundup Biactive is recommended to be used. Maximum concentrations should not exceed a 1:1 dilution with water, and 1:2 dilutions with water under hot, dry conditions (Monsanto, 2000).

Alternative methods

An interesting method, which is not widely used so far, is high-voltage electricity control. Tests by Diprose *et al.* (1980) provided data that high voltages around 5 kV RMS (0.5–1 A) destroy annual beets in pots.

Under field conditions, about 20 kW are necessary to control the plants. A tractor-driven system producing 8 kV RMS removed 75% of the bolting and weed beet. Krouský (2001) recommended the use of individual practices according to the density of weed beet and bolters. Hand selection or cutting is possible and sufficient for fields with 100 or less bolters or weed beets per hectare. Fields comprising from 100 to 1000 weed beets per hectare require 2–3 inter-row hoeings with effectiveness of about 70%, and the rest should be hand selected. Fields with 1000–10,000 weed beets per hectare should be hoed two or three times at 14-day intervals starting at the four-leaf stage of the weed beet. Then, a chemical control should be applied using a non-selective herbicide. The following crop (winter wheat) should be drilled using direct sowing 3–4 weeks after the beet harvest. In cereal stands, weed beet can be easily managed by post-emergence herbicides. Where weed beet infestation is at uncontrollable and highly damaging levels, then the above measures are not adequate and will not give an economic return. Farmers will recultivate and re-sow these fields with another crop just after the sugar (and also weed) beet emergence. The results of Soukup *et al.* (2003) demonstrated that an integration of preventive and direct control methods is needed, because none of the individual control measures (hoeing, glyphosate treatment) is effective alone.

Agricultural Consequences for Weed Beet Management Using GM Crops

Underestimation of the weed beet occurrence in initial phases of its introduction and spread led to rapid infestations of weed beet on arable land. As a result, the numbers of infested localities and weed beet abundance increase annually. On farms where no attention was paid to the control of weed beet, the situation is so serious that it will be impossible to grow sugarbeet for many years. Current control systems do not provide sufficient weed beet suppression in sugarbeet stands. New management tools for the effective control of weed beet populations are expected with the introduction of GM herbicide-tolerant sugarbeet cultivars. Non-tolerant weed beets would be well controlled by non-selective herbicides. Schäufele and Pfleiderer (2000) found that glufosinate and glyphosate herbicides were effective against the weed beet in their field trials with GM sugarbeet varieties (Liberty Link® and Roundup Ready®) in all tested timing and application doses. The advantage of this method is the ability to control weed beet at early growth stages before their seeds are formed. However, because of the longevity of weed beet seed in the soil, it will take many years to deplete the seed bank populations significantly. In addition, careful management is required to secure the long-term efficacy of this method. This includes: (i) prevention of the tolerance gene transfer into a weed beet population; and (ii) prevention of a soil infestation by new seeds carrying a herbicide tolerance from their mother plants.

Populations of weed beet could become very persistent reservoirs of various transgenes and important elements of gene flow inside and outside the agrosystem. Sugarbeet seed represents an entrance for the introduction of transgenes into adventitious beet populations. According to data obtained in the Czech Republic in 1999–2002, about 10–20% of seed lots are contaminated with weed beet every year (Table 16.4). The choice of seed production areas and the breeding systems can regulate the seed contamination with genotypes containing the *B* allele. If the transgene is built into the paternal line, bolters do not have a herbicide tolerance and can easily be managed in herbicide-tolerant crop stands. However, transgenes are more likely to flow into the wild/weed beet populations occurring in the seed production areas. When the transgene is built into the male-sterile maternal line (seed bearer), all mother plants pollinated by wild beet are likely to produce transgenic bolters.

Desplanque *et al.* (2002) discussed four factors influencing the spread of a transgene for herbicide tolerance during the seed production process: (i) use of male-sterile seed bearers; (ii) transgenic tetraploid pollinators; (iii) transgenic diploid pollinators; and (iv) cytoplasmic herbicide tolerance. The adventitious formation of transgenic weed beets is unavoidable, whatever scenario is adopted. Incorporation of the transgene into the tetraploid pollinator line is the best strategy to limit transgene flow in this respect – each contamination by wild pollen carrying the *B* allele leads to a non-tolerant, easy-to-manage weed beet hybrids. A different ploidy level can also decrease the fertility of bolters. It is generally agreed that pollen from triploid beets has low germination ability and the seeds of triploid varieties are less viable than those from diploid varieties.

Outlook

Intensive sugarbeet crop production systems seem to favour the introduction and spread of weed beet in general. In the future, it is necessary to study the influence of cropping systems and agronomic actions on dynamics in the cultural–weed beet complex to quantify and limit gene flow from transgenic beet to weed beets. A model comparison of cropping systems and a proposal of optimal cultivation practices has been suggested by Sester *et al.* (2003) (see also Lavigne *et al.*, Chapter 26, this volume) to minimize the likelihood of gene flow from transgenic beet to weed beets. For a validation of models, it is important to have precise input data of reproduction biology and population dynamics of weed beet under different climatic and soil conditions. Interdisciplinary cooperation including the research areas of plant genetics and breeding, ecology, ecosystem modelling, weed science and agronomy are necessary to manage the problem of weed beet and to ensure appropriate stewardship for the safe introduction of GM beet cultivars.

Acknowledgements

The research activities of the authors in this case are funded by the National Grant Agency for Agricultural Research, Grant Nos QD1360 and MSM 412100002.

References

Bartsch, D. and Schmitz, G. (2002) Recent experience with biosafety research and post-marked environmental monitoring in risk management of plant biotechnology derived crops. In: Thomas, J.A. and Fuchs, R.L. (eds) *Biotechnology and Safety Assessment*, 3rd edn. Elsevier Science, New York, pp. 13–37.

Bartsch, D., Lehnen, M., Clegg, J., Pohl-Orf, M., Schuphan, I. and Ellstrand, N.C. (1999) Impact of gene flow from cultivated beet on genetic diversity of wild sea beet populations. *Molecular Ecology* 8, 1733–1741.

Bartsch, D., Brand, U., Morak, C., Pohl-Orf, M., Schuphan, I. and Ellstrand, N.C. (2001) Biosafety of hybrids between transgenic virus-resistant sugar beet and Swiss chard. *Ecological Applications* 11, 142–147.

Boudry, P., Mörchen, M., Saumitou-Laprade, P., Vernet, P. and Van Dijk, H. (1993) The origin and evolution of weed beets: consequences for the breeding and release of herbicide-resistant transgenic sugar beets. *Theoretical and Applied Genetics* 87, 471–478.

CTM (2003) CTM Weed Beet Surfer. Available at: http://www.ctmharpley.freeuk.com

Danovská, A. (1996) Spôsoby ničenia burinnej repy v porastoch cukrovej repy. *Listy Cukrovarnické a řepařské* 112, 295–296.

Desplanque, B., Boudry, P., Broomberg, K., Saumitou-Laprade, P., Cuguen, J. and Van Dijk, H. (1999) Genetic diversity and gene flow between wild, cultivated and weedy forms of *Beta vulgaris* L. (Chenopodiaceae), assessed by RFLP and microsatelite markers. *Theoretical and Applied Genetics* 98, 1194–1201.

Desplanque, B., Hautekèete, N. and Van Dijk, H. (2002) Transgenic weed beets: possible, probable, avoidable? *Journal of Applied Ecology* 39, 561–571.

Diprose, M.F., Benson, F.A. and Hackam, R. (1980) Electrothermal control of weed beet and bolting sugar beet. *Weed Research* 20, 311–322.

Jassem, M., Olszevska, D. and Dabrowska, B. (1997) Fenologiczna i morfologiczna charakterystyka burakochwastów. *Biuletyn Instytutu Hodovli i Aklimatyzacji Roślin* 202, 213–219.

Kohout, V. (1996) *Kulturní rostliny jako plevel následných plodin*. Studijní informace ÚZPI, Prague.

Krouský, J. (2001) Plevelná řepa, staronový nepřítel. *Listy Cukrovarnické a řepařské* 117, 208–210.

Longden, P.C. (1989) Effects of increasing weed-beet density on sugarbeet yield and quality. *Annals of Applied Biology* 114, 527–532.

Maughan, G.L. (1984) Survey of weed beet in sugar beet in England 1978–81. *Crop Protection* 3, 315–325.

Monsanto (2000) *Roundup Biactive Product Guide*. Monsanto Agricultural Sector, Cambridge.

Mücher, T., Hesse, P., Pohl-Orf, M., Ellstrand, N.C. and Bartsch, D. (2000) Characterization of weed beet in Germany and Italy. *Journal of Sugar Beet Research* 37, 19–38.

Pohl-Orf, M., Morak, C., Wehres, U., Saeglitz, C., Driessen, S., Lehnen, M., Hesse, P., Mücher, T., von Soosten, C., Schuphan, I. and Bartsch, D. (2000) The environmental impact of gene flow from sugar beet to wild beet – an ecological comparison of transgenic and natural virus tolerance genes. In: Fairbairn, C., Scoles, G. and McHughen, A. (eds) *Proceedings of the 6th International Symposium on the Biosafety of Genetically Modified Organisms*, Saskatoon, Canada, pp. 51–55.

Rotowiper GmbH (2003) Rotowiper GmbH. Available at: http://www.rotowiper.de

Schäufele, W.R. and Pfleiderer, U.-E. (2000) Ansätze zur Bekämpfung von Ausfallraps, Kartoffeldurchwuchs und Unkrautrüben in Herbizidresistenten Zuckerrüben – erste Ergebnisse. *Zeitschrift für Pflanzenkrankheiten und Pflanzenschutz* 18, 403–409.

Sester, M., Colbach, N. and Darmency, H. (2003) Modelling gene flow from transgenic sugar beet to weed beet. In: *Programme and Abstract Book ESF Conference Introgression from Genetically Modified Plants into Wild Relatives and its Consequences*, 21–24 January , Universiteit of Amsterdam, Poster Abstract, p. 153.

Soukup, J. and Holec, J. (2002) Monitoring výskytu a diverzity plevelné řepy. In: *Proceedings of Conference Řepařství*. Czech University of Agriculture, Prague, pp. 55–58.

Soukup, J., Holec, J., Vejl, P., Skupinová, S. and Sedlák, P. (2002) Diversity and distribution of weed beet in the Czech Republic. *Journal of Plant Diseases and Protection*, Special Issue 18, 67–74.

Soukup, J., Holec, J. and Nováková, K. (2003) Supression of weed beet populations on arable land. In: *Programme and Abstract Book ESF Conference Introgression from Genetically Modified Plants into Wild Relatives and its Consequences*, 21–24 January, Universiteit of Amsterdam, Poster Abstract, p. 153.

Srba, V. (2002) Přehled odrőd cukrovky 2002. ÚKZÚZ Brno.

ÚKZÚZ (1990) Metodika prőzkumu a zjišťování výskytu plevelných řep. Správa ochrany rostlin, Prague.

Vášová, Z. (1995) Plevelná řepa v cukrovce. *Úroda* 7, 36–37.

17 A Protocol for Evaluating the Ecological Risks Associated with Gene Flow from Transgenic Crops into Their Wild Relatives: the Case of Cultivated Sunflower and Wild *Helianthus annuus*

DIANA PILSON[1], ALLISON A. SNOW[2], LOREN H. RIESEBERG[3] AND HELEN M. ALEXANDER[4]

[1]School of Biological Sciences, 348 Manter Hall, University of Nebraska, Lincoln, NE 68588–0118, USA; [2]Department of Evolution, Ecology, and Organismal Biology, 1735 Neil Ave., Ohio State University, Columbus, OH 43210–1293, USA; [3]Department of Biology, Jordan Hall, 1001 E. Third St. Indiana University, Bloomington, IN 47405–3700, USA; [4]Department of Ecology and Evolutionary Biology, 1200 Sunnyside Ave., University of Kansas, Lawrence, KS 66045–7534, USA; E-mail: dpilson1@unl.edu

Abstract

A widely acknowledged risk associated with transgenic crops is the possibility that hybridization with wild relatives will cause fitness-related transgenes to persist in wild populations. If wild plants acquire transgenes coding for resistance to herbivory, disease and/or environmental stress, they could become more abundant in their natural habitats or invade previously unsuitable habitats. In addition, wild populations containing transgenes that provide resistance to herbivores or disease may have additional effects on native insect and pathogen populations. However, little is known about whether these concerns are justified. For example, many weedy species are affected by herbivores, yet the impact of insect damage on population densities and invasiveness has rarely been examined. In order to determine if a transgene poses a risk to wild populations, or to the species with which the wild plant interacts, three questions must be addressed. These are: (i) are there genetic or geographical barriers to the escape of the transgene from the crop into wild populations? (ii) is the transgene expected to increase in frequency in wild populations? and (iii) what are the ecological consequences of escape? We are investigating these questions in the cultivated/wild sunflower system. In answer to the first two questions, we have found: few barriers to gene flow between the crop and wild population, and decreased lepidopteran herbivory and increased seed production in wild plants containing a Bt toxin gene specific to lepidopterans. In answer to the third question, increased seed production in individual plants (caused by, for example, a *Bt* gene) is likely to lead to an increase in the size of wild populations in western Nebraska, USA, but not in eastern Kansas, USA. We currently are carrying out experimental and modelling work to determine if larger local populations will lead to larger or more persistent metapopulations.

Introduction

Evaluating the ecological risks of transgenic crops has become increasingly important as more varieties are released, and as more acres are planted in these varieties. A widely acknowledged risk associated with transgenic crops is the possibility that hybridization with weedy relatives will cause fitness-related transgenes to persist in wild populations. Despite this understanding, there exist almost no data appropriate for evaluating the ecological effects of the escape of transgenes from crops into wild populations. We argue here that assessment of the risks associated with the escape of transgenes must sequentially address three questions.

The first of these questions is: is the transgenic variety of the cultivated plant sexually compatible with wild relatives? Clearly, if cultivated and wild populations are not sexually compatible, there exists no opportunity for the escape of a transgene, and risk assessment need go no further. However, many crop species, including rice, sorghum, canola (oilseed rape), sugarbeet, oats, squash, carrot, radish, strawberry, clover, sunflower and others, are known to hybridize with their wild relatives (Snow and Morán Palma, 1997; Zemetra *et al.*, 1998). The fecundity of crop–wild hybrids is often found to be lower than the fecundity of purely wild types (e.g. Snow *et al.*, 1998). However, fecundity in F_1s must be zero to prevent the transfer of beneficial alleles into wild populations.

If the crop can hybridize successfully with its wild relative, it is necessary to answer the second question: does the transgene confer a fitness benefit on the wild plant? If the transgene is of no benefit (or is always costly) to the wild plant, genetic drift (or purifying selection) will determine its fate in the wild population. For example, it may be that transgenes controlling traits important to harvesting the crop and shelf-life of the product, such as those associated with fruit ripening, will be neutral or costly in wild populations. Consequently, these traits are not expected to increase in frequency, and may pose little ecological risk. In contrast, transgenes for characters such as insect or pathogen resistance and drought tolerance may be beneficial to wild populations, and for this reason they may increase in frequency in wild populations by natural selection. Transgenes for herbicide tolerance are unlikely to increase fitness in purely wild populations, but weedy populations containing these genes may be more difficult to control with herbicides. Very little is known about the fitness effects of transgenes in wild populations; studies of the fitness effects of transgenes in wild relatives have been performed for no commercially released transgenes.

Finally, if a transgene confers a fitness benefit on a wild relative, it is necessary to answer a third question: what are the ecological consequences of the escape of the transgene into a wild population? Specifically, it is necessary to determine if the transgene alters interactions between the wild plant and its biotic and abiotic environment. A transgene that increases in frequency in wild plants does so, by definition, because it increases survival or fecundity, and one ecological risk that has been discussed in the literature is the effect of this increased individual fitness on the population size, dynamics and habitat use in the wild plant. In addition, transgenes that confer resistance to herbivores and pathogens will have effects on native species using the wild plant as a host. Clearly, these questions must be the crux of any ecological risk assessment. However, virtually no work has been done in these areas.

We have been addressing these three questions in using *Helianthus annuus*, which is the wild progenitor of cultivated sunflower, and a *Bt* transgene that confers resistance to lepidopteran herbivores. Here we summarize the work we have done towards answering each of these questions.

Sunflower and its Insect Herbivores as a Model System

Wild *H. annuus* is a native, self-incompatible, annual plant that is widespread throughout much of the USA, reaching its greatest abundance in midwestern states (Heiser, 1954), where most cultivated sunflower is grown (Schneiter, 1997). Wild sunflower is a disturbance specialist, and populations are typically patchy and ephemeral, relying on the soil seed bank and long-distance dispersal for opportunities to become established in new areas. In the absence of tilling or other types of disturbance, population size declines. In agricultural areas, however, repeated tilling allows wild sunflower populations to persist for many years, especially along field margins.

Wild sunflower is host to many insect herbivores, and many of these species are also pests in the crop. Some herbivores have negative effects on fitness in wild populations (Pilson, 2000), and they can also substantially reduce yield in some years and locations (Charlet *et al.*, 1997). The most damaging insect pests of cultivated sunflower are those that infest developing seed heads (weevil, moth and midge larvae) and those that transmit disease (e.g. stem weevils that transmit *Phoma* black stem) (Schneiter, 1997). In wild *H. annuus*, insect resistance is typically polygenic, and efforts to introgress strong resistance into the crop have been unsuccessful (Seiler, 1992). For these reasons, cultivated lines with transgenic resistance conferred by Bt toxins are being developed by a number of seed companies, and several field trials have been approved (http://www.isb.vt.edu). Different Bt toxins are specific to different groups of insects, including Lepidoptera, Coleoptera and Diptera. Bt-induced resistance to Coleoptera was first field tested in the USA in 1996, and resistance to Lepidoptera was approved for field testing in 1999, although none has been commercialized to date. Additional field trials have taken place in The Netherlands and Argentina (http://www.isb.vt.edu; http://siiap.sagyp.mecon.ar/http-hsi/english/conabia/liuk4.htm). Broad-spectrum resistance involving multiple *Bt* genes and other genes for insect resistance may also be developed in the future.

Are Wild and Cultivated Sunflower Sexually Compatible?

Cultivated sunflower, which is primarily planted in North and South Dakota, Nebraska, Kansas and eastern Colorado, is nearly always sympatric with wild *H. annuus* (Heiser, 1954; Schneiter, 1997). Field experiments have shown that pollinators can transfer crop pollen to wild plants as far as 1000 m away, with the frequency of hybrid seeds being greatest (up to 42%) at the crop margin (Arias and Rieseberg, 1994; Whitton *et al.*, 1997). Studies have also shown that F_1 hybrids usually produce fewer seeds per plant than their wild counterparts, but the magnitude of this difference varies a great deal among plants, regions and growing conditions (Snow and Morán Palma, 1997; Snow *et al.*, 1998). Furthermore, selectively neutral crop markers have persisted for many generations in wild plants sampled in California, Kansas, North Dakota and Canada (Whitton *et al.*, 1997; Linder *et al.*, 1998). These studies demonstrate that introgression of neutral or beneficial crop genes into wild gene pools can be an ongoing process wherever these taxa occur sympatrically. These studies and the sympatry of crop and wild plants (Heiser, 1954; Schneiter, 1997) demonstrate that both genetic and geographical barriers to gene flow from crop to wild sunflower are minimal. Thus, the answer to the first question is yes, and further studies of pre-commercial transgenes in the wild genetic background are necessary.

What is the Fitness Effect of the *Bt* Transgene in the Wild Background?

The next step is to determine if the transgene(s) increases the fitness of wild plants. In the case of a *Bt* transgene, we need to know if plants carrying the gene are resistant to herbivory (by the species specifically targeted by the transgene) and, further, whether this reduction in damage leads to an increase in fitness. We have addressed these questions for a *Bt* transgene that is specific to Lepidoptera (Snow *et al.*, 2003). Our study involved the Bt protein Cry1Ac, which is toxic to many lepidopteran species but is not expected to affect other insect taxa (Estruch *et al.*, 1997). Ingesting a very small amount of Bt toxin (e.g. parts per billion) typically causes susceptible insects to stop feeding and die within a few days (Estruch *et al.*, 1997), or move to a non-toxic host plant (Davis and Onstad, 2000).

Determining the ecological effects of pre-commercial transgenes is inherently difficult due to biosafety and regulatory concerns. Uncaged plants must be exposed to natural levels of insect damage and cross-pollination, yet dispersal and persistence of the transgene(s) must be prevented. Our solution to this difficult problem was to use male-sterile plants for the field experiments so that the possibility of transgene escape through pollen could be eliminated.

To simulate the effects of introgression of a *Bt* transgene from the crop, male-sterile wild plants from a population near the Cedar Point Biological Station (near Ogallala) in western Nebraska were bred with transgenic cultivars to create BC_1 and BC_3 progeny that segregated for both the *Bt* transgene (*Bt+* or *Bt−*) and for male sterility (male-sterile or male-fertile). However, to prevent the accidental escape of the transgene, we did not use *Bt+*/male-fertile plants in the field. BC_1 progeny were planted in the field in 1999 at the Cedar Point Biological Station and in an agricultural field near Burlington, in eastern Colorado, and BC_3 progeny were planted at the Nebraska site in 2000. The effect of the transgene was examined by comparing insect damage and fecundity between *Bt+*/male-sterile and *Bt−*/male-sterile plants. We also compared *Bt−*/male-sterile and *Bt−*/male-fertile plants to determine the effects of male sterility on herbivory and seed production (some seed predators feed on pollen (Delisle *et al.*, 1989; Korman and Oseto, 1989) and might avoid male-sterile plants).

The *Bt* transgene led to reduced lepidopteran damage at both field sites and in both years (see Fig. 17.1 for the data from the Nebraska 1999

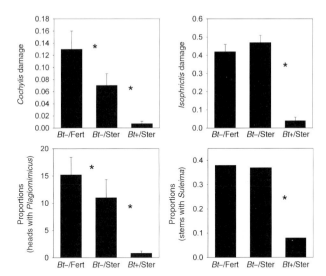

Fig. 17.1. Lepidopteran damage in Nebraska in 1999. *Cochylis hospes* (Chochylidae) and *Isophrictis similiella* (Gelechiidae) damage were categorized as 0, 1 or 2 (0, 1–30 or > 30 seeds eaten) for each inflorescence, and the mean value over all inflorescences on each plant was analysed by ANOVA. The proportion of heads on each plant attacked by *Plagiomimicus spumosum* (Noctuidae) was analysed by ANOVA. The proportion of plants with stem damage by *Suleima helianthana* (Tortricidae) was analysed by categorical ANOVA. In all analyses, we made planned contrasts between *Bt−*/male-sterile and *Bt−*/male-fertile plants, and between *Bt−*/male-sterile and *Bt+*/male-sterile plants, and significant contrasts (at *P* < 0.05) are indicated by asterisks.

experiment). However, the Bt toxin had no effect on amounts of damage caused by four non-lepidopteran species (see Fig. 17.2 for Nebraska 1999; Colorado 1999 and Nebraska 2000 show similar patterns). Although some of the non-lepidopteran species are negatively affected by competition with lepidopterans (M. Paulsen and D. Pilson, unpublished results), they did not cause more damage on *Bt+* plants than on *Bt–* controls. As expected, damage by some herbivores was reduced on male-sterile plants relative to male-fertile plants (Figs 17.1 and 17.2). *Bt+* plants produced an average of 55 and 23% more seeds per plant than *Bt–* plants in Nebraska in 1999 and 2000, and 14% more seeds per plant in Colorado in 1999 (Fig. 17.3). The reduction in herbivory on male-sterile plants suggests that using male-sterile plants to test for *Bt* effects may underestimate the fecundity advantage associated with the transgene. Had we been able to use pollen-producing *Bt* plants in the field, we might have documented more dramatic fecundity benefits of the transgene.

In any study of a single transformation event, it is not clear whether phenotypic effects (e.g. greater fecundity) are caused by the transgenic construct or by other mechanisms, such as position effects, pleiotropy, or close physical linkage with other crop genes. Thus, it is useful to determine whether effects associated with the *Bt* transgene can occur in the absence of lepidopteran herbivores. We performed a greenhouse experiment using BC_1 plants to examine this possibility, while recognizing there are many biotic and abiotic differences between field and greenhouse conditions. The *Bt* transgene had no effect on the number of inflorescences or seeds per plant in the greenhouse, regardless of whether the plants were grown under low

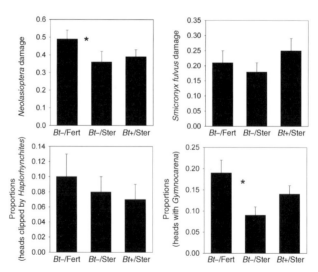

Fig. 17.2. Damage by non-lepidopteran herbivores in Nebraska in 1999. Damage by *Neolasioptera helianthi* (Cecidomyidae) and *Smicronyx fulvus* (Curculionidae) was quantified categorically (as described in Fig. 17.1). These data, as well as data for *Haplorhynchites aeneus* (Curculionidae) and *Gymnocarena diffusa* (Tephritidae), were analysed by ANOVA with planned contrasts (as described in Fig. 17.1).

water, low nutrient or control conditions, and regardless of whether they were male-fertile or male-sterile (Snow *et al.*, 2003; Fig. 17.4). This suggests that the transgene was not associated with an inherent fitness cost or benefit. It would be preferable to employ a wider range of growing conditions and several transgenic events in this type of study, but our results suggest that the fecundity advantage of transgenic plants in the field was due to protection from lepidopteran herbivores.

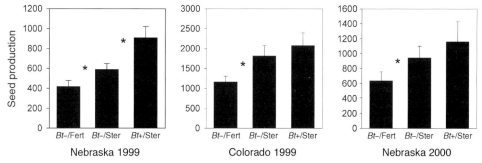

Fig. 17.3. Effects of the *Bt* transgene and male sterility on seed production per plant in Nebraska and Colorado in 1999 and Nebraska in 2000. Untransformed means and 1 SE are shown; *n* = 58–60 in Nebraska in 1999 and 47–49 in Colorado in 1999 and Nebraska in 2000. Data were analysed by ANOVA followed by planned contrasts as described in Fig. 17.1. When both years and sites were combined in a single ANOVA, the planned contrast between *Bt*+/male-sterile and *Bt*–/male-sterile plants was significant at *P* < 0.0206. The selection coefficients favouring the *Bt* gene were 0.35 in Nebraska in 1999, 0.13 in Colorado in 1999, and 0.19 in Nebraska in 2000.

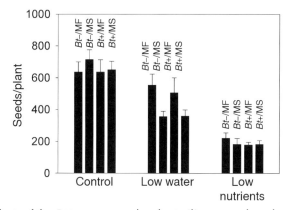

Fig. 17.4. Effects of the *Bt* transgene and male sterility on seed production of plants grown under three conditions (Control, Low water, Low nutrients) in a greenhouse. Means and 1 SE are shown. The only significant effects in a three-way ANOVA were growing condition (*P* < 0.0001) and the interaction between growing condition and male-fertility (*P* < 0.0042). Sample sizes from left to right were Control: 19, 24, 16, and 27 plants per treatment; Low water: 17, 27, 12 and 24; Low nutrients: 14, 11, 14, and 12.

This study shows that selection favouring an increase in the frequency of a *Bt* transgene has the potential to be quite strong. If herbivores cause more damage to F_1 (Cummings *et al.*, 1999) and BC_1 sunflowers than to wild genotypes, it is possible that the fecundity advantage associated with *Bt* would diminish with subsequent generations of backcrossing. On the other hand, because male-sterile plants had less damage from lepidopterans than those with pollen, we may have underestimated the fecundity advantage of *Bt* in this study. In addition, we have observed higher levels of lepidopteran damage on wild plants in other years (Pilson, 2000; D. Pilson and M. Paulsen, unpublished results). Therefore, we expect that subsequent generations of *Bt* wild plants would produce more seeds per plant than non-transgenic individuals in many locations and growing seasons. If so, the transgene is expected to spread quickly and kill susceptible, native lepidopterans that feed on wild sunflower.

What are the Ecological Consequences of the Escape of a *Bt* Gene into Wild Populations?

Although the escape of a transgene into a wild population, and its subsequent frequency increase by natural selection, are necessary, they are not sufficient to predict the environmental consequences of escape. Specifically, these processes are only important to the extent that they lead to the alteration of existing ecological interactions between the wild plant and its biotic and abiotic environment. Arguably, it is these potential ecological effects that should be the focus of attention in risk assessments.

In ongoing work, we are examining two types of ecological consequences of the escape of a *Bt* transgene into wild populations. These are: (i) the effect on the population dynamics of wild sunflower; and (ii) the effect on the community of native herbivores that feed on wild sunflower, and we present preliminary results from these studies here.

The effect of a *Bt* transgene on the population dynamics of wild sunflower

We have shown that a *Bt* gene increases seed production in wild plants, and is therefore likely to increase in frequency in wild populations. However, increased seed production by individual plants will only lead to larger populations, or more populations, if wild sunflower is currently seed limited. At present, little is known about processes controlling the population dynamics of wild sunflower. Thus, it is currently unclear what effect, if any, a *Bt* transgene will have on the dynamics of wild populations.

Surprisingly, there are few data with which to evaluate the importance of seed limitation on the population dynamics of any plant species. Silvertown and Franco (1993) and Silvertown *et al.* (1993) compared the sensitivities of demographic transitions in herbs and woody plants and found that population growth rates were more affected by changes in fecundity in semelparous species, suggesting that annual species, such as sunflower, are more likely than other plants to be seed limited. Louda and Potvin (1995) found that elimination of herbivores by application of insecticide increased not only individual fitness but also local population size in a native thistle. Crawley and Brown (1995) found that weedy roadway populations of oilseed rape were seed limited at the landscape scale. In contrast, Bergelson (1994) found that population size was not seed limited in experimental plantings of a diminutive cress, *Arabidopsis thaliana*, a result that is likely to be due to insufficient open space and competition from surrounding vegetation.

To evaluate the effect of increased seed production on the population dynamics of wild sunflower, we are using the following approach. First, in western Nebraska (in 2000 and 2001) and eastern Kansas (in 2000), we established experimental populations with varying amounts of seed production, and we are examining the effect of seed production in one year on population size and seed production in the following years. Secondly, we are using these experimental populations to derive parameters for spatially explicit metapopulation models. We will use these models to make predictions about the effect of seed production on metapopulation dynamics at our two study sites. Thirdly, we will validate our metapopulation models by comparison with empirical observations of metapopulation dynamics in both western Nebraska and eastern Kansas. Our two study sites are in very different environments (historically mixed short grass prairie in western Nebraska, and tall grass prairie in Kansas), and we expect that the dynamics will also be different in these habitats. This multi-year research programme is in its early stages, so here we only present data documenting the effect of seed production in one year on population size the following year.

In 2000, we established 12 experimental populations in each of four spatial blocks (48 populations in total) at each of our study sites (at the Cedar Point Biological Station, in western Nebraska, and at the Kansas Ecological Reserves, near Lawrence, in eastern Kansas). Each experimental population consisted of either 16 ($n = 32$) or 21 ($n = 16$) experimental plants. In each population, the experimental plants were surrounded by either 45 or 40 plants that were not allowed to disperse seed, and which served to maintain a similar competitive environment for all experimental plants. All of these plants were sown in a central (rototilled) 2 m $\times$ 2 m square (experimental plants in the central 1 m $\times$ 1 m) located in the centre of a larger 10 m $\times$ 10 m square. Half of the experimental populations with 16 plants were sprayed with a broad-spectrum insecticide to reduce herbivory (and increase seed production). Because sunflower is a disturbance specialist, and requires a recent disturbance for germination, we rototilled a 2 m $\times$ 4 m strip in one cardinal direction from each of the 48 plots (i.e. 12 tilled North, 12 tilled East, 12 tilled South and 12 tilled West). The 45 or 40 competitor plants were removed before dispersing any seeds, and experimental plants were allowed to disperse their seeds naturally. Probably due to drought during the 2000 growing season, many plants performed very poorly, and the treatments (number of experimental plants and spraying) had no effect on seed production at either study site. None the less, seed production (estimated by the number of heads dispersing seeds and an estimate of mean head area) varied by approximately an order of magnitude at both study sites (2016–23,381 seeds/population in Nebraska, and 5505–25,667 seeds/population in Kansas).

In 2001, we counted the number of seedlings emerging in each experimental population (in both the original square and the tilled strip). (Presumably, seeds were dispersed in all directions. However, with few exceptions, seeds only germinated in the tilled squares and strips.) We also counted the number of seedlings that survived to reproductive age and the number of inflorescences produced by each plant that survived to reproductive age. In Nebraska, we measured the inflorescence diameter of ten randomly chosen inflorescences in each experimental population, and estimated seed production in 2001 by estimating the total inflorescence area produced by each population.

We used regression methods (SAS Institute, 1989) to evaluate the effect of seed production in 2000 on the number of seedlings, number of reproductive plants, number of mature inflorescences and the estimated number of seeds produced in 2001. Because the direction in which the strips were tilled had a significant effect on all of these response variables (probably due to a consistent prevailing wind direction during seed dispersal in 2000), we included direction, as well as spatial block, as class variables in our analyses (PROC GLM in SAS). Before analysis, all variables were standardized to a mean of one and unit variance. Thus, these analyses provided the standardized partial regression coefficients evaluating the effect of seed production in 2000 on each of the separate response variables. Standardized coefficients are in units of standard deviation, and thus they can be directly compared among analyses and among sites.

While it may seem redundant to evaluate the effect of seed production in 2000 on all of these population responses in 2001, it is only redundant to the extent that there are no density-dependent processes affecting seed production. For example, it might be the case that many seeds germinate, but that intraspecific competition in high-density populations reduces the number of plants surviving to reproductive age to the same as seen in an originally lower density population. Similarly, many plants might survive, but plants might produce fewer inflorescences in higher density populations.

Our analyses suggest that in Nebraska, increasing seed production in one year increases both population size and the number of seeds dispersed the following year (Fig. 17.5). Although the pattern is similar in Kansas, there is no significant overall effect of seed production in 2000 on population size in 2001 (Fig. 17.5). These results suggest that density-dependent effects are more important in Kansas. In fact, in both Nebraska and Kansas, there appear to be density-dependent effects on seedling survival and the number of inflorescences on surviving plants (results not presented). However, these effects are stronger in Kansas, and it is these stronger effects that led to the non-significant effect of seed production on the next year's population size. From these data, we tentatively conclude that some sunflower populations

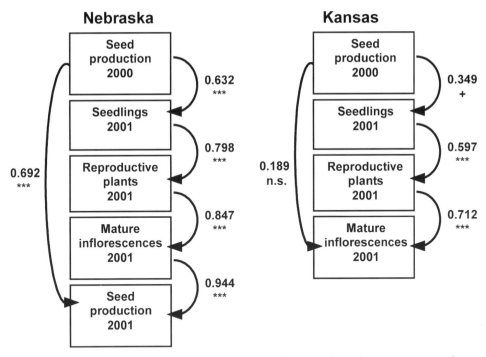

Fig. 17.5. The effect of the number of seeds dispersed in 2000 from the 48 experimental populations in Nebraska and Kansas on seedlings emerging, plants surviving to reproductive age, number of inflorescences with seeds, and estimated seed production (Nebraska only) in 2001. Each arrow indicates the standardized partial regression coefficient from a separate analysis that included spatial block and tilling direction, as well as the independent and response variables indicated. Standardized partial regression coefficients are in standard deviation units, and thus are comparable across analyses and sites. ***, $P < 0.0001$; +, $P < 0.10$; n.s., not significant.

are seed limited, and that increased seed production may lead to increased local population size. Additional analyses of these data, and of the second set of Nebraska populations, are necessary before we can draw firm conclusions.

We are also using these experimental populations (and additional manipulations of these populations in later years) to derive parameters for our metapopulation models. Specifically, we are evaluating the effect of seed production on patterns of seed dispersal, the establishment and decay of the seed bank, and the longevity of our experimental populations (both with and without additional disturbance). These empirically derived parameters, measured on the scale of our 2 m × 2 m populations, will be used in a spatially explicit metapopulation model. Metapopulation models were originally developed by Levins (1969), and in recent years have been a major focus of ecological inquiry (Hanski and Gilpin, 1991, 1997; Tilman and Kareiva, 1997). These models are often used in studies of organisms with very distinct habitat patches (e.g. frogs in ponds; Sjögren, 1994), but Antonovics *et al.* (1994), Freckleton and Watkinson (2002) and Higgins and

Cain (2002) have discussed how empirical metapopulation studies can be conducted with organisms that are patchily distributed but have broad habitat requirements. Our model will be written at the 2 m scale, because that is the scale at which we are deriving our parameters, but it can be scaled up from 2 to 80 m and beyond, enabling us to make predictions at larger spatial scales.

We will validate the results of our metapopulation model by comparison of model predictions with empirical observations of metapopulation dynamics. We have established metapopulation census routes along roadsides in western Nebraska (17.5 miles) and eastern Kansas (15.4 miles), near our local population study sites. Each autumn, when sunflower is in peak bloom, we count the number of flowering plants in each 0.05 mile section on both sides of the road. In addition, in either 17 (Nebraska) or 15 (Kansas) 80 m (~0.05 mile) segments, we count the number of flowering plants in each 2 m segment. In this way, we can connect observed dynamics at the scale of 2 m, 80 m and beyond, with the dynamics predicted by our model.

If the escape of a *Bt* gene into wild sunflower populations leads to an increase in the size or number of wild populations, there will probably be additional ecological consequences. For example, to the extent that sunflower increases in abundance, plant species that commonly co-occur with sunflower will probably be affected, and may decrease in relative frequency. These community-level effects of the escape of a transgene are currently beyond the scope of our work.

The effect of a *Bt* transgene on herbivore community structure

Clearly, if a *Bt* transgene escapes into wild sunflower populations, it will have a negative impact on the suite of susceptible native herbivores. Our data indicate that native herbivores will be killed by the Bt protein (Fig. 17.1) and, as the frequency of a *Bt* gene increases in wild sunflower, the negative impact on these native herbivores will also increase. In addition to the taxa shown to be affected in the present study, wild sunflower at our study sites is also attacked by two other native lepidopterans (*Homoeosoma electellum* and *Eucosma womonana*), and these species are also likely to be negatively affected by a *Bt* gene. Moreover, these species are specialists, feeding on only *H. annuus*, or on only *H. annuus* and other species in the genus *Helianthus*, and for this reason they cannot easily find food on other host plants. Although these species may evolve resistance to *Bt*, at least initially the suite of native lepidopterans will be severely affected by the escape of a lepidopteran-specific transgene into wild populations.

The effect of a lepidopteran-specific *Bt* gene on the non-lepidopteran members of the native herbivore community depends on existing interactions between these suites of herbivores. If the lepidopterans have negative competitive effects on any non-lepidopteran species, the release from competition provided by a *Bt* gene might allow these non-lepidopterans to increase in abundance. Experimental manipulations of damage by three of the lepidopterans (*Plagiomimicus*, *Homoeosoma* and *Cochylis*) indicate that only *Plagiomimicus* competes with other herbivores. Damage by *Neolasioptera helianthi* (sunflower seed midge) and *Smicronyx fulvus* (red sunflower seed weevil) is lower in inflorescences with *Plagiomimicus* than in inflorescences without this species (M. Paulsen and D. Pilson, unpublished results). However, because *Plagiomimicus* abundance is typically low, these interactions are probably only important every 5–10 years when *Plagiomimicus* reaches high abundance during population outbreaks (M. Paulsen and D. Pilson, unpublished results).

Summary

We have argued that assessing the ecological risks of the escape of transgenes into wild populations must sequentially address three questions. With respect to the question of cross-compatibility, we have found few barriers to gene flow between crop and wild sunflower. We also provided an unambiguous affirmative answer to the question of whether the transgene is positively selected in the wild. Wild plants containing a *Bt* toxin gene specific to lepidopterans exhibited decreased lepidopteran herbivory and increased seed production. Thus, we expect that *Bt* transgenes released in commercial sunflower will escape into and increase in frequency in wild populations (at least in locations and in years in which susceptible herbivores are abundant). Next it is necessary to evaluate the ecological consequences of transgenes in wild populations; however, virtually no work has been done in this area. Preliminary analyses suggest that increased seed production in individual plants (caused by, for example, a *Bt* gene) will lead to an increase in the size of wild populations in at least some locations. We are currently doing experimental and modelling work to determine if larger local populations will lead to larger or more persistent metapopulations.

Acknowledgements

We thank Jennifer Moody-Weis, Matt Paulsen, Nick Pleskac, Mike Reagon, Sarena Selbo and Diana Wolf for important contributions to the work presented here. Dow AgroSciences and Pioneer HiBred provided the transgenic cultivated line, and C. Scelonge at Pioneer HiBred performed the ELISA and PCR analyses. This work was funded by several grants from the US Department of Agriculture and by Dow AgroSciences and Pioneer HiBred. The data, analysis, interpretations and conclusions are solely those of the authors and not of Dow AgroSciences, L.L.C. or Pioneer HiBred International, Inc.

References

Antonovics, J., Thrall, P., Jarosz, A. and Stratton, D. (1994) Ecological genetics of metapopulations: the *Silene–Ustilago* plant–pathogen system. In: Real, L. (ed.)

Ecological Genetics. Princeton University Press, Princeton, New Jersey, pp. 146–170.

Arias, D.M. and Rieseberg, L.H. (1994) Gene flow between cultivated and wild sunflower. *Theoretical and Applied Genetics* 89, 655–660.

Bergelson, J. (1994) Changes in fecundity do not predict invasiveness: a model study of transgenic plants. *Ecology* 75, 249–252.

Charlet, L., Brewer, G.J. and Franzmann, B.A. (1997) Sunflower insects. In: Schneiter, A.A. (ed.) *Sunflower Technology and Production*. Publication No. 35, American Agronomy Society of America, Madison, Wisconsin, pp. 183–261.

Crawley, M.J. and Brown, S.L. (1995) Seed limitation and the dynamics of feral oilseed rape on the M25 motorway. *Proceedings of the Royal Society of London, Series B* 259, 49–54.

Cummings, C., Alexander, H.M. and Snow, A.A. (1999) Increased pre-dispersal seed predation in sunflower wild–crop hybrids. *Oecologia* 121, 330–338.

Davis, P.M. and Onstad, D.W. (2000) Seed mixtures as a resistance management strategy for European corn borers (Lepidoptera: Crambidae) infesting transgenic corn expressing Cry1Ab protein. *Journal of Economic Entomology* 93, 937–948.

Delisle, J., McNeil, J.N., Underhill, E.W. and Barton, D. (1989) *Helianthus annuus* pollen, an oviposition stimulant for the sunflower moth, *Homoeosoma electellum*. *Entomologia Experimentalis et Applicata* 50, 53–60.

Estruch, J.J., Carozzi, N.B., Desai, N., Duck, N.B., Warren, G.W. and Koziel, M.G. (1997) Transgenic plants: an emerging approach to pest control. *Nature Biotechnology* 15, 137–141.

Freckleton, R.P. and Watkinson, A.R. (2002) Large-scale spatial dynamics of plants: metapopulations, regional ensembles and patchy populations. *Journal of Ecology* 90, 419–434.

Hanski, I. and Gilpin, M.E. (1991) Metapopulation dynamics: brief history and conceptual domain. *Biological Journal of the Linnean Society* 42, 3–16.

Hanski, I. and Gilpin, M.E. (1997) *Metapopulation Dynamics: Ecology, Genetics, and Evolution*. Academic Press, New York.

Heiser, C.B. (1954) Variation and subspeciation in the common sunflower, *Helianthus annuus*. *American Midland Naturalist* 51, 287–305.

Higgins, S.I. and Cain, M.L. (2002) Spatially realistic plant metapopulation models and the colonization–competition trade-off. *Journal of Ecology* 90, 616–626.

Korman, A.K. and Oseto, C.Y. (1989) Structure of the female reproductive system and maturation of oocytes in *Smicronyx fulvus* (Coleoptera: Curculionidae). *Annals of the Entomological Society of America* 82, 94–99.

Levins, R. (1969) Some demographic and genetic consequences of environmental heterogeneity for biological control. *Bulletin of the Entomological Society of America* 15, 237–240.

Linder, C.R., Taha, I., Seiler, G.J., Snow, A.A. and Rieseberg, L.H. (1998) Long-term introgression of crop genes into wild sunflower populations. *Theoretical and Applied Genetics* 96, 339–347.

Louda, S.M. and Potvin, M.A. (1995) Effect of inflorescence feeding insects on the demography and lifetime fitness of a native plant. *Ecology* 76, 229–245.

Pilson, D. (2000) Flowering phenology and resistance to herbivory in wild sunflower, *Helianthus annuus*. *Oecologia* 122, 72–82.

SAS Institute (1989) *STAT User's Guide*, version 6, 4th edn. SAS Institute, Cary, North Carolina.

Schneiter, A.A. (ed.) (1997) *Sunflower Technology and Production*. Agronomy Monographs No. 35, Soil Science Society of America, Madison, Wisconsin.

Seiler, G.J. (1992) Utilization of wild sunflower species for the improvement of cultivated sunflower. *Field Crops Research* 30, 195–230.

Sjögren, P. (1994) Distribution and extinction patterns within a northern metapopulation of the pool frog, *Rana lessonae. Ecology* 75, 1357–1367.

Silvertown, J. and Franco, M. (1993) Plant demography and habitat: a comparative approach. *Plant Species Biology* 8, 67–73.

Silvertown, J., Franco, M., Pisanty, I. and Mendoza, A. (1993) Comparative plant demography – relative importance of life-cycle components to the finite rate of increase in woody and herbaceous species. *Journal of Ecology* 81, 465–476.

Snow, A.A. and Morán Palma, P. (1997) Commercialization of transgenic plants: potential ecological risks. *BioScience* 47, 421–431.

Snow, A.A., Morán Palma, P., Rieseberg, L.H., Wszcelaki, A. and Seiler, G. (1998) Fecundity, phenology, and seed dormancy of F_1 wild–crop hybrids in sunflower (*Helianthus annuus*, Asteraceae). *American Journal of Botany* 85, 794–801.

Snow, A.A., Pilson, D., Rieseberg, L.H., Paulsen, M.J., Pleskac, N., Reagon, M.R., Wolf, D.E. and Selbo, S. (2003) A Bt transgene reduces herbivory and enhances fecundity in wild sunflowers. *Ecological Applications* 13, 279–286.

Tilman, D. and Kareiva, P. (eds) (1997) *Spatial Ecology: the Role of Space in Population Dynamics and Interspecific Interactions.* Princeton University Press, Princeton, New Jersey.

Whitton, J., Wolf, D.E., Arias, D.M., Snow, A.A. and Rieseberg, L.H. (1997) The persistence of cultivar alleles in wild populations of sunflowers five generations after hybridization. *Theoretical and Applied Genetics* 95, 33–40.

Zemetra, R.S., Hansen, J. and Mallory-Smith, C.A. (1998) Potential for gene transfer between wheat (*Triticum aestivum*) and jointed goatgrass (*Aegilops cylindrica*). *Weed Science* 46, 313–317.

18 A Review on Interspecific Gene Flow from Oilseed Rape to Wild Relatives

ANNE-MARIE CHÈVRE[1], HENRIETTE AMMITZBØLL[2], BRODER BRECKLING[3], ANTJE DIETZ-PFEILSTETTER[4], FRÉDÉRIQUE EBER[1], AGNÈS FARGUE[1], CÉSAR GOMEZ-CAMPO[5], ERIC JENCZEWSKI[1], RIKKE JØRGENSEN[2], CLAIRE LAVIGNE[6], MATTHIAS S. MEIER[7], HANS C.M. DEN NIJS[8], KATHRIN PASCHER[9], GINETTE SEGUIN-SWARTZ[10], JEREMY SWEET[11], C. NEAL STEWART JR[12] AND SUZANNE WARWICK[13]

[1]INRA-Station Amélioration des Plantes, PO Box 35327, Domaine de la Motte, F-35653 Le Rheu, cedex, France; [2]RISØ National Laboratory, DK-4000 Roskilde, Denmark; [3]Center for Environmental Research Technology (UFT), University of Bremen, PO Box 33 04 40, D-28334 Bremen, Germany; [4]Federal Biological Research Centre for Agriculture and Forestry, Messeweg 11–12, D-38104 Braunschweig, Germany; [5]Escuela T.S. Ing. Agrónomos, Universidad Politécnica, 28040 Madrid, Spain; [6]Laboratoire Ecologie, Systématique et Evolution, UPRESA 8079, CNRS/Université Paris-Sud, Centre d'Orsay, Bâtiment 362, F-91405 Orsay Cedex, France; [7]Swiss Federal Institute of Technology, Geobotanical Institute, Zürichbergstrasse 38, CH 8044 Zürich, Switzerland; [8]IBED, University of Amsterdam, PO Box 94062, 1090 GB Amsterdam, The Netherlands; [9]Institute of Ecology and Conservation Biology, Department of Conservation Biology, Vegetation and Landscape Ecology, University of Vienna, Althanstraße 14, A-1090 Vienna, Austria; [10]Agriculture and Agri-Food Canada, Saskatoon, Saskatchewan SN7 0X2, Canada; [11]NIAB, Huntingdon Road, Cambridge CB3 0LE, UK; [12]Department of Plant Sciences, University of Tennessee, Knoxville, TN 37996–4561, USA; [13]Agriculture and Agri-Food Canada, Ottawa, Ontario K1A 0C6, Canada; E-mail: Anne-Marie. Chevre@rennes.inra.fr

Abstract

One of the main concerns about the commercial release of transgenic crops is the likelihood of transgene spread from cultivated species into wild relatives. This question is relevant for oilseed rape/canola (*Brassica napus*, AACC, $2n = 38$), as this species is partially allogamous with several wild relatives that are often sympatric with oilseed rape production. A workshop sponsored by the European Science Foundation (11–13 June 2001, Rennes, France) was held: (i) to identify the main weeds present in European and North American countries; (ii) to review results on the ability of oilseed rape to hybridize and backcross with wild relatives; (iii) to review the usefulness and limitations of the tools available for monitoring interspecific hybridization and gene introgression; and (iv) to provide recent results on modelling of gene flow.

Introduction

Various issues must be addressed when assessing the likelihood of gene flow from transgenic crops to wild relatives: (i) the existence of close relatives growing sympatrically with the crop and the similarity between the crop/wild relative phenologies, resulting in overlapping flowering in time and space; (ii) the production of fertile interspecific F_1 hybrids and their survival; (iii) the transmission of transgenes from the crop through the successive backcross generations; (iv) stable gene introgression through recombination between the genomes of the crop and the wild relative; and (v) maintenance of the introgressed genes within natural populations and potential stabilization of a new weed. Most of these questions have been examined for oilseed rape and for some of its wild relatives. At a workshop sponsored by the European Science Foundation (ESF; 11–13 June 2001, Rennes, France), participants presented information on the status of gene flow research in various countries and discussed approaches, methods and tools to assess the occurrence of gene flow between oilseed rape and its main wild relatives. A consensus was reached as to the main wild relatives requiring monitoring, on the basis of their relative importance as weeds and their potential for interspecific hybridization and gene introgression with oilseed rape. The first modelling results were presented, as well as research perspectives.

Oilseed Rape, its Main Wild Relatives and their Biogeographies

Oilseed rape (*Brassica napus* L., *AACC*, $2n = 38$) is a natural hybrid between the diploid species, *B. rapa* L. (*AA*, $2n = 20$) and *B. oleracea* L. (*CC*, $2n = 18$). In spite of the presence of wild populations of the progenitors, wild oilseed rape populations have not been reported. However, feral populations of *B. napus* occur frequently in anthropogenically influenced landscapes, including agricultural landscapes, and can be found at ruderal sites and in disturbed urban areas. Among the 235 species belonging to the Brassicaceae tribe (Warwick *et al.*, 2000), to which oilseed rape belongs, 14 species are found in areas where oilseed rape is cultivated in Europe and/or North America. The distribution of the species and their abundance are reported in Table 18.1.

The first seven related species (Table 18.1) are the most abundant from an agricultural perspective. More detailed information can be found on the

Table 18.1. Distribution and abundance of the main relatives of oilseed rape in Europe and North America.

	Austria	Czech Rep.	Denmark	France	Germany	Italy	Netherlands	Spain	Switzerland	UK	Canada	USA
Brassica rapa L. (2n = 20)	+++	+	+++	+	++	+	++++	+	+++	++	+++	+++
Brassica nigra (L.) Koch (2n = 16)	+		++	+++	+	+++	+++	+	+		+	+++
Raphanus raphanistrum L. (2n = 16)	++++	++++	+++	++++	++	++++	++	++++	++++	+++	+++	+++
Raphanus sativus L. (2n = 18)	++*		+	+	+	+	++	+	+++	+++	+++	+++
Sinapis arvensis L. (2n = 18)	++++	++++	++++	++++	++++	++++	++++	++++	++++	++++	++++	++++
Sinapis alba L. (2n = 24)	++		++	+++	+*	+++	+	++	+	+++	+*	+
Hirschfeldia incana (L.) Lagrèze-Fossat (2n = 14)	++	+++	++	+++	++	+++	++	++++	+	++		++
Brassica oleracea L. (2n = 18)	++*		++	+	+*	+*	+*	+	+*	+	+*	+*
Brassica juncea (L.) Czern.[a] (2n = 36)	++*	+	++						+		+	+
Eruca vesicaria (L.) Cav. ssp. sativa (2n = 22)	++		++	+		+	++	+++	+			+
Diplotaxis erucoides (L.) DC. (2n = 14)				++		+		+				
Diplotaxis muralis (L.) DC.[a] (2n = 42)	++	+	+++	+	++	+	++	+	+	++	++	++
Diplotaxis tenuifolia (L.) DC. (2n = 22)	+++	+	++	+	++	++++	+++	+	+	++	+	++
Erucastrum gallicum (Willd.) O.E. Schulz[a] (2n = 30)	++		++	+			++		++		++	++

The first seven species occur most frequently as weeds. Occurrence of the species in these countries: (+) indicates that the relative is present in the country, (++), (+++) and (++++) indicate that the relative is rare, less abundant or abundant, respectively, in cultivated areas, including the areas where oilseed rape is currently grown or could potentially be grown.
[a]Species that are primarily selfers; the others are allogamous; *feral populations. For *B. oleracea*, (+) autochtonous (native in the mainland) or (+*) feral populations.

following websites for maps and ecological data on Canadian crucifers (Warwick *et al.*, 1999), and for geographical distribution on all species in the Brassicaceae (HYPPA, 2000; Warwick *et al.*, 2000).

Assessment of Interspecific Hybridization, Gene Introgression and Fitness

Tools for the characterization of interspecific hybrids and gene introgression

Various tools have been used to assess the occurrence of hybridization and gene introgression between oilseed rape and its major wild relatives. Their relative usefulness and limitations will be discussed.

Morphological markers

The morphology of leaves, flowers and pods has been used successfully to distinguish species of the Brassicaceae (Gomez-Campo, 1980). However, morphological markers are largely insufficient to identify interspecific F_1 hybrids unambiguously between oilseed rape and wild relatives because the morphology of the hybrids is often similar to the female parent rather than being intermediate. In cases where oilseed rape was the female parent, morphological criteria are useful in the backcross generations to the wild relative, when the hybrids recover part of the weedy morphology.

Cytological studies

Weedy relatives of oilseed rape possess chromosome numbers ranging from $2n = 14$ to 42 (Table 18.1). When interspecific F_1 hybrids are produced from normal gametes, they possess half of the chromosomes of each parental species. Because of the small chromosome size in the Brassicaceae, chromosome counting is difficult. However, a linear regression can be established between chromosome number and DNA content, allowing the estimation of chromosome number by flow cytometry. Flow cytometry is therefore useful for studying large numbers of plants even if the method does not provide an exact estimate of the number of chromosomes (± 1 or 2 chromosomes; Eber *et al.*, 1997). Interspecific F_1 hybrids can also be produced from unreduced gametes and thus carry one or both parental genomes at the diploid stage; these hybrids sometimes have a chromosome number so close to that of the parental species that differentiation by flow cytometry is virtually impossible, for instance, *B. napus–Raphanus raphanistrum* hybrids produced from unreduced gametes of *R. raphanistrum* (genome *ACRrRr*) have $2n = 37$ chromosomes, compared with $2n = 38$ of oilseed rape. In ambiguous cases, it is necessary to analyse the meiotic behaviour of the hybrids and/or their genomic constitution by other methods, such as genomic *in situ* hybridization (GISH), provided that the species do not share a common or a closely related genome, as in the case of *B. rapa* and *B. napus*. In the following generations, chromosome numbers can be determined by chromosome counting or flow cytometry, but molecular markers and/or GISH analyses are needed to demonstrate that plants with the same chromosome number as the wild relative contain introgressed segments of the crop species.

Transgenes

In gene flow from transgenic crops to wild relatives, hybrids can be distinguished by screening large populations using the characters conferred by the transgenes, e.g. herbicide resistance (Warwick *et al.*, 2003), green fluorescent protein (GFP) (Halfhill *et al.*, 2001, 2002), etc. However, this approach does not distinguish interspecific hybrids from transgenic oilseed rape volunteers that have escaped from cultivation.

Molecular markers

A large number of molecular markers are now available for the Brassicaceae. These markers can be used to detect interspecific hybridization and genetic introgression, notably when the markers are species specific. In most cases, species-specific markers have to be developed for each species combination. Such markers are usually difficult to identify in *Brassica* species that share a common genome, but relatively easier to identify in other species combinations in the Brassicaceae. Screening of the intraspecific variability in the wild relatives is needed to ensure the absence of crop-specific markers prior to gene flow analyses. Several DNA bulks of plants per parental species can be used to assess the intra- and interspecific variability of a species.

Cytoplasmic markers have been developed for determining the maternal parentage of hybrids. In order to identify chloroplast regions suitable for species discrimination, Fluch and co-workers (unpublished data presented at the workshop) screened an equivalent of 16,600 bp of the tobacco chloroplast genome in various Brassicaceae species with a total of 34 polymerase chain reaction (PCR) primers in various combinations. Further investigations limited the primer set to four primer pair combinations, which proved to be applicable for species discrimination and so for maternal hybrid detection in the Brassicaceae.

Various nuclear markers have been used to distinguish species of the Brassicaceae. Random amplified polymorphic DNA (RAPD) and amplified fragment length polymorphism (AFLP) are directly useful for detecting interspecific hybrids because these markers do not require preliminary sequence information. Both types of markers are based on PCR technology using genomic DNA and yield highly informative fingerprints with several dominant markers. Dominance can be a limitation when parental species are heterozygous. While the presence of a band specific to a species is a clear indication that this species has contributed to the initial hybrid production, the absence of the band is uninformative. AFLP markers have been used to confirm hybrids of oilseed rape and wild relatives (Hansen *et al.*, 2001; Warwick *et al.*, 2003) and to quantify the proportion of the crop genome remaining in subsequent backcross generations to the wild species (Halfhill *et al.*, 2003). The presence of several species-specific markers is therefore needed or, alternatively, species-specific monomorphic markers could be used. Bands of interest (species-specific amplified fragments) may also be cloned, sequenced and transformed into sequence characterized amplified regions (SCARs) by designing specific primers. Microsatellites, simple sequence repeats (SSRs), have been developed recently among different Brassicaceae species and could provide highly informative tools (Lowe *et al.*, 2002).

A new type of molecular marker based on the presence or absence of retroelements (i.e. SINEs, LINEs and retrotransposons) at different loci has been utilized in Brassicaceae phylogenetic studies (Lenoir *et al.*, 1997) and is being developed for use in gene flow studies. The very low probability of convergent insertions, the knowledge of the ancestral state and the high stability of retroelement insertions make these markers unique and particularly well suited for monitoring gene flow. Once a given retroelement insertion has been phylogenetically assigned to a group of species, the detection of that particular retroelement-containing locus in a plant of a species not belonging to that group is clear evidence of interspecific genetic exchange. This is because the likelihood of the transposition of a retroelement in distantly related species to the same location as the ancestral site is virtually impossible. Such oilseed rape-specific markers from SINE sequences are under development to monitor gene flow between oilseed rape and in populations of *R. raphanistrum* (Prieto and Deragon, unpublished data presented at the workshop).

Assessment of interspecific hybridization, gene introgression and fitness: baseline data available for each *B.napus*–wild relative combination

Two methods can be used to assess the occurrence of gene flow between oilseed rape and its wild relatives: (i) successive analysis of the different steps in the process of gene dispersal (hybridization, backcrossing, selfing, etc.) and establishment (introgression, persistence, fitness); and (ii) screening for the presence of oilseed rape markers within populations of weedy relatives. Until recently, the first approach has been used more frequently, because the second approach raises questions about distinguishing introgressive markers from markers jointly inherited from a common ancestor and about assessing the large intraspecific variability of wild species. The potential for *B. napus*–wild relative hybridization (Table 18.2) has been assessed in controlled environments, either by hand pollination or by paired parental plants grown in isolation. It has also been assessed in experimental field trials and in commercial oilseed rape fields where the wild relative occurs. The production of subsequent generations has been undertaken for some of the wild relatives as well as the analysis of gene introgression in controlled experiments or in natural wild populations. Finally, the fitness of crop–wild relative hybrids has been evaluated for some combinations. Detailed information is reviewed below for each *B. napus*–wild relative combination.

Table 18.2. Current baseline available concerning the different combinations.

Female–Male	Interspecific hybrid production		Production of subsequent generations	Introgression in the wild relative genome	Introgression in spontaneous wild populations
	Hand pollination	Field conditions			
B. napus–B. rapa	Yes	Yes	Yes	Yes	Yes
B. rapa–B. napus	Yes	Yes	Yes	Yes	Yes
B. napus–B. oleracea	Yes	?	?	?	?
B. oleracea–B. napus	Yes	?	?	?	?
B. napus–B. nigra	Yes	?	Yes	?	?
B. nigra–B. napus	Yes	?	?	?	?
B. napus–R. raphanistrum	Yes	Yes	Yes	No	?
R. raphanistrum–B. napus	Yes	Yes	No	No	No
B. napus–S. arvensis	Yes	Yes*	?	?	?
S. arvensis–B. napus	Yes	No	No	No	No
B. napus–H. incana	Yes	Yes*	Yes	No	?
H. incana–B. napus	Yes	Yes	Yes	No	?
B. napus–E. gallicum	Yes	No	Yes	No	?
E. gallicum–B. napus	No	No	No	No	No

Yes, data provided; No, data provided but negative results; *, hybrid obtained only on male-sterile oilseed rape; ?, no experimental data.

B. napus–B. rapa

B. rapa is one of the progenitors of oilseed rape, and the occurrence of natural gene flow between these species has been demonstrated in several European countries. In Denmark, Hansen *et al.* (2001) found extensive transfer of nuclear as well as plastid DNA from oilseed rape into a self-maintained weedy population of *B. rapa*. The *B. rapa* plants had a proportional decrease in oilseed rape-specific markers, suggesting that the markers were introgressed beyond the stage of backcross hybrids. The results also showed that transplastomic oilseed rape, i.e. plants carrying transgenes engineered into chloroplast DNA, will not prevent dispersal of these genes to wild relatives. In several other mixed populations of oilseed rape and *B. rapa*, introgression was less pronounced. The extent of gene flow from the crop depends on agricultural and weed control practices and, when the species co-exist over several growth seasons, introgression is likely to take place (Perl *et al.*, 2002; Hauser *et al.*, 2003). Snow *et al.* (1999) showed that there are no apparent fitness differences between introgressed glufosinate-resistant and wild *B. rapa*. In the UK, *B. rapa* occurs both as a weed and as a wild plant, particularly along riverbanks. The extent of hybridization and historical introgression between oilseed rape and weedy *B. rapa* in a natural field situation in the Humberside area has been examined (Wilkinson *et al.*, 2003). AFLP analysis of the *B. rapa* population showed evidence of introgression in the samples from the seed bank, but no evidence of introgression in samples from mature plants in the field, indicating that there may be selection pressure against backcrossed individuals. Scott and Wilkinson (1998) found a very low frequency of oilseed rape-specific markers in wild populations of *B. rapa* in the UK, although few populations were sympatric with oilseed rape crops.

In Canada, *B. napus* and *B. rapa* currently make up approximately 90 and 10%, respectively, of the oilseed rape (canola) crop in western Canada, with only *B. napus* being grown in eastern Canada. Both *B. napus* and *B. rapa* occur as weedy volunteer (feral) populations. In addition to recent oilseed rape escapes, *B. rapa* also occurs as separate naturalized weed populations in eastern Canada, where its range extends into both canola-growing and non-canola-growing areas. In a field plot trial, a mean hybridization frequency of 6.6% (range 0–36%) was obtained in plants of *B. rapa* (naturalized Québec genotype) that were grown at a density of 1 plant/m^2 with glyphosate-resistant *B. napus* (Warwick *et al.*, 2003). Hybridization was also detected in two field populations of *B. rapa* growing in commercial fields of glyphosate-resistant *B. napus* in Québec (Warwick *et al.*, 2003). Hybrids were confirmed by the presence of the herbicide resistance marker, the 3x ploidy level and evidence of species-specific AFLP markers from both species. Persistence and spread of the transgene currently are being monitored (S.I. Warwick, Ottawa, 2003, personal communication).

B. napus–B. oleracea

Although interspecific *B. napus* × *B. oleracea* hybrids have been obtained after hand pollination, spontaneous hybridization is unlikely because wild populations of *B. oleracea* naturally grow in coastal cliffs and are therefore rarely sympatric with oilseed rape crops. In the UK, one sympatric location was identified by Wilkinson *et al.* (2000), but interspecific hybridization was not detected.

B. napus–B. nigra

Interspecific hybrids have been produced, mainly by hand pollination, with larger numbers of hybrids obtained using oilseed rape as the female parent, the number of hybrids ranging from 0.9 to 3.4 per 100 pollinated flowers (Bing *et al.*, 1991; Kerlan *et al.*, 1992). Eight plants were produced from reciprocal F$_1$ hybrids either by backcrossing to *B. napus* or by open pollination. The hybrids yielded few seeds upon open pollination (Bing *et al.*, 1991).

B. napus–R. raphanistrum

Reciprocal hybrids were obtained after hand pollination (Kerlan *et al.*, 1992) and under optimal conditions using either male-sterile oilseed rape (Baranger *et al.*, 1995) or isolated plants of *R. raphanistrum* (Darmency *et al.*, 1998) as the female parent. Field trials performed under agronomic conditions were conducted in France (Chèvre *et al.*, 2000) and Australia (Rieger *et al.*, 2001) with a herbicide-resistant oilseed rape variety and various densities of *R. raphanistrum*. In France, the frequency of interspecific F_1 hybrids on either parent was low, ranging from 10^{-5} to 10^{-7}. Five successive generations (G1–G5) were derived in field test plots where an equal number of herbicide-resistant hybrids and *R. raphanistrum* had been planted. The chromosome number of the hybrids decreased during the G3, G4 (Chèvre *et al.*, 1997) and G5 generations. The percentage of herbicide-resistant plants decreased, while male and female fertility increased. Some herbicide-resistant plants had a female fertility equivalent to that of *R. raphanistrum*, but none had 18 chromosomes as in *R. raphanistrum*, suggesting that the transgene had not been introgressed through recombination into the *R. raphanistrum* genome. In the G5 generation, hybrids carrying the oilseed rape cytoplasm, and exhibiting chlorophyll deficiency, had a fitness value 100 times lower than hybrids with the *R. raphanistrum* cytoplasm (Guéritaine *et al.*, 2002). Herbicide-resistant hybrids with the *R. raphanistrum* cytoplasm always carried at least one additional oilseed rape chromosome and had a growth pattern similar to that of *R. raphanistrum*, but with male and female fitness values twice lower than that of *R. raphanistrum* (Guéritaine *et al.*, 2002).

In the UK, surveys conducted over 6 years of *R. raphanistrum* growing in close proximity to transgenic oilseed rape fields have not detected the presence of the transgene (Eastham and Sweet, 2002).

In Canada, *R. raphanistrum* occurs in both agricultural and ruderal habitats. The species also occurs in naturally disturbed habitats, such as seashores, particularly in Eastern Canada. It only co-occurs with oilseed rape in eastern Canada and a restricted area in western Canada (one region in Alberta). Canadian genotypes of *R. raphanistrum* from eastern and western regions of Canada (Alberta (AB), Prince Edward Island (PEI) and Québec (QC)) and one from France (FR) were compared in: (i) paired-crossing experiments with *B. napus* in the greenhouse (individual flowering plants of each species were bagged together, shaken daily and allowed to cross naturally); and (ii) in experimental field trials where plants of *R. raphanistrum* (AB, PEI, QC and FR genotypes) were grown at a density of 1 plant/m^2 with glyphosate-resistant *B. napus* or with *B. napus* expressing GFP. There was evidence for both *R. raphanistrum* genotype and individual plant differences in dormancy, vernalization requirements and seed production via unreduced gametes and/or 'selfing'. A single F_1 hybrid (PEI genotype) was detected in the field plot experiment utilizing the *B. napus* herbicide-resistant marker (Warwick *et al.*, 2003). The PEI hybrid had reduced fertility (< 1%), an uneven proportion of AFLP makers (100 and 20% of the *B. napus* and *R. raphanistrum* markers, respectively) and a chromosome number of approximately $2n = 35$–37. GISH data confirmed a genomic structure of *RrRrAC*, i.e. an unreduced gamete from *R. raphanistrum* (18 chromosomes) and a reduced gamete from *B. napus* (19 chromosomes). Surveys to detect hybridization between glyphosate-resistant *B. napus* oilseed rape crops and *R. raphanistrum* under commercial production were conducted in western (Alberta) and eastern (Québec) Canada in 2000 and 2001. A total of 19 populations were screened and no interspecific hybrids were detected (Warwick *et al.*, 2003).

B. napus–Sinapis arvensis

Sexual hybrids were obtained under greenhouse conditions with *S. arvensis* as the female parent at a frequency of ≤ 0.0015% of the potential seed output (Moyes *et al.*, 2002). Interspecific hybrids have been produced with oilseed rape as the female parent in hand pollination studies, with hybridization rates averaging 2.2 hybrids per 100 pollinated flowers (Kerlan *et al.*, 1992) and from 0 to 0.0049% of the total seed potential (Moyes *et al.*, 2002). F_1 hybrids have also been produced after cultivation of male-sterile oilseed rape with *S. arvensis* as the pollinator with 0.18 hybrid seeds per 100 flowers (Chèvre *et al.*, 1996); few seeds were obtained on these hybrids after backcrossing to *S. arvensis* and open pollination (Chèvre *et al.*, 1996). In contrast, no hybrids have been detected under field conditions (Bing *et al.*, 1991, 1996; Lefol *et al.*, 1996a; Moyes *et al.*, 2002).

In the UK, surveys conducted over 6 years of *S. arvensis* growing near fields of transgenic oilseed rape have not detected the presence of the transgene (Eastham and Sweet, 2002).

In Canada, *S. arvensis* is found in agricultural and ruderal habitats and is the major weed in oilseed rape-growing areas in western North America. A 2-year field survey to detect hybridization between *S. arvensis* and glyphosate-resistant *B. napus* crops was conducted in Saskatchewan. A total of 42 and 37 populations of *S. arvensis* were sampled in 1999 and 2000, respectively; no interspecific hybrids were detected (Warwick *et al.*, 2003).

B. napus–Hirschfeldia incana

Interspecific hybridization between oilseed rape and *H. incana* was successful after hand pollination with 1.3 or 3.1 hybrids per 100 flowers from crosses with *H. incana* used as female or as male, respectively (Kerlan *et al.*, 1992), and after field experiments using male-sterile oilseed rape as the female parent with 1.9 hybrids per 100 flowers (Eber *et al.*, 1994). When *H. incana* was used as the female parent, Lefol *et al.* (1996b) showed that interspecific hybridization occurred spontaneously in field experiments. On average over 3 years, 0.6 hybrids were produced per isolated plant of *H. incana*, which represented 0.4% of the seed. No gene introgression from oilseed rape into *H. incana* was detected after five generations of backcrossing to *H. incana* (Darmency and Fleury, 2000).

B. napus–Erucastrum gallicum

E. gallicum occupies agricultural and ruderal habitats throughout Canada and is a common weed in Europe. Large populations of *E. gallicum* in oilseed rape fields are reported for a limited region in Saskatchewan, Canada. In 1993, manually performed hybridization between *B. napus* and an *E. gallicum* population from Alberta produced 0 or 0.11 hybrid seeds per 100 pollinated flowers using *E. gallicum* as female or as male, respectively (Lefol *et al.*, 1997). The single hybrid obtained on the *B. napus* parent was fertile, self-compatible and possessed RAPD markers from both parents. In 1994, 20 BC_1 progeny plants were transplanted to the field and self/sib progeny resembling the *E. gallicum* parent were collected in subsequent years (1995–1999). Plants from the 1995 population had the chromosome number ($2n = 30$), glucosinolate profile and RAPD profile of *E. gallicum*. Genomic characterization of the BC_1 progeny and the 1995–1999 populations using AFLP markers is in progress. A 2-year field survey to detect hybridization between *E. gallicum* and glyphosate-resistant *B. napus* crops was conducted in Saskatchewan. A total of three and 35 populations of *E. gallicum* were sampled in 1999 and 2000, respectively; no interspecific hybrids were detected (Warwick *et al.*, 2003).

The Fate of Hybrids and Transgenes

Several interspecific hybrids have been described, but gene introgression in a wild relative has only been confirmed for *B. napus*–*B. rapa* hybrids. It is likely that the probability of intergenomic recombination varies with the structural differentiation of chromosomes and/or the presence of genes controlling intergenomic pairing/recombination. Regulation of transgene expression can also be expected in a new genetic background.

Analysis of genetic mechanisms of introgression

The production of viable, unreduced gametes has been reported in both the parental species (Chèvre *et al.*, 2000; Rieger *et al.*, 2001; Warwick *et al.*, 2003) and interspecific hybrids (Chèvre *et al.*, 1998; Warwick *et al.*, 2003). Their influence on genomic structure and on the fertility of the subsequent generations is under study.

In addition, it has been shown that homologous recombination depends on the genome carrying the transgene (i.e. in oilseed rape, a transgene integrated into the C genome is less likely to be transmitted to *B. rapa*; Metz *et al.*, 1997; Lu *et al.*, 2002; but see Halfhill *et al.*, Chapter 20, this volume), but also on the distribution of the genes along the chromosome arms. To test the latter hypothesis, Jenczewski and co-workers (unpublished data presented at the workshop) proposed a strategy to analyse the rate of introgression of oilseed rape markers into the genome of *R. raphanistrum*. Advanced backcross generations of interspecific hybrids between male-sterile oilseed rape (seven independent herbicide-resistant transgenic F_1 hybrids obtained from the same parental lines) and *R. raphanistrum* as the recurrent parent were produced. The chromosomal structure of the BC_1 populations (each population originating from one of the seven transgenic F_1 hybrids) displayed a different pattern. These differences were less pronounced in the BC_2 generation, whether the hybrids were treated with herbicide or not. Only one line, designated 235.3, remained different due to the fact that most of the plants had a higher chromosome number. In the BC_3 generation, the reduction in the number of chromosomes continued in the different hybrids and was more pronounced in the herbicide-untreated populations. Line

235.3 still displayed a significantly higher number of chromosomes. These advanced generations are now being screened with specific oilseed rape markers evenly distributed over the oilseed rape genome.

Regulation and stability of transgene expression

When assessing the environmental fate of transgenic plants in the field, one must consider expression, exposure and persistence of transgenic products in the interspecific hybrids. Regulation of transgene expression is complex and modulated by epigenetic events, chromatin structure, interactions and, at the individual gene level, by promoters, enhancers and terminators (reviewed by Meyer, 2000). Recent activity in plant molecular biology has been devoted to understanding homology-dependent gene silencing in transgenic plants. For example, rather than enhanced expression, increased gene dosage in plants can result in silencing of both transgenes and homologous endogenous genes. This phenomenon is termed co-suppression (reviewed by Turner and Schuch, 2000). Transgene, viral and endogenous gene silencing can occur either at the transcriptional level or after transcription has occurred (Kooter *et al.*, 1999).

In addition to gene silencing, the most important factor controlling gene expression is upstream regulator element activity (e.g. by promoters and enhancers). Commercial transgenic plants in the future will rely on the judicious use of tissue-specific and inducible promoters, along with appropriate enhancers to effect high gene expression only when and where desired. For example, if *Bt* expression had been better tuned in crops, much of the potential side effects of *Bt* exposure (e.g. the Monarch butterfly debacle) could have been avoided.

While many observational data have been recorded on transgene silencing, it seems to be transgenic event specific, and there has been no explanation why one transgenic event is silenced, and another with an identical transgene construct is not. Halfhill *et al.* (2002) analysed the expression of transgenes in transgenic oilseed rape–*B. rapa* hybrids, and in BC_1 and BC_2 generations. High expressing oilseed rape events led to high expression in the weed hybrid and BC genetic backgrounds for a *Bt cry1Ac* transgene. There is no reason to suspect that introgression into wild relatives would trigger a non-silenced event to be silenced.

Modelling

In the modelling field, three main steps are usually being considered concerning gene flow from a transgenic crop to a wild species: (i) effective pollen dispersal, resulting in the arrival of a pollen grain on a pistil; (ii) hybridization and introgression; and (iii) fate of the transgene in the new population. To our knowledge, no published model considers the three steps in detail. Most models only consider steps (i) and (iii). More details of these models and corresponding references can be found in Lavigne *et al.* (Chapter 26, this volume). Models have also been developed to predict gene flow from cultivated oilseed rape to feral populations of oilseed rape. Several models such as GENESYS (Colbach *et al.*, 2001a, b) and the GenEERA project study gene flow at the plant or landscape level.

These models could constitute a basis for the development of models studying the gene flow between oilseed rape and wild relatives at the landscape level. For example, the GENESYS model takes into account the different flowering dynamics of the oilseed rape populations according to their environment to calculate hybridization rates between populations. It also simulates the effects of some morphology traits on pollen and seed dispersal. The advantage of this model is that it takes into account field pattern, crop and management of field margins. The project GenEERA is currently developing a model that represents feral oilseed and wild relatives on an individual basis. This model will be used to examine the implication of the local processes of spread and persistence when extrapolated to the landscape level.

A model was also developed by Thompson *et al.* (2003) to study the mechanisms of introgression of a transgene from a cultivated crop to wild relatives but, in order to do this, the authors had to simplify the biology of the plants and created a virtual *Brassica* species.

A first model for predicting gene flow from oilseed rape to *B. rapa* with and without selection pressure was developed by Lu and Kato (2001) and showed the importance of the fitness value at various generations, and of the location of the transgene (Lu *et al.*, 2002). More general models consider the likelihood of hybridization depending on the genetic system determining species incompatibility.

The final step would be to combine all the phenomena implied in transgene escape (pollen and seed dispersal, hybridization, selection, genetic drift, dormancy, human management of fields and wild environments, etc.) into models incorporating the spatial scale of the landscape, with interactions among fields, feral or volunteer populations and wild relative populations.

Conclusions

From the data presented, we showed that a combination of different tools is needed to assess transgene escape. Monitoring an interspecific gene flow event is a difficult and labour-intensive activity, particularly when hybridization frequencies are low and the pollen donor is unknown. In order to detect such an event, large numbers of individuals must be screened and suitable marker systems must be available for efficient and effective screening procedures. Using herbicide resistance markers allows the screening of a large number of samples and has been used frequently in experimental interspecific gene flow studies in oilseed rape, but does not distinguish between hybrids and the transgenic crop. Further analyses using cytogenetic and molecular markers are needed to identify the hybrids.

Immediate detection of a rare gene flow event in nature is unlikely. In the case of the transmission of a herbicide resistance trait, the event is more likely to be detected as part of a monitoring programme for naturally arising herbicide-resistant weed mutants or biotypes. Detection requires sufficient time for a large enough build-up of the resistant weed population before a producer notices that the weeds are not being controlled by the recommended herbicide application. Determination of the origin of the herbicide resistance is then a straightforward procedure, i.e. molecular identification of the transgene in the weed population to confirm that its origin is due to gene flow. Increased support for such ecological monitoring of herbicide-resistant weeds is an integral part of monitoring for gene flow after the release of herbicide-resistant crops (see also den Nijs and Bartsch, Chapter 27, this volume).

To support such work, national programmes have been developed. In 1999, the French Ministry of Education, Research and Technology invited the submission of grant proposals for research on the impact of genetically modified organisms (GMOs). The objective was to bolster studies on the identification, assessment and management of GMO-related impacts for the environment and food safety. In Germany, within the framework of the BMBF (Ministry of Education and Research), a nationally funded research programme on biosafety and monitoring, a joint project on the potential effects of the cultivation of transgenic oilseed rape has been initiated recently. The project intends cooperation with another BMBF-funded project on oilseed rape, GenEERA, with respect to modelling of the dispersal potential of oilseed rape.

The ESF workshop held in Rennes was an important step in bringing together the current knowledge on gene flow between oilseed rape and wild relatives. The workshop contributions demonstrated that hybridization of oilseed rape is possible with a number of related species. While gene flow between *B. napus* and *B. rapa* is common, it is not fully understood how frequently other hybridization events occur under natural conditions. The ongoing research, especially the molecular analysis of hybrids and subsequent generations and modelling at the farm scale and landscape level, will bring further insight into mitigating the likelihood of transgene escape into the environment.

References

Baranger, A., Chèvre, A.M., Eber, F. and Renard, M. (1995) Effect of oilseed rape genotype on the spontaneous hybridization rate with a weedy species: an assessment of transgene dispersal. *Theoretical and Applied Genetics* 91, 956–963.

Bing, D.J., Downey, R.K. and Rakow, G.F.W. (1991) Potential of gene transfer among oilseed *Brassica* and their weedy relatives. *GCIRC Rapeseed Congress*, 9–11 July, Saskatoon, Canada, pp. 1022–1027.

Bing, D.J., Downey, R.K. and Rakow, G.F.W. (1996) Hybridizations among *Brassica napus*, *B. rapa* and *B. juncea* and their two weedy relatives *B. nigra* and *Sinapis arvensis* under open pollination conditions in the field. *Plant Breeding* 115, 470–473.

Chèvre, A.M., Eber, F., Baranger, A., Kerlan, M.C., Barret, P., Vallee, P. and Renard, M. (1996) Interspecific gene flow as a component of risk assessment for transgenic Brassicas. *Acta Horticulturae* 407, 169–179.

Chèvre, A.M., Eber, F., Baranger, A. and Renard, M. (1997) Gene flow from transgenic crops. *Nature* 389, 924.

Chèvre, A.M., Eber, F., Baranger, A., Hureau, G., Barret, P., Picault, H. and Renard, M. (1998) Characterisation of backcross generations obtained under field conditions from oilseed rape–wild radish F$_1$ interspecific hybrids: an assessment of transgene dispersal. *Theoretical and Applied Genetics* 97, 90–98.

Chèvre, A.M., Eber, F., Darmency, H., Fleury, A., Picault, H., Letanneur, J.C. and Renard, M. (2000) Assessment of interspecific hybridization between transgenic oilseed rape and wild radish under normal agronomic conditions. *Theoretical and Applied Genetics* 100, 1233–1239.

Colbach, N., Clermont Dauphin, C. and Meynard, J.M. (2001a) GENESYS: a model of the influence of cropping system on gene escape from herbicide tolerant rapeseed crops to rape volunteers – I. Temporal evolution of a population of rapeseed volunteers in a field. *Agriculture Ecosystems and Environment* 83, 235–253.

Colbach, N., Clermont Dauphin, C. and Meynard, J.M. (2001b) GENESYS: a model of the influence of cropping system on gene escape from herbicide tolerant rapeseed crops to rape volunteers – II. Genetic exchange among volunteer and cropped populations in a small region. *Agriculture Ecosystems and Environment* 83, 255–270.

Darmency, H. and Fleury, A. (2000) Mating system in *Hirschfeldia incana* and hybridization to oilseed rape. *Weed Research* 40, 231–238.

Darmency, H., Lefol, E. and Fleury, A. (1998) Spontaneous hybridization between oilseed rape and wild radish. *Molecular Ecology* 7, 1467–1473.

Eastham, K. and Sweet, J.B. (2002) Genetically modified organisms: the significance of gene flow through pollen transfer. *European Environment Agency, Environmental Issue Report* No 28.

Eber, F., Chèvre, A.M., Baranger, A., Vallée, P., Tanguy, X. and Renard, M. (1994) Spontaneous hybridization between a male-sterile oilseed rape and two weeds. *Theoretical and Applied Genetics* 88, 362–368.

Eber, F., Letanneur, J.C. and Chèvre, A.M. (1997) Chromosome number of oilseed rape (*Brassica napus*)–wild radish (*Raphanus raphanistrum*) spontaneous hybrids and of their progeny estimated by flow cytometry. *Cruciferae Newsletter* 19, 17–18.

Gomez-Campo, C. (1980) Morphology and morpho-taxonomy in the tribe *Brassiceae*. In: Tsunoda, S., Hinata, K. and Gomez-Campo, C. (eds) Brassica *Crops and Wild Allies*. Japan Scientific Societies Press, Tokyo, Japan, pp. 3–31.

Guéritaine, G., Sester, M., Eber, F., Chèvre, A.M. and Darmency, H. (2002) Fitness of backcross six of hybrids between transgenic oilseed rape (*Brassica napus*) and wild radish (*Raphanus raphanistrum*). *Molecular Ecology* 11, 1419–1426.

Halfhill, M.D., Richards, H.A., Mabon, S.A. and Stewart, C.N. Jr (2001) Expression of GFP and Bt transgenes in *Brassica napus* and hybridization with *Brassica rapa*. *Theoretical and Applied Genetics* 103, 659–667.

Halfhill, M.D., Millwood, R.J., Raymer, P.L. and Stewart, C.N. Jr (2002) Bt-transgenic oilseed rape hybridization with its weedy relative, *Brassica rapa*. *Environmental Biosafety Research* 1, 19–28.

Halfhill, M.D., Millwood, R.J., Weissinger, A.K., Warwick, S.I. and Stewart, C.N. Jr (2003) Additive transgene expression and genetic introgression in multiple green fluorescent protein transgenic crop × weed hybrid generations. *Theoretical and Applied Genetics* 107, 1533–1540.

Hansen, B.L., Siegismund, H.R. and Jørgensen, R.B. (2001) Introgression between oilseed rape (*Brassica napus* L.) and its weedy relative *B. rapa* in a natural population. *Genetic Resources and Crop Evolution* 48, 621–627.

Hauser, T.P., Damgaard, C. and Jørgensen, R.B. (2003) Frequency dependent fitness of hybrids between oilseed rape (*Brassica napus*) and weedy *B. rapa* (Brassicaceae). *American Journal of Botany* 90, 571–578.

HYPPA (2000) HYpermedia for Plant Protection – Weeds. Available at: http://www.inra.fr/Internet/Centres/Dijon/malherbo/hyppa/

Kerlan, M.C., Chèvre, A.M., Eber, F., Baranger, A. and Renard, M. (1992) Risk assessment of outcrossing of transgenic rapeseed to related species: I. Interspecific hybrid production under optimal conditions with emphasis on pollination and fertilization. *Euphytica* 62, 145–153.

Kooter, J.M., Matzke, M.A. and Meyer, P. (1999) Listening to the silent genes: transgene silencing, gene regulation and pathogen control. *Trends in Plant Science* 4, 340–347.

Lefol, E., Danielou, V. and Darmency, H. (1996a) Predicting hybridization between transgenic oilseed rape and wild mustard. *Field Crops Research* 45, 153–161.

Lefol, E., Fleury, A. and Darmency, H. (1996b) Gene dispersal from transgenic crops II. Hybridization between oilseed rape and the wild hoary mustard. *Sexual Plant Reproduction* 9, 189–196.

Lefol, E.B., Séguin-Swartz, G. and Downey, R.K. (1997) Sexual hybridisation in crosses of cultivated *Brassica* species with the crucifers *Erucastrum gallicum* and *Raphanus raphanistrum*: potential for gene introgression. *Euphytica* 95, 127–139.

Lenoir, A., Cournoyer, B., Warwick, S.I., Picard, G. and Deragon, J.M. (1997) Evolution of SINE S1 retrotransposons in Cruciferae plant species. *Molecular Biology and Evolution* 14, 934–941.

Lowe, A.J., Jones, A.E., Raybould, A.F., Trick, M., Moule, C.L. and Edwards, K.J. (2002) Transferability and genome specificity of a new set of microsatellite primers among *Brassica* species of the U triangle. *Molecular Ecology Notes* 2, 7–11.

Lu, C.M. and Kato, M. (2001) Fertilization fitness and relation to chromosome number in interspecific progeny between *Brassica napus* and *Brassica rapa*: a comparative study using natural and resynthesized *B. napus*. *Breeding Science* 51, 73–81.

Lu, C.M., Kato, M. and Kakihara, F. (2002) Destiny of a transgene escape from *Brassica napus* into *Brassica rapa*. *Theoretical and Applied Genetics* 105, 78–84.

Metz, P.L.J., Jacobsen, E., Nap, J.-P., Pereira, A. and Stiekema, W.J. (1997) The impact of biosafety of the phosphinothricin-tolerance transgene in inter-specific *B. rapa* × *B. napus* hybrids and their successive backcrosses. *Theoretical and Applied Genetics* 95, 442–450.

Meyer, P. (2000) Transcriptional transgene silencing and chromatin components. *Plant Molecular Biology* 43, 221–234.

Moyes, C.L., Lilley, J.M., Casais, C.A., Cole, S.G., Haeger, P.D. and Dale, P.J. (2002) Barriers to gene flow from oilseed rape (*Brassica napus*) into populations of *Sinapis arvensis*. *Molecular Ecology* 11, 103–112.

Perl, M., Hauser, T.P., Damgaard, C. and Jørgensen, R.B. (2002) Male fitness of oilseed rape (*Brassica napus*), weedy *B. rapa* and their F_1 hybrids when pollinating *B. rapa* seeds. *Heredity* 89, 212–218.

Rieger, M.A., Potter, T.D., Preston, C. and Powles, S.B. (2001) Hybridisation between *Brassica napus* L. and *Raphanus raphanistrum* L. under agronomic field conditions. *Theoretical and Applied Genetics* 103, 555–560.

Scott, S.E. and Wilkinson, M.J. (1998) Transgene risk is low. *Nature* 393, 320.

Snow, A.A., Andersen, B. and Jørgensen, R.B. (1999) Costs of transgenic herbicide resistance introgressed from *Brassica* napus into weedy *B. rapa. Molecular Ecology* 8, 605–615.

Thompson, C.J., Thompson, B.J.P., Ades, P.K., Cousens, R., Garnier-Gere, P., Landman, K., Newbigin, E. and Burgman, M.A. (2003) Model-based analysis of the likelihood of gene introgression from genetically modified crops into wild relatives. *Ecological Modelling* 162, 199–209.

Turner, M. and Schuch, W. (2000) Post-transcriptional gene-silencing and RNA interference: genetic immunity, mechanisms and applications. *Journal of Chemical Technology and Biotechnology* 75, 869–882.

Warwick, S.I., Francis, A. and Mulligan, G.A. (1999) *Brassicaceae of Canada*: Agriculture and Agri-Food Research Branch Technical Publication, ECORC, Ottawa, Canada Contribution No. 981317.1255. Available at: http://cbif.gc.ca/

Warwick, S.I., Francis, A. and La Fleche, J. (2000) *Guide to Wild Germplasm of* Brassica *and Allied Crops (Tribe Brassiceae, Brassicaceae)*, 2nd edn. Agriculture and Agri-food Research Branch Publication, ECORC Ottawa, Canada. Contribution No. 991475. Available at: http://www.brassica.info

Warwick, S.I., Simard, M.-J., Légère, A., Beckie, H.J., Braun, L., Zhu, B., Mason, P., Séguin-Swartz, G. and Stewart, C.N. Jr (2003) Hybridization between transgenic *Brassica napus* L. and its wild relatives: *B. rapa* L., *Raphanus raphanistrum* L., *Sinapis arvensis* L., and *Erucastrum gallicum* (Willd.) O.E. Schulz. *Theoretical and Applied Genetics* 107, 528–539.

Wilkinson, M.J., Davenport, I.J., Charters, Y.M., Jones, A.E., Allainguillaume, J., Butler, H.T., Mason, D.C. and Raybould, A.F. (2000) A direct regional scale estimate of transgene movement from genetically modified oilseed rape to its wild progenitors. *Molecular Ecology* 9, 983–991.

Wilkinson, M.J., Elliot, L., Allainguillaume, J., Shaw, M.W., Norris, C., Welters, R., Alexander, M., Sweet, J. and Mason, D.C. (2003) Hybridization between rapeseed and *B. rapa* in the UK: a first step towards national-scale risk assessment of GM rapeseed. *Science* 302, 457–459.

19 Gene Introgression and Consequences in *Brassica*

RIKKE BAGGER JØRGENSEN, HENRIETTE AMMITZBØLL, LISE B. HANSEN, MARINA JOHANNESSEN, BENTE ANDERSEN AND THURE P. HAUSER

Plant Research Department, PLE-309, Risø National Laboratory, DK-4000 Roskilde, Denmark; E-mail: rikke.bagger.jorgensen@risoe.dk

Abstract

Transgenes may be transferred from genetically modified (GM) crops to the wider environment through crosses with compatible wild or weedy relatives. For oilseed rape (*Brassica napus*), we found extensive transfer of nuclear as well as plastid DNA (cpDNA) to *Brassica rapa* in an environment with poor weed control. Some of the plants with markers from both species were apparently introgressed beyond the stage of the BC_1 generation. In conventionally managed fields with oilseed rape as crop and the wild recipient as a weed, the introgression was insignificant or not detected, so apparently the extent of gene flow from the crop depended on the agricultural management or other environmental effects. Our results also showed that oilseed rape plastids were introgressed to *B. rapa* under field conditions. Field and laboratory experiments revealed that fitness of interspecific F_1 hybrids and backcross plants with *B. rapa* were variable but could be as high as and even higher than the fitness of the parental species.

We present results that show the importance of genotype and environment (i.e. agronomic practice and density/proportion of plant types) in the introgression of oilseed rape genes to *Brassica* and *Raphanus* species. In the light of our findings, we discuss the perspectives of releasing GM oilseed rape.

Introduction

Flow of transgenes by hybridization between genetically modified (GM) crops and wild relatives is not necessarily a risk to the environment. The consequences to natural and cultivated ecosystems depend on a wide range of factors. Therefore, the effects are addressed most appropriately by a targeted risk analysis of the transgenic plant in its proper environment. However, baseline knowledge about the extent of gene flow between the crop and related recipients as well as survival of the resulting progeny is always requested in the risk assessment process. In this chapter, we present some of our results on introgression of oilseed rape genetic material (*Brassica napus*, $2n = 38$, genomes *AACC*) to related species in the *Brassica* and *Raphanus* genus and fitness analysis of the introgressed plants. In Denmark, the wild relatives of oilseed rape, *B. rapa* and *R. raphanistrum*, are rather common as weeds in agricultural fields. They are found mostly in oilseed rape fields but also in other types of crops where they can form weedy populations together with volunteer oilseed rape. *B. rapa* and *R. raphanistrum* can also occur at ruderal sites but are rarely found in more natural habitats. *B. rapa* and *R. raphanistrum* are likely recipients of oilseed rape genes and, therefore, quantification of the crop–wild gene flow is relevant. Other, less abundant relatives, i.e. *B. juncea*, might also have potential for spontaneous gene exchange with oilseed rape.

Gene Flow from Oilseed Rape to *Brassica rapa*

Frequency of F₁ hybrids

B. rapa ($2n = 20$, genomes *AA*) is one of the parental species of oilseed rape. *B. rapa* is a common annual weed in agricultural fields worldwide in the temperate zone. Outside the field, *B. rapa* populations are ephemeral, as seeds will only germinate when the soil is turned. Harberd (1975) reported the spontaneous occurrence of the *B. napus* × *B. rapa* hybrid (*B.* × *harmsiana*) in oilseed rape fields. Frequencies of F₁ hybrids between oilseed rape and the weedy *B. rapa* have been reported from field experiments and survey of natural populations of the wild species. Jørgensen and Andersen (1994), Jørgensen (1996), Landbo *et al.* (1996), Jørgensen *et al.* (1998), Scott and Wilkinson (1998) and Pertl *et al.* (2002) found hybrid frequencies of between 0 and 69% of the seeds depending on parental genotypes, density and proportions of plants and agricultural practice. Generally, *B. rapa* produces more interspecific hybrids than oilseed rape when the two species grow together under natural conditions (Jørgensen and Andersen, 1994; Hauser *et al.*, 1997; Jørgensen *et al.*, 1998).

Frequency of backcrossing in a field experiment

The F₁ hybrids have reduced pollen fertility (Jørgensen and Andersen, 1994; Pertl *et al.*, 2002), but spontaneous backcrossing does take place. *B. rapa* and interspecific hybrids with a transgene providing Basta resistance were sown together in field experiments to assess the extent of backcrossing (Mikkelsen *et al.*, 1996) to the weedy parent. Seed set per pod on interspecific hybrids was low (~2.5) compared with seed set on the parental species (16–23). An average of 67% of the plants developed from seeds harvested on 32 interspecific hybrids were herbicide resistant. Among 865 offspring, plants with a *B. rapa*-like morphology were selected for further analysis. A few

(0.5%) were almost identical to *B. rapa* (chromosome number $2n = 20$, high pollen fertility) and set a normal number of seeds in crosses with genuine *B. rapa* (Mikkelsen *et al.*, 1996). The reciprocal cross *B. rapa* × hybrid was not observed among more than 2000 offspring from 30 *B. rapa* plants.

Introgression between oilseed rape and *B. rapa* in a natural environment with poor weed control

In an organic field of barley and legumes in eastern Denmark, some morphologically deviating *Brassica* plants were observed together with weedy oilseed rape and *B. rapa*. At flowering, all *Brassica* plants were collected from a 3 m² plot. Leaf material from a total of 102 plants was analysed using amplified fragment length polymorphism (AFLP) markers specific to *B. napus* or *B. rapa*. The development of these species-specific markers was described in Hansen *et al.* (2001). Among the AFLP markers used in the analysis, three were specific to *B. rapa* (one monomorphic and two polymorphic markers; due to the homology between the *A* genome of *B. rapa* and *B. napus*, only a few *B. rapa*-specific markers were identified) and 21 to oilseed rape (17 monomorphic and four polymorphic markers), 18 of which were positioned on the *C* genome. In parallel with the analysis of the natural population, F_1 interspecific hybrids and first and second backcross generations with *B. rapa* were produced and the inheritance of the same markers was studied in offspring from these controlled crosses: *B. rapa* was the female in the controlled crosses and, as all AFLP markers specific to oilseed rape were transferred to offspring plants, these markers were judged to be nuclear.

We revealed a pronounced introgression in the natural weedy population (Fig. 19.1; Hansen *et al.*, 2001): 45 plants were introgressed having both oilseed rape- and *B. rapa*-specific markers. Among the remaining 57 plants, seven had only *B. napus*-specific markers and 50 plants had only *B. rapa* markers. Figure 19.1 gives the distribution of oilseed rape markers in the 102 plants. The monomorphic markers showed that there was only one first-generation hybrid (F_1) among the analysed plants. The infrequency of F_1 hybrids could be due to a small number of oilseed rape plants compared with *B. rapa*. We assume that the natural population had maintained itself since 1987, the last time oilseed rape was cultivated in this field. The proportion of oilseed rape had probably decreased since then as *B. rapa* has a better survival over time due to pronounced seed dormancy (Landbo and Jørgensen, 1997). The long existence of the mixed population suggests that some of the plants with DNA markers from both species were advanced generations of introgressed plants. This was confirmed by comparing the marker distribution in the natural population with distribution of the very same markers in the BC_1 and BC_2 generation from the controlled crosses (Fig. 19.1; Hansen *et al.*, 2001). As the marker distribution in BC_2 plants resembled the marker complement in the natural population, we tentatively conclude that the latter was introgressed beyond the BC_1 generation. Progression of introgression was studied in offspring from the parental plants in the field

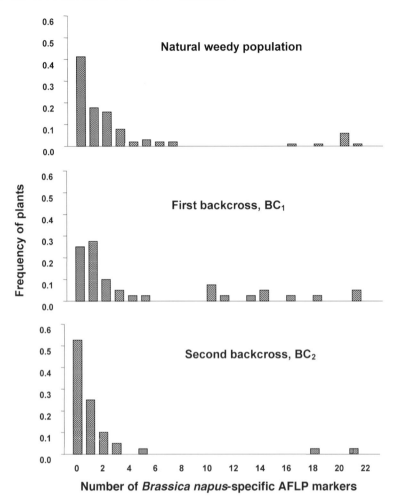

Fig. 19.1. Distribution of oilseed rape-specific AFLP markers in the weedy population of *B. rapa* found together with oilseed rape in an organic field (top), in the BC$_1$ generation (middle) and BC$_2$ generation (bottom) from control crosses.

(Hansen *et al.*, 2003), and the marker analysis showed that *B. rapa* most often functioned as the maternal plant in the introgression process, and that the amount of oilseed rape DNA was diminished in the majority of offspring compared with their introgressed maternal plants. However, we found that introgression brought about both incorporation of *B. napus* C genome DNA into the *B. rapa* genome and exchange of chloroplast DNA, producing *B. rapa*-like plants with oilseed rape chloroplasts. The chromosome number was counted in 15 offspring plants from five introgressed females in the natural population. Offspring chromosome numbers were 20–26, and they had from one to 12 C genome-specific markers. The presence of plants with C genome markers and 20 chromosomes suggests that recombination had taken place between the C genome of oilseed rape and the A genome of *B. rapa*.

Spontaneous introgression between oilseed rape and *B. rapa* in a conventionally managed field

The inefficient weed control in the organic field probably accounts for the high frequency of introgressed plants. In accordance with this, the frequency of introgressed plants from conventionally grown oilseed rape fields was much lower. In 2450 plants from seeds harvested in eight populations of weedy *B. rapa* found in conventional managed fields, we only detected two plants introgressed beyond the F_1 stage and 81 F_1 hybrids. The introgressed plants had more than six *B. napus*-specific AFLP markers in addition to the *B. rapa* markers. For the majority of these plants, the results were obtained from random amplification of polymorphic DNA (RAPD) and isoenzyme data, and only a few hundred plants were analysed by AFLP using the same markers as in the organic field. As the AFLP provided more markers than the other marker techniques, the frequency of introgressed plants might have been underestimated. On the other hand, the frequency of introgressed plants is probably overestimated compared with the normal field situation as most of the plants analysed were reared in growth chambers from seeds harvested in *B. rapa* populations. In nature, the survival of introgressed plants to the adult and reproductive stage is probably lower. Figure 19.2 compares the data on introgressed plants from the organic and the conventional fields. For UK environments, a low frequency of transgene dispersal from oilseed rape to *B. rapa* was predicted from findings of a low number of F_1 hybrids germinated from seed harvested on *B. rapa* in natural populations found along rivers (Scott and Wilkinson, 1998). Hybridization between the true wild *B. rapa* and oilseed rape in the UK very much depends on the sympatry of these populations with the crop. Weedy populations of *B. rapa* have now also been found in the UK. These populations have a higher frequency of hybridization than the riverbank populations because they

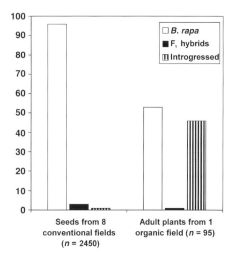

Fig. 19.2. Frequency of *B. rapa* and introgressed plants from mixed populations of oilseed rape and *B. rapa*.

occur as a weed within the oilseed rape crop. AFLP analysis of seed bank material from one of these weedy UK populations has revealed introgression; however, the mature plants studied were not introgressed, indicating that fitness of introgressed plants could be substantially decreased (see Norris *et al.*, Chapter 9, this volume). Our data from the Danish populations demonstrate that introgression between oilseed rape and *B. rapa* can be considerable when the two species are found in long existing mixed populations.

Fitness of introgressed plants

We analysed the fitness of different generations of introgressed plants in field and growth chamber experiments. Hauser *et al.* (2003) showed that the seed set of F_1 hybrids between *B. napus* and *B. rapa* was dependent on the environmental conditions. The F_1 hybrids could produce more seeds than *B. rapa* (maximum seed set/plant: *B. rapa* 4850; F_1 hybrids 6700), but also fewer seeds, dependent on the number and densities of parents and backcross plants in the population. As to the F_2 and BC_1 with *B. rapa*, Hauser *et al.* (1998) found large variation in fitness, with a smaller fraction of these plants being just as fit as *B. rapa* and *B. napus*. In a further advanced backcross programme, BC_1 plants were selected for being *B. rapa*-like and these plants backcrossed to *B. rapa* to obtain BC_3 plants. The reproductive fitness of this BC_3 generation was as great as that for *B. rapa* (Snow *et al.*, 1999). The BC_3 plants segregated 1:1 for a *bar* transgene providing glyphosinate resistance. There was no difference in the seed set and survival of the GM siblings compared with the non-GM siblings when the herbicide was not applied, indicating that cost associated with the transgene was negligible. The overall conclusion from our fitness data is that in some environments, some of the introgressed plants will be as fit as their wild parent and occasionally as fit as the crop. These results support our finding of persistent populations of introgressed plants occurring spontaneously in the agroecosystem (Hansen *et al.*, 2001, 2003).

Transfer of plastid-encoded genes from oilseed rape to *B. rapa*

In *Brassica*, like most other angiosperms, chloroplasts are not transmitted by the pollen (Corriveau and Coleman, 1988). Therefore, transplastomic oilseed rape, where the transgene is engineered into the plastid genome, can only disperse the novel genes through the seed (Daniell *et al.*, 1998; Scott and Wilkinson, 1999). The origin of the chloroplast DNA was analysed in 91 of the plants from the natural population in the organic field mentioned above (Hansen *et al.*, 2003). The analysis was performed by polymerase chain reaction (PCR) using primers to non-coding regions of the cpDNA (Taberlet *et al.*, 1991). The primers amplify species-specific markers for oilseed rape and for *B. rapa*. The analysis assigned the chloroplast of the plants to either oilseed rape or *B. rapa* type, and this information was compared with their nuclear AFLP fingerprint. The cpDNA analysis showed that besides the seven *B. napus*-like plants, one *B. rapa*-like plant and two introgressed plants carried the oilseed rape chloroplasts. In a huge and persisting weedy *B. rapa* population in a conventionally managed field, we found that nine of the 23 plants analysed carried oilseed rape chloroplasts. The AFLP analysis of these plants failed to detect any oilseed rape-specific nuclear markers, but an AFLP marker segregating together with the *B. rapa* cytoplasm was also missing, supporting that chloroplast introgression had taken place in these plants. Our results from these long-lasting populations of oilseed rape and *B. rapa* indicate that transgenes positioned in the chloroplast DNA of *B. napus* will be captured by *B. rapa*. If oilseed rape and subsequently the F_1 hybrid are females in the crosses, fully fertile *B. rapa*-like plants with 20 chromosomes and oilseed rape chloroplasts can be produced after just two generations (Mikkelsen *et al.*, 1996). This could be a pathway for production of transplastomic *B. rapa*. Scott and Wilkinson (1999) analysed the chloroplast inheritance in 47 F_1 hybrids harvested on wild *B. rapa* and sired by oilseed rape. They only found plants with a *B. rapa* cytoplasm and concluded that transgene introgression from transplastomics would occur extremely rarely in populations of *B. rapa* found along riverbanks in the UK.

Gene Flow from Oilseed Rape to *B. juncea*

In northern Europe, *B. juncea* ($2n = 36$, genomes *AABB*) is rarely cultivated, but it may occasionally occur as a weed or as a ruderal plant. Spontaneous hybridization between oilseed rape and *B. juncea* has been reported (Frello *et al.*, 1995; Jørgensen *et al.*, 1998). Depending on the proportions of the parental species, up to 3% of the offspring harvested on *B. juncea* were hybrids. Production of hybrids with *B. napus* as female was less successful (Jørgensen *et al.*, 1998). Pollen fertility of the hybrids was rather low, 0–28%. In a study of marker transfer from oilseed rape to the first backcross generation with *B. juncea*, 20 *B. napus*-specific RAPD markers were all transferred and most of them in the expected frequencies (Frello *et al.*, 1995).

Gene Flow from Oilseed Rape to *R. raphanistrum*

R. raphanistrum ($2n = 18$, genomes *RR*) is a weed in agricultural fields and also occurs as a ruderal plant. In separate pollen-tight growth cabinets, plants from wild populations of *R. raphanistrum* were mixed with a male-sterile oilseed rape line and bumble bees were used as pollinators. A Danish, a Swiss and a French population of *R. raphanistrum* were used as paternal parents and seeds were harvested on the male-sterile oilseed rape. The offspring was typed by species-specific inter-simple sequence repeat (ISSR) markers and morphology. The results showed large differences in hybridization potential between the different populations of *R. raphanistrum*, the Danish population being the least likely father of F_1 hybrids as only 0.02 F_1 hybrids were produced per pod compared with the Swiss population that produced many times more hybrids per pod (0.64). Such a difference in the hybridization potential among populations or varieties of Brassicaceae may be common (see Guéritaine *et al.*, 2003) and may influence the frequency of gene introgression and result in regional variations in risk associated with GM oilseed. Even though F_1 hybrids between oilseed rape and *R. raphanistrum* can be formed, the recombination of oilseed rape genetic material into the genome of *R. raphanistrum* seems to be difficult. Chèvre *et al.* (Chapter 18, this volume) did not observe genomic recombination despite recurrent backcrosses to the wild parent.

Effects of Growing Genetically Modified Oilseed Rape

Our gene flow analysis is based on DNA markers of the oilseed rape nucleus and plastids. Most of these markers are supposed to be selectively neutral. There is no reason to believe that transgenes will be introgressed differently from these endogenous genes. However, the rate of transgene transfer may be increased due to selection in favour of the transgenic traits, which, for example, will be the case for herbicide tolerance in the agroecosystem (Snow *et al.*, 1999). In conclusion, given the right environment, oilseed rape is apparently a potent donor of genes to *B. rapa*, and introgressed plants survive and reproduce in the natural populations. Experiments in non-selective environments suggest that the fitness of some of the backcross plants will be equivalent to that of the wild parent. The reproductive fitness of interspecific F_1 hybrids can be even greater than that of both parents. We have also shown that integration of genes in the plastid DNA or on C chromosomes of oilseed rape will not provide safeguards towards natural introgression. However, the frequency of gene transfer can be reduced by efficient control of related weeds, e.g. by efficient herbicide control. Therefore, the development of herbicide-tolerant *B. rapa* will probably be slow in the conventionally managed fields and herbicide-tolerant cultivars will provide easy and secure management of *B. rapa* over many years. As many farmers convert to agricultural practices with no or low herbicide usage, weed problems may increase. The newly started organic production of oilseed rape may promote the presence of mixed weedy populations of oilseed rape and *B. rapa* in succeeding crops. Transgenes from neighbouring GM oilseed rape may be transferred to *B. rapa* and *B. napus* in these populations. This scenario may give rise to different transgenes becoming stacked in these reservoir populations and problems in relation to threshold values of GM in the organic production. Transfer of transgenes from oilseed rape to wild relatives other than *B. rapa* seems to be less likely.

Acknowledgements

The Danish Research Agency (Centre for Bioethics and Risk Assessment), the SMP II programme and the EU (CONFLOW, FW 5) are acknowledged for providing funding for this project.

References

Corriveau, J.L. and Coleman, A.W. (1988) Rapid screening method to detect potential biparental inheritance of plastid DNA and results for over 200 angiosperm species. *American Journal of Botany* 75, 1443–1458.

Daniell, H., Datta, R., Varma, S., Gray, S. and Lee, S.-B. (1998) Containment of herbicide resistance through genetic engineering of the chloroplast genome. *Nature Biotechnology* 16, 345–348.

Frello, S., Hansen, K.R., Jensen, J. and Jørgensen, R.B. (1995) Inheritance of rapeseed (*Brassica napus*) specific RAPD markers and a transgene in the cross *B. juncea* × (*B. juncea* × *B. napus*). *Theoretical and Applied Genetics* 91, 236–241.

Guéritaine, G., Bonavent, J.F. and Darmency, H. (2003) Variation of prezygotic barriers in the interspecific hybridization between oilseed rape and wild radish. *Euphytica* 130, 349–353.

Hansen, L.B., Siegismund, H.R. and Jørgensen, R.B. (2001) Introgression between oilseed rape (*Brassica napus* L.) and its weedy relative *B. rapa* L. in a natural population. *Genetic Resources and Crop Evolution* 48, 621–627.

Hansen, L.B., Siegismund, H.R. and Jørgensen, R.B. (2003) Progressive introgression between *Brassica napus* (oilseed rape) and *B. rapa*. *Heredity* 19, 276–283.

Harberd, D.J. (1975) *Brassica* L. In: Stace, C.A. (ed.) *Hybridization and the Flora of the British Isles*. Academic Press, London, pp. 137–139.

Hauser, T.P., Jørgensen, R.B. and Østergard, H. (1997) Preferential exclusion of hybrids in mixed pollinations between oilseed rape (*Brassica napus*) and weedy *B. campestris* (Brassicaceae). *American Journal of Botany* 84, 756–762.

Hauser, T.P., Jørgensen, R.B. and Østergård, H. (1998) Fitness of backcross and F$_2$ hybrids between weedy *Brassica rapa* and oilseed rape (*B. napus*). *Heredity* 81, 436–443.

Hauser, T.P., Damgaard, C. and Jørgensen, R.B. (2003) Frequency dependent fitness of hybrids between oilseed rape (*Brassica napus*) and weedy *B. rapa* (Brassicaceae). *American Journal of Botany* 90, 571–578.

Jørgensen, R.B. (1996) Spontaneous hybridization between oilseed rape (*Brassica napus*) and weedy relatives. *Acta Horticulturae* 407, 193–200.

Jørgensen, R.B. and Andersen, B. (1994) Spontaneous hybridization between oilseed rape (*Brassica napus*) and weedy *B. campestris* (Brassicaceae). *American Journal of Botany* 81, 1620–1626.

Jørgensen, R.B., Andersen, B., Hauser, T.P., Landbo, L., Mikkelsen, T. and Østergård, H. (1998) Introgression of crop genes from oilseed rape (*Brassica napus*) to related wild species – an avenue for the escape of engineered genes. *Acta Horticulturae* 459, 211–217.

Landbo, L. and Jørgensen, R.B. (1997) Seed germination in weedy *Brassica campestris* and its hybrids with *B. napus*: implications for risk assessment of transgenic oilseed rape. *Euphytica* 97, 209–216.

Landbo, L., Andersen, B. and Jørgensen, R.B. (1996) Natural hybridization between oilseed rape and a wild relative: hybrids among seeds from weedy *B. campestris*. *Hereditas* 125, 89–91.

Mikkelsen, T.R., Andersen, B. and Jørgensen, R.B. (1996) The risk of crop transgene spread. *Nature* 380, 31.

Pertl, M., Hauser, T.P., Damgaard, C. and Jørgensen, R.B. (2002) Male fitness of oilseed rape (*Brassica napus*), weedy *B. rapa*, and their F_1 hybrids when pollinating *B. rapa* seeds. *Heredity* 88, 212–218.

Scott, S.E. and Wilkinson, M.J. (1998) Transgene risk is low. *Nature* 393, 320.

Scott, S.E. and Wilkinson, M.J. (1999) Low probability of chloroplast movement from oilseed rape (*Brassica napus*) into wild *Brassica rapa*. *Nature Biotechnology* 17, 390–392.

Snow, A.A., Andersen, B. and Jørgensen, R.B. (1999) Cost of transgenic herbicide resistance introgressed from *Brassica napus* into weedy *B. rapa*. *Molecular Ecology* 8, 605–615.

Taberlet, P., Gielly, L., Pautou, G. and Bouvet, J. (1991) Universal primers for amplification of three non-coding regions of chloroplast DNA. *Plant Molecular Biology* 17, 1105–1109.

20

Transgene Expression and Genetic Introgression Associated with the Hybridization of GFP Transgenic Canola (*Brassica napus* L.) and Wild Accessions of Bird Rape (*Brassica rapa* L.)

MATTHEW D. HALFHILL[1,3], SUZANNE I. WARWICK[2] AND C. NEAL STEWART JR[3]

[1]*North Carolina State University, Crop Science Department, Raleigh, NC 27695, USA;* [2]*Agriculture and Agri-Food Canada (AAFC), Ottawa, ON, Canada K1A 0C6;* [3]*University of Tennessee, Department of Plant Sciences, Knoxville, TN 37996, USA; E-mail: nealstewart@utk.edu*

Abstract

The level of transgene expression in crop × weed hybrids and the degree to which crop-specific genes are integrated into hybrid populations are important factors in assessing the potential ecological and agricultural risks of gene flow associated with genetic engineering. The average transgene zygosity and genetic structure of transgenic hybrid populations change with the progression of generations toward weed species, and the green fluorescent protein (GFP) transgene is an ideal marker to quantify transgene expression in advancing populations. Several hybrid generations (BC_1F_1 and BC_2F_1) were produced by backcrossing various GFP/*Bt* transgenic canola (*Brassica napus* cv. Westar) and bird rape (*B. rapa*) hybrid generations onto *B. rapa*. The average fluorescence of each successive hybrid generation was analysed, and homozygous canola lines and hybrid populations that contained individuals homozygous for GFP (BC_2F_2 Bulk) demonstrated significantly higher fluorescence than hemizygous hybrid generations (F_1, BC_1F_1 and BC_2F_1). These data demonstrate that the expression of the GFP gene was additive, and fluorescence could be used to determine zygosity status. The formation of homozygous individuals within hybrid populations also increases the average level of transgene expression as generations progress. Amplified fragment length polymorphism (AFLP) DNA analysis was used to quantify the degree of *B. napus* introgression into multiple backcross hybrid generations with *B. rapa*. The F_1 hybrid generations contained 95–97% of the *B. napus*-specific AFLP markers, and each successive backcross generation demonstrated a reduction of markers resulting in 15–29% presence in the BC_2F_2 Bulk population. Multiple backcrosses to a weedy genetic background may result in the ploidy level of the wild relative, but a significant number of crop-specific genes will be co-introgressed into that background. The consequences of the crop-specific genetic load associated with transgene flow must, in future studies, be assessed in field-level experiments, as it is conceivable that backcrossed hybrids may be more crop-like in their physiology and ecology when compared with the wild-type weed.

Introduction

Transgenic crops are rapidly becoming a staple of modern agriculture. Since 1992, the US Department of Agriculture (USDA) has deregulated 58 events of transgenic crops for commercial field release (Animal and Plant Health Inspection Service, APHIS, Permits 2003, www.isb.vt.edu). Deregulated crop species include beet (*Beta vulgaris* L.), canola (*Brassica napus* L.), maize (*Zea mays* L.), cotton (*Gossypium hirsutum* L.), flax (*Linum usitatissimum* L.), papaya (*Carica papaya* L.), potato (*Solanum tuberosum* L.), rice (*Oryza sativa* L.), soybean (*Glycine max* Merr.), squash (*Cucurbita pepo* L.), tobacco (*Nicotiana tabacum* L.) and tomato (*Lycopersicon esculentum* Mill.). Despite the fact that over 3.5 trillion transgenic plants have been grown in the USA, there have been few detectable ecological consequences (Stewart *et al.*, 2000). This is not to say that growing transgenic plants is completely risk free.

Of the widely acknowledged risks of transgenic plants (Stewart *et al.*, 2000), the one that we have focused on in our USDA-funded research since 1994 has been gene flow and ecological consequence. Transgene movement via hybridization between closely related species can be a mode of transgene flow directly into wild populations (Raybould and Gray, 1993), so that a transgene might move from crop to weed. If expressed in the genetic background of a weed species, a transgene could conceivably change the fitness of the weed in nature. In the worst-case scenario, the weed could become more invasive and competitive.

Gene flow from transgenic canola (*Brassica napus*, AACC, $2n = 38$) to different canola varieties and weedy relatives is a potential and realistic risk of the implementation of biotechnology on an agricultural scale. Hybridization between transgenic canola varieties with congeneric and confamilial wild relatives represents a path for transgenic phenotypes to be acquired by natural populations (Warwick *et al.*, 1999). Canola cultivation results in significant volunteer populations during subsequent years, and transgenic volunteer populations, particularly herbicide-tolerant volunteers, present additional management concerns (Légère *et al.*, 2001; Simard *et al.*, 2002). Several studies have demonstrated hybridization between canola and wild relatives in crosses with both closely related species (*B. rapa* L., AA, $2n = 20$) and distantly related (*Raphanus raphanistrum* L., RrRr, $2n = 18$) crosses under agricultural conditions (Jørgensen and Anderson, 1994; Scott and Wilkinson, 1998; Chèvre *et al.*, 2000; Rieger *et al.*, 2001; Halfhill *et al.*, 2002; Warwick *et al.*, 2003). Hybridization experiments involving transgenic plants have shown that transgenes will be expressed in hybrid plants, and that transgenic phenotypes should be expected in agricultural fields where crops may cross-pollinate with weedy species (Metz *et al.*, 1997; Halfhill *et al.*, 2001, 2002; Warwick *et al.*, 2003).

Previous reports of transgene expression in hybrid populations have focused on qualitative assessments of transgenic phenotypes, such as herbicide tolerance, insect resistance or marker genes (Mikkelsen *et al.*, 1996; Metz *et al.*, 1997; Harper *et al.*, 1999; Snow *et al.*, 1999; Chèvre *et al.*, 2000; Halfhill *et al.*, 2001). As more is understood about the importance of transgene copy number and population structure, quantitative assessments at the population level will become necessary to evaluate potential risk effectively. There have been mixed results when investigating the relationship between transgene zygosity and expression. Several studies have demonstrated additive transgene expression between homozygous and hemizygous progeny from an independent transformation event (Hobbs *et al.*, 1990; Stewart *et al.*, 1996; Niwa *et al.*, 1999; James *et al.*, 2002), while other studies demonstrate no differences in expression based on zygosity (Hobbs *et al.*, 1990; Caligari *et al.*, 1993; Scott *et al.*, 1998; James *et al.*, 2002). Interspecific hybridization initially generates hemizygous F_1 individuals with one copy of the transgene locus. Over evolutionary time, the average zygosity of a hybrid population equilibrates as backcrossing and intermating occur, resulting, in the absence of selection, in Hardy–Weinberg equilibrium for diploidized plants. If differential transgene expression occurs in a mixed population composed of hemizygous and homozygous individuals, the average transgenic phenotype for a given population may change in subsequent generations.

The canola × *B. rapa* model system has been used to demonstrate that a weedy, *B. rapa*-like phenotype can be recovered after several backcross generations. Previous studies have shown that the F_1 hybrid generation is triploid (AAC, $2n = 29$) and, after multiple generations of backcrossing, the ploidy level of these generations is reduced to the original diploid level of the weedy parent, *B. rapa* (Metz *et al.*, 1997; Halfhill *et al.*, 2002). Quantification of the genetic introgression of the canola genome into backcrossed generations has been reported using several DNA marker systems, such as restriction fragment length polymorphism (RFLP) (Jørgensen and Andersen, 1994; Mikkelsen *et al.*, 1996), inter-simple sequence repeat (SSR) (Scott and Wilkinson, 1998) and amplified fragent length polymorphism (AFLP) analysis (Hansen *et al.*, 2001; Warwick *et al.*, 2003). Hybridization may significantly change the genetic composition of weedy populations, and understanding the degree of crop gene introgression may help predict how introgressed populations will interact in agricultural and natural environments.

Assuming that transgenes will persist in backcrossed populations, the next step in evaluating the biosafety of transgenic hybrid populations is to determine what factors control the levels of transgene expression and to quantify the degree of *B. napus* genetic introgression in hybrid generations. We report results that determine transgene expression within various hybrid generations using the green fluorescent protein (GFP), which has been shown to be a quantitative marker for studying transgene expression

(Harper *et al.*, 1999; Halfhill *et al.*, 2001; Stewart, 2001; Millwood *et al.*, 2003; Richards *et al.*, 2003). GFP analysis allows for the estimation of transgene expression levels through non-destructive, *in situ* measurements with a fluorescence spectrophotometer. The results from these experiments describe how zygosity variation correlates with transgene expression and protein synthesis within a hybrid population, and how AFLP analysis was used to quantify the degree of *B. napus* introgression into multiple backcross hybrid generations with *B. rapa*.

Green Fluorescent Protein Transgene Tagging

An essential component of our gene flow research has included tagging fitness-enhancing genes, such as a *Bacillus thuringiensis* (*Bt*) *cry1Ac* gene that encodes an endotoxin protein for caterpillar control, with a transgene encoding GFP (*Aequorea victoria* mGFP5-ER variant green fluorescent protein). GFP can be visualized in mature plant tissues macroscopically by illuminating the plants in the dark with a long-wave UV or blue lamp. Therefore, for the first time, transgene segregation and expression can be assessed simultaneously in all plant tissues in real time (Stewart, 1996, 2001). GFP could also be used to mark seeds, flowers or other organs. For example, pollen could be painted with GFP to study gene flow and pollination behaviour, especially long-distance gene flow (Hudson *et al.*, 2001). There is no fitness cost with expressing GFP in plants either cytoplasmically or when targeted to the endoplasmic reticulum (ER). However, ER targeting yields better GFP expression in the field (Harper *et al.*, 1999). Furthermore, we can assess the zygosity status of transgenic plants, in either the initial recipient species or hybrids with wild relatives, by simple observation or using more sophisticated fluorescence spectrometry (Halfhill *et al.*, 2003).

Hybridization Experiments

The potential for gene flow between multiple GFP, GFP/*Bt* and *Bt* canola has been quantified under both laboratory and field experiments (Halfhill *et al.*, 2001, 2002). In a laboratory-based crossing experiment, nine T_2 GFP/*Bt* events (GT1–9) and three GFP events (GFP1–3) were hand crossed with three lines of *B. rapa* (CA from Irvine, California, courtesy of Art Weiss; 2974 from Milby and 2975 Waterville, Québec, Canada). Plants in six of the canola lines (GT1, GT5, GT7, GT8, GFP1 and GFP2) were homozygous for GFP as detected by fluorescence analysis, while the other six lines were still segregating (3:1). The zygosity status of the lines was analysed by selfed progeny analysis. Plants were classified as homozygous if all selfed progeny were GFP fluorescent, and hemizygous if any non-fluorescent progeny were recorded. Parental canola lines were germinated on moist filter paper, and fluorescent individuals were selected for the hybridization experiment. The *B. rapa* lines were used as the pollen recipients. Both species were allowed to flower, and hand crossing was performed by removing a canola flower and pollinating the *B. rapa* plants. The hand crossing continued as long as both plants continued to flower. All seeds were collected from the *B. rapa* parent, and were germinated on moist filter paper and screened by visual assay for GFP fluorescence. Plants that expressed GFP were backcrossed in the same fashion as above. The hybrids were used as pollen donors to produce backcrosses with their respective *B. rapa* accession.

All GT and GFP events generated F_1 transgenic hybrids through the hand-crossing experiment (Table 20.1). The hybrids were characterized in the same manner as the parental canola events, and demonstrated GFP macroscopic fluorescence patterns identical to the parent canola events. The hybrid lines backcrossed with *B. rapa* resulted in a backcrossed generation (BC_1) for each line crossed. The frequency of transgenic hybrids (F_1) recovered from each cross ranged from 25 to 86%, and the percentage of BC_1 plants recovered ranged from 15 to 34%. The BC_1 hybridization frequency was similar between all crossing types, with the hemizygous F_1 hybrid donating pollen to the maternal weedy parent.

The genomic location of transgene integration into canola, whether on the *A* or *C* genome, has been suggested to play a role in the ability of transgenic events to pass fitness-enhancing transgenes to *B. rapa* (Metz *et al.*, 1997). In this model, hybrid plants were putatively triploid (*AAC*), and the

Table 20.1. Percentage recovery of transgenic progeny from canola × *B. rapa* hybridizations and hybrids backcrossed into *B. rapa*.

Transgenic event	F_1 hybrids				BC_1 hybrids			
	CA	2974	2975	Total	CA	2974	2975	Total
GT1	69	81	38	62	34	25	41	33
GT2	38	88	81	77	23	35	31	30
GT3	81	50	63	65	24	10	30	20
GT4	38	56	56	50	7	30	36	26
GT5	81	75	81	79	39	17	39	31
GT6	50	50	54	51	26	12	26	21
GT7	31	75	63	56	30	19	31	26
GT8	56	75	69	67	22	22	21	22
GT9	81	31	31	48	27	28	23	26
GFP1	50	88	75	71	18	33	32	27
GFP2	69	88	100	86	26	20	57	34
GFP3	19	38	19	25	10	22	11	15

The numbers indicate the percentage of progeny recovered from the hand crosses that screened positive for GFP fluorescence under UV light. ANOVA indicates that the GFP3 line produced a significantly lower percentage ($P < 0.05$) of transgenic BC_1 hybrids than four other lines (GT1, GT2, GT5 and GFP1); however, the other 11 lines were not significantly different from one another. GT (canola events with GFP and *Bt*), GFP (canola events with GFP only), CA, 2974 and 2975 (California, Milby and Waterville, Québec populations of *B. rapa*, respectively).

chromosomes on the C genome were unstably passed or lost during meiotic divisions. If the transgene is on the C genome, the gene could be lost to the next generation, leading to no transgenic backcrosses. The location of the transgene insertion would result in different backcross frequencies between transgenic events, and certain lines would be 'safer' with regard to gene flow and integration. However, because the A and C genomes share a significant degree of homology, recombination rates may be high and allow for increased rates of transgene integration into *B. rapa* (Tomiuk *et al.*, 2000). Our finding (Halfhill *et al.*, 2001) that 12 independent canola events generate backcrosses at similar rates supports the hypothesis that there are probably few 'safe' locations in the canola genome with regards to gene flow. This is in contrast to Metz *et al.* (1997), in which two independent herbicide-tolerant canola events produced BC_1 plants at significantly different rates. A sample size of 12 transgenic events is the largest analysed to date, and adds data to an argument that had been historically theoretical.

Transgene Copy Number and Average Whole Plant Fluorescence

The effect of transgene copy number on transgene expression in GFP canola and transgenic hybrid generations was quantified using GFP fluorescence (Halfhill *et al.*, 2003). The identity of homozygous and hemizygous individuals was determined by progeny analysis as described previously. GFP fluorescence of the T_1 canola plants within each transformation event was correlated to the zygosity of each individual (Fig. 20.1). Homozygous individuals demonstrated significantly higher fluorescence at 508 nm compared with hemizygous individuals ($P < 0.05$). After standardization, the homozygous and hemizygous fluorescence profiles differed only in the magnitude of the GFP peak (480–540 nm) (Fig. 20.1). For example, GFP2 homozygous plants exhibited an average fluorescence of 10.5 ± 0.4 (all units in 10^5 counts per second (c.p.s.) $\pm$ SD) compared with hemizygous plants at 7.6 ± 1.4 c.p.s. and non-transformed canola at 5.3 ± 0.7 c.p.s. When the latter background level of fluorescence was removed from each sample, homozygous individuals fluoresced twice as much as hemizygous individuals. The difference between zygosity states was consistent and statistically significant within all ten transgenic lines analysed in the study (data not shown).

All hybrid generations within each crossing line were analysed for average GFP fluorescence (Fig. 20.2). In each crossing line, the homozygous canola plants exhibited significantly higher average GFP fluorescence when compared with the hemizygous hybrid generations (F_1, BC_1F_1 and BC_2F_1) and wild-type *B. rapa* parent (analysis of variance (ANOVA) Fisher's protected least squares difference (PLSD), $P < 0.05$). The average magnitude of fluorescence of GT1 was 10.1 ± 0.8 (10^5 c.p.s. $\pm$ SD at 508 nm), and the hemizygous generations ranged from 7.6 ± 0.3 to 8.2 ± 0.5 c.p.s. The hemizygous hybrid lines were not significantly different from one another, and always exhibited significantly greater fluorescence than the *B. rapa* parent. The BC_2F_2 Bulk generation was composed of a mixture of homozygous and

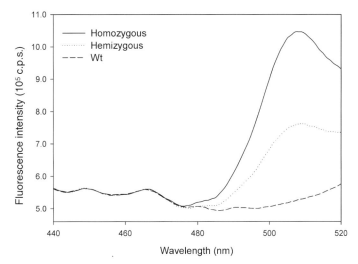

Fig. 20.1. Average scanning fluorescence emission of homozygous versus hemizygous individuals of T_1 GFP2 canola when excited with 385 nm UV light. Non-transformed canola (Westar, Wt) was used as a control. Homozygous individuals exhibited an average fluorescence at 508 nm of 10.5 ± 0.4 (all units in 10^5 c.p.s. $\pm$ SD) compared with hemizygous individuals that had an average fluorescence of 7.6 ± 1.4. Wt exhibited an average fluorescence of 5.3 ± 0.7.

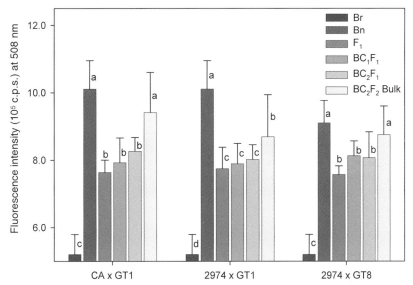

Fig. 20.2. Average GFP fluorescence of parental transgenic canola (Bn) and F_1, BC_1F_1, BC_2F_1 and BC_2F_2 Bulk hybrid generations with *B. rapa* (Br). Leaves were excited with 385 nm UV light and fluorescence intensity was measured at 508 nm (c.p.s.). Fluorescence averages ($\pm$ SD) per generation were compared by ANOVA within each crossing line (CA $\times$ GT1, 2974 $\times$ GT1 and 2974 $\times$ GT8), and letters indicate significant differences between generations (Fisher's PLSD, $P < 0.05$).

hemizygous individuals, and in all cases exhibited greater fluorescence than the hemizygous generations (Fig. 20.2). In crossing lines CA × GT1 and 2974 × GT8, average GFP fluorescence of the BC_2F_2 Bulk generation was not significantly different from the original homozygous canola event.

Additive Transgene Expression

The GFP gene demonstrated additive transgene expression in ten independent transformation events of canola. In all canola lines, homozygous individuals that contained two copies of the transgene locus fluoresced twice as much as hemizygous individuals above the background level of fluorescence. In previous studies (Hobbs *et al.*, 1990; Caligari *et al.*, 1993; Scott *et al.*, 1998; Allen *et al.*, 2000; James *et al.*, 2002), independent transformation events modified with the same plasmid have been shown to exhibit a wide range of expression levels and stability through generations. Allen *et al.* (2000) and James *et al.* (2002) established that some transgenic events show transgene silencing, while others are consistently expressed, and matrix attachment regions (MARs) were shown in each case to limit transgene silencing of a β-glucuronidase (GUS) transgene in tobacco and rice, respectively. In the canola lines analysed in this study, no evidence of transgene silencing was found, and this may be due to the inclusion of a single tobacco RB7 MAR on the pSAM12 plasmid that was used for the GT canola transformations (Harper *et al.*, 1999). The transgenic canola lines were also generated by positive selection for the GFP transgenic phenotype in tissue culture (Stewart *et al.*, 2002). This type of selection could possibly have removed low-expressing transgenic events or those with the tendency to silence, and skewed the T_0 population towards events that were resistant to transgene silencing. The combination of positive selection and the inclusion of a MAR on the plasmid has produced a population of transgenic canola events that demonstrate additive transgene expression and resistance to transgene silencing.

Additive transgene expression of GFP has also been shown in two other plant models, tobacco (*Nicotiana tabacum*) and *Arabidopsis thaliana* (L.) Heynh. (Niwa *et al.*, 1999; Molinier *et al.*, 2000). In the tobacco model, classes based on fluorescence intensity (high, low and no fluorescence) were used to predict the zygosity status of segregating T_1 progeny. Our study could also categorize individuals based on a class level (data not shown). The numerical quantification provided by fluorescence spectrophotometry allowed for a precise, quantifiable measure of the GFP phenotype and was therefore useful in models estimating recombinant protein per unit area based on fluorescence (Richards *et al.*, 2003). In these models, GFP fluorescence has been shown to correlate with the total amount of GFP present in water and plant extracts. In the construction of fusion proteins, in which GFP is fused to another recombinant protein, simple fluorescence measurements could predict the yield or phenotype of an otherwise immeasurable transgenic character. Vain *et al.* (2000) quantified GFP fluorescence in transgenic *A. thaliana* lines using blue laser-based spectrophotometry on a FluorImager imaging system (FluorImager SI, Molecular Dynamics). Although this system was also quantitative, the instrument required small plant samples and, therefore, if one desired to measure fluorescence of larger plants, leaves would have to be harvested and placed in the instrument. With the Fluoromax-2 fluorescence spectrophotometer (Jobin Yvon & Glen Spectra, Edison, New Jersey), sampling was conducted on large plants with the use of a fibre optic cable directly on intact plant tissues. Millwood *et al.* (2003) recently reported the development of a portable fluorescence spectrophotometer (GFPMeter, OptiScience, Tyngsboro, Massachusetts) that is highly accurate and can be used under field conditions. This instrument allows for future applications of GFP in large-scale agriculture, and may expand the uses of GFP that have been limited by the requirement for laboratory-based quantitative systems.

GFP Hybrid Generations

The GFP transgenic phenotype was qualitatively useful in the selection of hybrid plants at the seedling stage, and allowed for the accurate selection of transgenic hybrids and introgression events in subsequent generations. Positive selection for the GFP phenotype is advantageous in comparative studies, because non-transgenic individuals can be germinated under identical conditions without the confounding effects of destructive collection required for molecular-based confirmation analyses, thus allowing for risk assessment studies to predict the competitive ability of transgenic individuals under field conditions.

In advanced hybrid generations, the average transgenic phenotype was shown to increase as inter-mating altered the degree of homozygosity within the population. The finding that later generations will reach the average transgenic phenotype of the crop variety must be included in risk management strategies. After hybridization occurs and hemizygous F_1 individuals are produced, the population should be expected to shift the degree of the transgenic phenotype based on the dynamics of breeding within the hybrid population and weedy plants. Future research must, therefore, quantify the frequency of backcrossing and selfing amongst transgenic individuals under field conditions.

Genetic Introgression

The number of genes originating from *B. napus* (i.e. crop-specific genes) introgressing into transgenic hybrid generations was quantified using AFLP analysis (Halfhill *et al.*, 2003). Five primer pair combinations were used to generate a total of 270 consistently amplified AFLP bands, 92 of which were *B. napus*-specific markers. Each hybrid generation was analysed for the presence/absence of these *B. napus*-specific AFLP markers, and it was determined that the number of markers decreased with each backcross generation

(Fig. 20.3). The F_1 generation contained between 95 and 97% of the *B. napus*-specific markers, and the BC_1F_1 generation contained between $62 \pm 12\%$ and $75 \pm 14\%$ ($\pm$ SD). The BC_2F_1 generation continued this general trend, but demonstrated wider variation in percentage crop-specific markers in each of the crossing lines. The CA $\times$ GT1 and 2974 $\times$ GT1 BC_2F_1 generations had $30 \pm 7\%$ and $25 \pm 3\%$ of the markers, respectively, in contrast to the 2974 $\times$ GT8 BC_2F_1 generation at $73 \pm 2\%$. The BC_2F_2 Bulk populations showed a further reduction of *B. napus*-specific markers, ranging from 15 ± 1 to $29 \pm 3\%$ per crossing line.

Consequences of Genetic Load from Linkage Disequilibrium

Transgenes will be expressed in hybrid and genetic backgrounds that are predominantly *B. rapa* (Halfhill *et al.*, 2002), and a significant amount of crop-originating genetic material will co-introgress into the hybrid generations. Until recently, the predominant ecological hypothesis has been that certain transgenes that enhance the fitness of the crop would also increase the fitness of the wild relative that harbours the fitness-enhancing transgene. For instance, a *Bt* transgene conferring insect resistance was shown to increase the fitness of the crop (Stewart *et al.*, 1997) when defoliating insects are present, and therefore the transgenic crop would be protected from herbivory relative to the non-transgenic crop. Would this pattern hold true

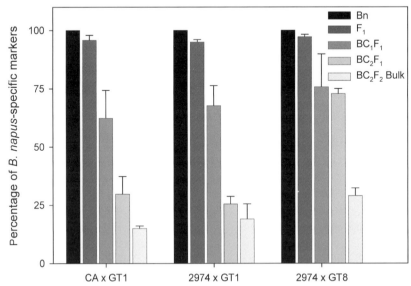

Fig. 20.3. AFLP analysis of canola (*B. napus*, Bn) and F_1, BC_1F_1, BC_2F_1 and BC_2F_2 Bulk hybrid generations. AFLP analysis with five specific primer sets yielded 92 *B. napus*-specific markers. Within each crossing line (CA $\times$ GT1, 2974 $\times$ GT1 and 2974 $\times$ GT8), the percentage of Bn-specific markers $\pm$ SD for each generation is shown (five plants per generation).

for crop × weed hybrids and, more importantly, introgressed backcrossed generation plants? What is the consequence of the genetic load from crop genes associated with transgenes during introgression with wild relatives? What is the relationship between the number of crop genetic markers (genes) that flank a transgene and are transmitted with it during introgression and the fitness and competitiveness of the resulting introgressed hybrid and weed? What is the effect of linkage disequilibrium during introgression?

Our approach for testing ecological performance has been to perform competition experiments between crops and transgenic weedy hybrids (studies in progress for GFP/*Bt* canola and *B. rapa*). In these experiments, winter wheat has been grown in competition with wild-type or *Bt B. rapa* hybrids under greenhouse and field conditions. If the transgenic wild relative plants are, indeed, more competitive, one should be able to observe a decrease in biomass or a yield penalty in the winter wheat crop. Results from our recent field weed–crop competition experiments have, therefore, caused us to put forth an alternative hypothesis. We hypothesize that genetic load of genes from crops hitchhiking with the linkage group of a transgene introgressed into wild relatives will moderate weediness potential of the transgenic 'weed'. Multiple backcrosses to a weedy genetic background may result in the ploidy level of the wild relative, but a significant number of crop genes in that background will make the backcrossed hybrid more crop-like when compared with the wild-type weed in its physiology and ecology. However, even if the transgene itself would confer a fitness increase, the genetic load of crop genes may have the opposite effect and cause the new transgenic weed to be slower growing and relatively less invasive or competitive than the wild-type weed. Indeed, the introgression process could make a transgenic weed a 'wimp' (Adam, 2003; Stewart *et al.*, 2003a). In essence, linkage disequilibrium hypothetically would be a more powerful evolutionary force in domesticating weeds compared with the selective advantage conferred by the transgene.

Consequence of Gene Flow

Transgenes have almost certainly been transferred from *B. napus* to free-living *B. rapa* plants into the F_1, BC_1 and BC_2 generations and beyond. This occurrence does not signify a risk in and of itself. Genes have been flowing back and forth between crops and weeds since crops have existed as such. Furthermore, there are much greater imminent risks in agricultural systems compared with gene flow and its results (e.g. pesticide residues, soil erosion, land use). So the appropriate question to be asked is not 'will gene flow happen?' It most certainly will, especially from *B. napus* into *B. rapa*. But will transgenes persist in unintended hosts and will there be any detrimental consequences? Thus far, all experiments have artificially selected for transgene persistence. We know nothing about transgene persistence via introgression in wild species, but the barriers to introgression may be significant (Stewart *et al.*, 2003b).

The most predominant transgenic trait in commercial production is herbicide tolerance (HT) that allows plants to survive otherwise lethal doses of glyphosate, glufosinate and bromoxynil. Most of the canola in Canada is planted to exploit herbicide tolerance (75% is transgenic HT, but 25% is non-transgenic imidazolinone HT), and therefore many opportunities exist for interspecific gene flow resulting in HT hybrids. However, the biggest problem associated with HT canola is volunteerism in follow-up years that will require alternative and/or additional control measures (Hall *et al.*, 2000; Légère *et al.*, 2001; Beckie *et al.*, 2003). Transgenic HT volunteer or feral canola plants would have no selective advantage outside of agriculture and/or other disturbed sites where the herbicide is not applied. However, HT volunteers can emerge from the seed bank over multiple years and serve as a potential pollen source for dispersal of transgenes to weedy relatives and canola crops that follow in rotation or are located in nearby fields. There does not appear to be an associated fitness cost of the HT trait in *B. napus* canola, and therefore the trait would not be selected against and could persist in habitats having no exposure to herbicide pressure (Simard *et al.*, 2003).

HT wild relatives would be another variation of this problem and, although initially less severe because of significantly fewer numbers of hybrids produced, these may increase in severity over time as a result of backcrossing and subsequent introgression into the wild population. Similarly to the HT volunteers, transgenic HT wild relatives are unlikely to have a selective advantage outside of agriculture and/or other disturbed sites where the herbicide is applied. Also, we may assume no fitness cost (as discussed above) in the acquisition of HT in a hybrid, and therefore no selection against these genes in the weedy populations in the absence of herbicide selection pressure. Snow *et al.* (1999) showed that fitness costs associated with transgenic glufosinate resistance introgressed from *B. napus* into weedy *B. rapa* were negligible.

There are several transgenic traits that must be assessed for ecological consequence; traits that might confer some fitness advantage to plant populations in both agricultural and non-agricultural settings. The trait that we have used is insect resistance conferred by *Bt cry1Ac*, which kills lepidopteran herbivores on crucifers such as the diamondback moth (*Plutella xylostella*). These plants have been field tested by commercial entities as well as academic laboratories. However, other insect resistance genes such as *Bt Vip3A*, *Photorhabdus luminescens* toxins, cholesterol oxidase (CO) from *Streptomyces* culture filtrate, proteinase inhibitors, lectins and chitinases represent the future of insect resistance traits (reviewed by Stewart, 1999). Other transgenic phenotypes such as disease resistance and tolerance to various stresses such as aluminium, salt and cold will also need special assessment. One approach for testing ecological performance is to perform competition experiments between crops and transgenic weedy hybrids (as described above for *Bt* canola and *B. rapa*). Future research can then expand from the current understanding that transgene flow will occur, to making predictions of the potential ecological and agricultural consequences of such gene flow.

References

Adam, D. (2003) Transgenic crop trial's gene flow turns weeds into wimps. *Nature* 421, 462.

Allen, G.C., Spiker, S. and Thompson, W.F. (2000) Use of matrix attachment regions (MARs) to minimize transgene silencing. *Plant Molecular Biology* 43, 361–379.

Beckie, H.J., Warwick, S.I., Nair, H. and Séguin-Swartz, G. (2003) Gene flow in commercial fields of herbicide-resistant canola (*Brassica napus*). *Ecological Applications* 13, 1276–1294.

Caligari, P.D.S., Yapabandara, Y.M.H.B., Paul, E.M., Perret, J., Roger, P. and Dunwell, J.M. (1993) Field performance of derived generations of transgenic tobacco. *Theoretical and Applied Genetics* 86, 875–879.

Chèvre, A.M., Eber, F., Darmency, H., Fleury, A., Picault, H., Letanneur, J.C. and Renard, M. (2000) Assessment of interspecific hybridization between transgenic oilseed rape and wild radish under agronomic conditions. *Theoretical and Applied Genetics* 100, 1233–1239.

Halfhill, M.D., Richards, H.A., Mabon, S.A. and Stewart, C.N. Jr (2001) Expression of GFP and Bt transgenes in *Brassica napus* and hybridization and introgression with *Brassica rapa*. *Theoretical and Applied Genetics* 103, 659–667.

Halfhill, M.D., Millwood, R.J., Raymer, P.L. and Stewart, C.N. Jr (2002) Bt-transgenic oilseed rape hybridization with its weedy relative, *Brassica rapa*. *Environmental Biosafety Research* 1, 19–28.

Halfhill, M.D., Millwood, R.J., Weissinger, A.K., Warwick, S.I. and Stewart, C.N. Jr (2003) Additive transgene expression and genetic introgression in multiple green fluorescent protein transgenic crop × weed hybrid generations. *Theoretical and Applied Genetics* 107, 1533–1540.

Hall, L., Topinka, K., Huffman, J., Davis, L. and Good, A. (2000) Pollen flow between herbicide-resistant *Brassica napus* is the cause of multiple-resistant *B. napus* volunteers. *Weed Science* 48, 688–694.

Hansen, L.B., Siegismund, H.R. and Jørgensen, R.B. (2001) Introgression between oilseed rape (*Brassica napus* L.) and its weedy relative *B. rapa* L. in a natural population. *Genetics Research and Crop Evolution* 48, 621–627.

Harper, B.K., Mabon, S.A., Leffel, S.M., Halfhill, M.D., Richards, H.A., Moyer, K.A. and Stewart, C.N. Jr (1999) Green fluorescent protein as a marker for expression of a second gene in transgenic plants. *Nature Biotechnology* 17, 1125–1129.

Hobbs, S.L.A., Kpodar, P. and Delong, C.M.O. (1990) The effect of T-DNA copy number, position and methylation on reporter gene expression in tobacco transformants. *Plant Molecular Biology* 15, 851–864.

Hudson, L.C., Chamberlain, D. and Stewart, C.N. Jr (2001) GFP-tagged pollen to monitor pollen flow in transgenic plants. *Molecular Ecology Notes* 1, 321–324.

James, V.A., Avart, C., Worland, B., Snape, J.W. and Vain, P. (2002) The relationship between homozygous and hemizygous transgene expression levels over generations in populations of transgenic rice. *Theoretical and Applied Genetics* 104, 553–561.

Jørgensen, R.B. and Andersen, B. (1994) Spontaneous hybridization between oilseed rape (*Brassica napus*) and weedy *B. campestris* (Brassicaceae): a risk of growing gentically modified oilseed rape. *American Journal of Botany* 81, 1620–1626.

Légère, A., Simard, M.-J., Thomas, A.G., Pageau, D., Lajeunesse, J., Warwick, S.I. and Derksen, D.A. (2001) Presence and persistence of volunteer canola in Canadian cropping systems. In: *Proceedings of the Brighton Crop Protection Conference – Weeds*. British Crop Protection Council, Farnham, UK, pp. 143–148.

Metz, P.L.J., Jacobsen, E., Nap, J.-P., Pereira, A. and Stiekema, W.J. (1997) The impact of biosafety of the phosphinothricin-tolerance transgene in inter-specific *B. rapa* × *B. napus* hybrids and their successive backcrosses. *Theoretical and Applied Genetics* 95, 442–450.

Mikkelsen, T.R., Andersen, B. and Jørgensen, R.B. (1996) The risk of crop transgene spread. *Nature* 380, 31.

Millwood, R.J., Halfhill, M.D., Harkins, D., Russotti, R. and Stewart, C.N. Jr (2003) Instrumentation and methodology for quantifying GFP fluorescence in intact plant organs. *BioTechniques* 34, 638–643.

Molinier, J., Himber, C. and Hahne, G. (2000) Use of green fluorescent protein for detection of transformed shoots and homozygous offspring. *Plant Cell Reports* 19, 219–223.

Niwa, Y., Hirano, T., Yoshimoto, K., Shimizu, M. and Kobayashi, H. (1999) Non-invasive quantitative detection and applications of non-toxic, S65T-type green fluorescent protein in living plants. *Plant Journal* 18, 455–463.

Raybould, A.F. and Gray, A.J. (1993) Genetically modified crops and hybridization with wild relatives: a UK perspective. *Journal of Applied Ecology* 30, 199–219.

Richards, H.A., Halfhill, M.D., Richards, R.J. and Stewart, C.N. Jr (2003) GFP fluorescence as an indicator of recombinant protein expression in transgenic plants. *Plant Cell Reports* 22, 117–121.

Rieger, M.A., Potter, T.D., Preston, C. and Powles, S.B. (2001) Hybridization between *Brassica napus* L. and *Raphanus raphanistrum* L. under agronomic field conditions. *Theoretical and Applied Genetics* 103, 555–560.

Scott, A., Woodfield, D. and White, D.W.R. (1998) Allelic composition and genetic background effects on transgene expression and inheritance in white clover. *Molecular Breeding* 4, 479–490.

Scott, S.E. and Wilkinson, M.J. (1998) Transgene risk is low. *Nature* 393, 320.

Simard, M.J., Légère, A., Pageau, D., Lajeunnesse, J. and Warwick, S.I. (2002) The frequency and persistence of canola (*Brassica napus*) volunteers in Québec cropping systems. *Weed Technology* 16, 433–439.

Simard, M.J., Légère, A., Séguin-Swartz, G., Nair, H. and Warwick, S.I. (2003) Fitness of double (imidazolinone + glufosinate) vs single herbicide resistant canola (*Brassica napus*). *Weed Science Society of America Abstracts* 43, 75–76.

Snow, A.A., Andersen, B. and Jørgensen, R.B. (1999) Costs of transgenic herbicide resistance introgressed from *Brassica napus* into weedy *B. rapa*. *Molecular Ecology* 8, 605–615.

Stewart, C.N. Jr (1996) Monitoring transgenic plants using *in vivo* markers. *Nature Biotechnology* 14, 682.

Stewart, C.N. Jr (1999) Insecticidal transgenes into nature: gene flow, ecological effects, relevancy, and monitoring. In: Lutman, P.W. (ed.) *Gene Flow and Agriculture: Relevance for Transgenic Crops.* Symposium Proceedings No. 72, British Crop Protection Council, Farnham, UK, pp. 179–190.

Stewart, C.N. Jr (2001) The utility of green fluorescent protein in transgenic plants. *Plant Cell Reports* 20, 376–382.

Stewart, C.N. Jr, Adang, M.J., All, J.N., Boerma, H.R., Cardineau, G., Tucker, D. and Parrott, W.A. (1996) Genetic transformation, recovery, and characterization of fertile soybean transgenic for a synthetic *Bacillus thuringiensis cryIAc* gene. *Plant Physiology* 112, 121–129.

Stewart, C.N. Jr, All, J.N., Raymer, P.L. and Ramachandran, S. (1997) Increased fitness of transgenic insecticidal rapeseed under insect selection pressure. *Molecular Ecology* 6, 773–779.

Stewart, C.N. Jr, Richards, H.A. and Halfhill, M.D. (2000) Transgenic plants and biosafety: science, misconceptions and public perceptions. *BioTechniques* 29, 832–843.

Stewart, C.N. Jr, Halfhill, M.D. and Millwood, R.J. (2002) GFP in transgenic plants: *Brassica* transformation. In: Hicks, B. (ed.) *Green Fluorescent Protein, Methods and Protocols, Methods in Molecular Biology.* Humana Press, Totowa, New Jersey, pp. 245–252.

Stewart, C.N. Jr, Halfhill, M.D., Raymer, P. and Warwick, S.I. (2003a) Transgene introgression and consequences in *Brassica*. In: *Introgression from Genetically Modified Plants into Wild Relatives and its Consequences. Programme and Abstract Book.* University of Amsterdam, The Netherlands, p. 59. Poster Abstract.

Stewart, C.N. Jr, Halfhill, M.D. and Warwick, S.I. (2003b) Transgene introgression from genetically modified crops to their wild relatives. *Nature Reviews Genetics* 4, 806–817.

Tomiuk, J., Hauser, T.P. and Jørgensen, R.B. (2000) A- or C-chromosomes, does it matter for the transfer of transgenes from *Brassica napus*? *Theoretical and Applied Genetics* 100, 750–754.

Vain, P., Worland, B., Kohli, A., Snape, J. and Christou, P. (2000) The green fluorescent protein (GFP) as a vital screenable marker in rice transformation. *Theoretical and Applied Genetics* 96, 164–169.

Warwick, S.I., Beckie, H. and Small, E. (1999) Transgenic crops: new weed problems for Canada? *Phytoprotection* 80, 71–84.

Warwick, S.I., Simard, M.J., Légère, A., Braun, L., Beckie, H.J., Mason, P., Zhu, B. and Stewart, C.N. Jr (2003) Hybridization between *Brassica napus* L. and its wild relatives: *B. rapa* L., *Raphanus raphanistrum* L. and *Sinapis arvensis* L., and *Erucastrum gallicum* (Willd.) O.E. Schulz. *Theoretical and Applied Genetics* 107, 528–539.

21 Insect-resistant Transgenic Plants and Their Environmental Impact

ROSIE S. HAILS[1] AND BEN RAYMOND[2]

[1]NERC Centre for Ecology and Hydrology, Mansfield Rd, Oxford OX1 3SR, UK, E-mail: RHA@ceh.ac.uk; [2]Imperial College, Silwood Park, Ascot, Berks SL5 7PY, UK

Abstract

The introduction of GM insect-resistant (GMIR) crops has both environmental and agronomic implications. Risk assessment has traditionally taken a 'three-tiered' approach prior to field-scale release, namely laboratory experiments, greenhouse experiments and small-scale field trials. In addition, with the commercialization of GM crops in other countries, evidence can also be obtained on certain aspects of their potential impact at the field scale. We review three areas of potential risk for GMIR plants: (i) potential impacts on non-target insects; (ii) the potential for enhanced ecological fitness as a consequence of introgression of the transgene into wild relatives; and (iii) changes in pesticide use as a result of GMIR crop use.

Introduction

Herbivorous natural enemies can reduce the survival and fecundity of plants in natural communities. In response to these selection pressures, plants have evolved numerous resistance mechanisms, from toxic secondary compounds to hairy leaves, to reduce the impact of their enemies. This, in turn, reduces the reproductive potential of the natural enemy and induces a second cycle of evolution to overcome these resistance mechanisms. This reciprocal evolutionary interaction is termed co-evolution (Janzen, 1980) and is thought to have generated much of the chemical and morphological diversity found in different plant species (Dixon, 2001). Conventional plant breeding follows a similar path, except that the initial phase, the selection for resistance to herbivores and pathogens of crops, is initiated by the plant breeder, drawing from the pool of genetic variation available in the crops' sexually compatible wild relatives. These measures are effective for a time, reducing the need for broad-spectrum insecticides in agroecosystems, but new cycles of co-evolution result in natural enemies circumventing these mechanisms.

The development of microbial insecticides, particularly the insecticidal proteins from the soil bacterium *Bacillus thuringiensis*, has allowed more targeted control against some lepidopteran and coleopteran pests. These proteins are produced by the bacterium as protoxins, which are then activated in the insect midgut. The activated toxins bind to midgut cells, causing severe damage, and eventually death. However, as with all insecticidal sprays, the influence of spray drift on non-target species remains, and stem borers and other cryptic pests are hard to reach. An alternative pest control strategy has been developed in the production of genetically modified insect-resistant (GMIR) plants. By expressing genes derived from *B. thuringiensis* (*Bt*), *Bt* plants express partially activated toxins, and this provides resistance to a range of lepidopteran and coleopteran pests. Other insecticidal constructs include lectins (e.g. the snowdrop lectin GNA) and proteinase inhibitors. In most cases, these transgenes are driven by constitutive promoters, so gene expression occurs in most tissues most of the time. *Bt* cotton and *Bt* maize were amongst the first major products of plant biotechnology, and have been widely adopted by farmers in the USA.

However, Europe has yet to adopt the commercial growing of GM crops. The reasons behind this are complex, but one contributory factor is likely to be an increasing awareness of the dependence of much wildlife on the management of European agricultural systems. Recent analysis of long-term data sets has illustrated the correlation between changes in agricultural practice over the latter half of the last century (including the increased use of fertilizers, pesticides and irrigation) and current biodiversity declines (e.g. Chamberlain *et al.*, 2000; Benton *et al.*, 2002; Robinson and Sutherland, 2002). These observations are underpinned by other studies investigating the mechanisms behind these correlations: for example, the switch from spring- to winter-sown crops has resulted in declines in weed biodiversity and potential food sources for birds at a crucial time of year (e.g. Rands, 1986; Hald, 1999; Brickle *et al.*, 2000). It is on to this stage of declining biodiversity, but increasing environmental awareness, that the future of GM crops in Europe is being considered.

There are concerns that such insect-resistant crops may have detrimental effects on non-target insects and, ultimately, biodiversity more generally. There is also the possibility that the introgression of genes from GMIR crops to their wild relatives may enhance survival or fecundity. Will this result in enhanced ecological fitness, and an increase in the abundance of some related plant species? It is also possible that the associated reductions in pesticides, and the incorporation of GMIR crops with a more integrated form of pest management, may provide considerable environmental benefits. We focus on these three areas, but emphasize that this is not a comprehensive review of all consequences of the commercialization of GMIR crops.

The Potential Consequences of GMIR Crops for Non-target Species

Identifying hazards: which non-target species are susceptible?

The identification of potential hazards, often conducted in the laboratory, is vital as it provides focus for further risk assessment, and many of the studies in the published literature fall into this category. Impacts may be direct, through toxicity of the *Bt* proteins to the natural enemy, or indirect, through reducing the quality of the prey. This would have implications for the compatibility of GM crops with biological control.

An example of subtle indirect effects is provided by a study on parasitoids of the potato aphid. Parasitoids (*Aphelinus abdominalis*) were exposed to aphids (*Myzus euphorbiae*) feeding on GNA-expressing or control potatoes. Both the weight and the sex ratio of the parasitoids were significantly reduced by the transgenic potato treatment (Couty *et al.*, 2001a). A second experiment, controlling for aphid size, caused the impacts on weight and sex ratio to disappear. However, a reduction in parasitoid fecundity remained, suggesting that there were also direct effects of GNA (Couty *et al.*, 2001b). There are a number of other studies in the literature that have demonstrated the potential for negative impacts on predators, parasitoids, non-target insects of economic or conservation value (e.g. Burgess *et al.*, 1996; Hilbeck *et al.*, 1998a, b; Losey *et al.*, 1999) and soil fauna (Saxena and Stotzky, 2000). Such studies provide the groundwork for a more comprehensive risk assessment.

From hazard identification to risk assessment

To interpret the significance of such findings, tritrophic interactions need to be embedded in a population and community context, in which behaviour, density dependence, prey choice and a host of other factors can play their role in determining the outcome (Schuler *et al.*, 1999a). This represents a considerable challenge to ecologists and entomologists, but there are a few notable examples of how this can be achieved.

Specialist parasitoids are thought to be one of the most vulnerable groups of natural enemies, as they have no alternative hosts to turn to when prey quality is low. The specialist parasitoid, *Cotesia plutella*, when forced to develop in a susceptible *Bt*-fed host (diamondback moth, *Plutella xylostella*), inevitably dies, as there is insufficient time to complete development. Schuler and colleagues investigated how this outcome might alter if the parasitoid was presented with the range of choices it could encounter in the field (Schuler *et al.*, 1999b). If parasitoids were given a choice between susceptible hosts feeding on *Bt* or wild-type oilseed rape, significantly more (> 80%) would choose those feeding on wild-type leaves, as they respond to the amount of damage caused by the larvae. Given the choice of

Bt-susceptible or -resistant hosts feeding on *Bt* oilseed rape, around 80% chose the latter (in which they can complete development) for precisely the same reason (Schuler *et al.*, 1999b). This illustrates that the behavioural choices made in the field will greatly limit the exposure of these parasitoids to this particular hazard. It also illustrates that the non-GM refugia required in GM crops provide ecological advantages beyond the control of the development of *Bt* resistance.

One of the best known studies in which hazard identification has been translated into a full risk assessment is the study of the monarch butterfly, a spectacular butterfly which migrates each year down the length of the USA. The larvae feed exclusively on milkweed plants, some of which are found in and around maize fields. The hazard identified was that if larvae were fed milkweed leaves dusted with pollen from *Bt* maize, then after 4 days, nearly 50% had died (Losey *et al.*, 1999). These results created considerable interest and the Environmental Protection Agency then put out a call for a full risk assessment, taking account of the degree to which the monarch larvae were likely to be exposed to the hazard. This led to a comprehensive risk assessment on this one species with respect to *Bt* maize. Estimation of the likely degree of exposure of monarch larvae to *Bt* pollen required the quantification of a number of different elements. These included: (i) the density of maize pollen found on milkweed plants, and the rate at which this declined with distance from the field; (ii) levels of expression of *Bt* in pollen; and (iii) degree of phenological overlap between larval feeding periods and pollen shed in maize. One of the key findings was that the expression of the *Bt* proteins in pollen varied considerably between maize varieties. Only one variety (*Bt*176) expressed levels that were sufficient and consistent enough to cause concern (Hellmich *et al.*, 2001; Stanley-Horn *et al.*, 2001). Maize pollen is relatively heavy – and densities drop to around 14 grains per leaf within 2 m of a maize field (Pleasants *et al.*, 2001). Finally, phenological overlap varied considerably across the USA, from 10 to 60% (Oberhauser *et al.*, 2001).

Sears and colleagues put all this information together in a quantitative risk assessment (Sears *et al.*, 2001). Overall risk (*R*) is the combined probability of exposure (P_e) and toxicity (P_t). This can be expressed as:

$$R = P_e \times P_t$$

If it is assumed that 5% of planted maize is *Bt*176, 15% is of other *Bt* varieties, and the remainder is non-GM maize, then an upper estimate for *R* is 1 in 250 larvae. *Bt*176 is now being phased out (throughout 2003), so a second estimate was made on the basis that *Bt*176 is no longer grown. If *Bt* maize is adopted until grown at its legal maximum of 80% (given current refuge requirements), and all future varieties have toxicities less than or equal to the current varieties (excluding 176), then *R* is estimated as 1 in 2000 larvae (Sears *et al.*, 2001). The consortium of ecologists concluded that the risk of *Bt* maize to monarch populations is currently, and should remain, low.

This case study illustrates how quantitative risk assessments can be made through adopting techniques developed in relation to pesticides and other environmental contaminants.

The Consequences of Gene Flow: Will Insect-resistant Plants Become Invasive?

Only a small proportion of introduced (non-native) species become invasive, but those few can be responsible for considerable economic loss and represent one of the greatest threats to biodiversity (Sala *et al.*, 2000). Crop plants hybridize with wild relatives, allowing gene flow from cultivars to wild populations (Scott and Wilkinson, 1998; Ellstrand *et al.*, 1999). Therefore, the possibility that the introgression of transgenes could alter the invasive potential of wild relatives of crops should be carefully considered. Here we discuss the possibility that the introgression of insect resistance genes could cause wild relatives to become invasive by considering two issues. First, we ask if any information can be obtained by considering the factors that cause non-native plants to become invasive? Is there any evidence that they escape regulation by invertebrate herbivores? Secondly, we ask what direct experimental evidence is available on changes in ecological fitness for GMIR crops or wild relatives.

Non-native invasive plants – do they escape regulation by insect herbivores?

The factors that cause some species to become invasive have been discussed extensively (for a recent review, see Kolar and Lodge, 2001). One hypothesis is that when removed from its native community of invertebrate herbivores and other natural enemies, a plant species benefits from enhanced survival and fecundity, resulting in an increased long-term population growth rate. This has been called herbivore escape, or the natural enemy release hypothesis (Elton, 1958; Williamson, 1996). The diversity of herbivore resistance and tolerance mechanisms in plants illustrates that herbivores have played an important role in evolutionary time (Dixon, 2001). However, as a broad generalization, invertebrate herbivores in their native range tend to have little impact on the abundance of their host plant and, consequently, the exclusion of insect herbivores may have negligible impact on plant densities (Crawley, 1983). While there is strong evidence for the role of vertebrate herbivores in the regulation of plant populations (Crawley, 1990), the evidence for insect herbivores is much weaker (Crawley, 1989). It is as if the co-evolutionary cycle is currently tipped in favour of the plant.

Nevertheless, insect herbivores can play a role in population regulation for plants with certain life history traits. Plants that rely on current seed production for recruitment to the next generation are vulnerable to herbivory that reduces fecundity (Maron and Vilà, 2001). Classic examples include monocarpic perennials that do not have long-lived seed banks, e.g. Platte thistle (*Cirsium canescens*). Recruitment to the next generation relies critically on seed production in the previous generation (Louda and Potvin, 1995). In contrast, long-lived plants and those that exhibit seed dormancy are much less likely to be limited by herbivores (Maron and Vilà, 2001).

However, if resources are limiting for some species, allocation of resources to defence or tolerance mechanisms leaves less for growth and competition. The consequences are that in an enemy-free environment, defended genotypes would be out-competed by undefended genotypes (e.g. Agrawal *et al.*, 1999; Redman *et al.*, 2001). Thus, the acquisition of 'cheap' herbivore resistance mechanisms could possibly replace more costly defence mechanisms, and allow reallocation of resources to competitive ability. Consequently, enemy release could be delayed rather than immediate, and difficult to detect in its early stages. Such scenarios, however, rely on the assumption that only one class of natural enemies (e.g. herbivores) are responsible for population regulation, whereas this may often be more complex, with more than one group of potential regulating factors operating at different densities, and on the assumption that expressing *Bt* toxins is metabolically quite 'cheap'.

The classical biological control of invasive non-native plants has often been used in support of the herbivore release hypothesis. The successful control of a non-native plant by an insect herbivore does illustrate the potential of herbivores to regulate plant populations (McEvoy *et al.*, 1991), but it does not follow that the same herbivore is responsible for regulating the plant in its native habitat (nor that resistance to that herbivore in its native habitat would cause herbivore release). Biological control involves a simplified version of an ecological community – in extreme cases simplified to just two players – the pest and the potential control agent (though in the competitive context of the invaded community). Other species, which may be responsible for controlling the herbivore itself, could be absent (Hosking, 1995).

Direct experimental evidence that native herbivores can significantly retard the invasion of non-native plants is scarce. Keane and Crawley (2002) emphasized that the role of herbivores should be considered in the context of plant competition. They suggested that a form of 'herbivore release' would occur if herbivore pressure was greater on natives than exotics, with the non-native species taking advantage of this situation. This interaction between the role of natural enemies and competition from other plant species was first exposed by Elton (1958) in his 'ecological resistance hypothesis'. Removal of all herbivores (or enemies) should then result in greater native competitive ability and a reduction in abundance of the invasive species, but few studies have followed this experimental model (Keane and Crawley, 2002).

Thus, to date, there is little direct experimental evidence that herbivore release has played a major role in the invasion of exotic plant species, but this may partly be because of a lack of appropriate experiments. Plants with specific life history traits (seed-limited recruitment, and lack of seed dormancy) can be limited by herbivores, and therefore have the potential to be released from herbivores in non-native habitats (or through the introduction of insect toxin genes). However, over evolutionary time, insect herbivores have had major fitness impacts on plants, resulting in the array of anti-herbivore defences we now observe.

Is there any evidence for changes in ecological fitness for GMIR crops or wild relatives?

Fitness may be defined as the proportional contribution of an individual to the next generation (Begon *et al.*, 1996). To compare invasiveness between two plant varieties, a direct measure is to compare the relative finite rates of increase, λ, defined as:

$$\lambda = N_t + 1/N_t$$

Thus, the finite rate of increase is a multiplication factor between successive generations. There are relatively few experiments addressing this question directly, one exception being a set of experiments conducted in the early 1990s, in which four crops (oilseed rape, sugarbeet, maize and potatoes) were introduced into a variety of habitats across the UK. For oilseed rape, sugarbeet and maize, the GM varieties were herbicide tolerant, but the GM potatoes expressed either *Bt* or pea lectin. In no case did the GM plants exhibit a greater finite rate of increase compared with the conventional plants, and transgenic plants never persisted significantly longer than conventional plants (Crawley *et al.*, 2001). These results illustrate that these crops are unlikely to persist for long outside of cultivation, and that the herbivore resistance genes are unlikely to enhance their survival.

Recent work on wild sunflowers has illustrated that the presence of a *Bt* transgene, designed to confer tolerance to insect herbivores, results in significant increases in fecundity (Snow *et al.*, 2003). Fecundity is sometimes used as a surrogate for fitness in the ecological literature and, all other things being equal, plants producing more seeds will make a greater proportional increase to the next generation. However, in the absence of herbivores, there may be fitness costs to carrying the transgene, just as there are costs to non-transgenic herbivore resistance genes. Such costs have been found to be more prevalent than previously supposed (Strauss *et al.*, 2002). The difference between costs and benefits will determine the overall fitness advantage, and this may depend upon the frequency and distribution of herbivore attack in the natural habitat. This could be resolved by comparative studies of *Bt* and wild-type wild sunflowers in their natural habitats over a number of seasons.

Will enhanced fitness lead to changes in population abundance? This will depend upon the rate at which herbivores may evolve to circumvent any novel resistance mechanisms, and the degree to which local population dynamics are seed limited. Rates of co-evolution may be quite rapid in response to *Bt* genes (Shelton *et al.*, 1993; Ferré and van Rie, 2002), in which case any increases in fitness may be transient. If wild sunflower populations are seed limited and herbivore populations remain susceptible, then enhanced fitness may well translate into an increase in abundance. A recent review of seed sowing experiments aimed to unravel the extent to which natural populations are seed limited (Turnbull *et al.*, 2000). Approximately 50% of seed augmentation experiments showed some evidence of seed limitation; this tends to occur in early successional habitats and with early successional species.

In many instances, little is known about the factors that regulate and limit populations of wild relatives, the potential recipients of transgenes. In particular, lack of knowledge of the relative regulatory roles of pathogens and herbivores, likely targets of transgenes, makes interpretation of their impact on vital rates, such as fecundity, difficult.

The Consequences of Adoption of GMIR Crops on Agricultural Practice

The adoption of GMIR crops can have important environmental consequences if they can affect synthetic insecticide usage and the abundance of non-target fauna in agroecosystems. A simple relationship between farm-level biodiversity and pesticide use is not easy to demonstrate, but broad-spectrum insecticides can cause considerable direct mortality to beneficial and non-target insects (Robinson and Sutherland, 2002). High-intensity farming and high insecticide use is usually associated with lower abundance and diversity of non-target fauna (e.g. Kenmore, 1991; Brown and Schmitt, 2001; Mäder et al., 2002; Wilby and Thomas, 2002). Many independent sources have shown that in general, there is less pesticide use in GMIR relative to conventional crops, and their adoption can have benefits for the abundance of non-targets. This overall picture is not clear-cut in every instance and it is best to examine the evidence in turn for each of the three commercial GMIR crop systems.

Cotton

Cotton, Gossypium hirsutum, engineered to express CryIAc B. thuringiensis toxins (hereafter Bt cotton) is now widely planted in Australia, China and the USA. In Australia, Bt cotton has reduced the need for insecticide sprays by 40–60% in the first few years of use (Pyke and Fitt, 1998). Research has not found any negative effects on beneficial or non-target insect species (Fitt et al., 1994; Fitt, 2000). While Helicoverpa armigera and H. punctigera, the main cotton pests in Australia, are much less susceptible to CryIAc than many other Lepidoptera and require some additional control with pesticides (Fitt, 2000), Australian farmers adopt Bt cotton partly to guarantee stable yields and partly because of concerns over pesticide contamination from conventional crops (Shelton et al., 2002). In China, the reduction in foliar pesticide usage associated with Bt cotton is 60–80% (Shelton et al., 2002).

In the USA, the primary cotton pests are the cotton bollworm, Helicoverpa zea, and the tobacco budworm, Heliothis virescens. The pink bollworm, Pectinophora gossypiella, is only a major pest of cotton in the western part of the USA. Bt cotton effectively controls all three species, and most secondary lepidopteran pests. Data compiled by the National Agricultural Statistics

Service for the US Department of Agriculture show that overall pesticide use in cotton has declined since 1995 (before *Bt* cotton was available) and there has been a marked decrease in the proportional application of the more toxic insecticides and an increasing reliance on beneficial insects and bio-insecticides since 1998. However, other factors, most notably the boll weevil eradication programme, have contributed to a reduction in pesticide use in US cotton. Nevertheless, in areas where damage from boll weevil is low, *Bt* cotton can effectively reduce the need for relatively toxic broad-spectrum insecticides (Environmental Protection Agency, 2001). States that historically have had high damage levels from Lepidoptera have been quick to adopt large acreages of *Bt* technology. Using data compiled by Williams (2002) for the National Cotton Council, it can be shown that the reduction in insecticides sprayed for Lepidoptera in each state since 1995 is significantly related to the proportion of *Bt* cotton planted in that region (Fig. 21.1; for the decrease in pesticide usage between 1995 and 1998, $F_{1,25} = 18.1$, $P < 0.001$; for the decrease between 1995 and 2001, $F_{1,23} = 6.65$, $P < 0.05$).

The decrease in total insecticide usage in one of the years examined (1998) was also significantly related to proportional adoption of *Bt* cotton ($F_{1,25} = 9.40$, $P < 0.01$), although not in 2001. This could be attributed to the fact that arthropod pest pressure was higher in 1998, when they caused an estimated 8% reduction in yield, than in 2001 when the arthropod yield reduction was 4.5% (Williams, 2002).

Survey data on pesticide use in the USA have been complemented by a number of field studies investigating the impact of *Bt* cotton on beneficial and non-target fauna. In a large-scale study in four cotton-growing regions, Head *et al.* (2001) found that increased numbers of beneficial arthropods in *Bt* relative to conventional cotton were found when insecticide applications

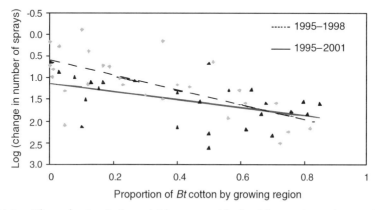

Fig. 21.1. The reduction in insecticides sprayed for Lepidoptera in each state since 1995 (on a log scale) against the proportion of *Bt* cotton planted in that region. For the decrease in pesticide usage between 1995 and 1998, $F_{1,25} = 18.1$, $P < 0.001$; for the decrease between 1995 and 2001, $F_{1,23} = 6.65$, $P < 0.05$. Data compiled by Williams (2002) for the National Cotton Council (http://www.msstate.edu/Entomology/Cotton.html).

needed to be made in conventional crops as a result of high pest numbers. Other studies have found either no difference between conventional and *Bt* cotton in numbers of non-targets (Wilson *et al.*, 1992; Flint *et al.*, 1995) or occasionally higher number of beneficials in *Bt* crops (van Tol and Lentz, 1998). Bird counts in cotton fields have increased markedly (by 15%) since 1995 and, in a state by state analysis, the magnitude of the reduction in pesticide use was significantly correlated with the increase in bird numbers (Environmental Protection Agency, 2001).

Maize

Field maize, *Zea mays*, has been engineered to express *B. thuringiensis* toxins (usually Cry1Ab or Cry1Ac) active against Lepidoptera. The main targets for these toxins are the European corn borer, *Ostrinia nubilalis*, and the Southwestern corn borer, *Diatraea grandiosella*. The larvae of the corn borers have a cryptic habit and are only susceptible to sprayed insecticides for a short period. Since only 5–8% of conventional field maize acreage was typically sprayed for corn borers, large changes in insecticide use were not expected to result from the use of *Bt* maize (Environmental Protection Agency, 2001). Nevertheless, the US Environmental Protection Agency estimates that an annual reduction in 1.6 Mha of insecticide treatments has occurred since *Bt* maize was introduced in 1995 (Environmental Protection Agency, 2001). On the other hand, Benbrook (2001) estimated that insecticide use targeting corn borers rose from 4% of field maize in 1995 to 5% in 2000. Carpenter and Gianessi (2001) assert that there was a 6% reduction in the use of all pesticides recommended for the control of corn borers between 1995 and 1999 and estimated that perhaps 1.5% of this reduction is attributable to *Bt* maize. However, insecticides in maize are not targeted against single pests, and the damage caused by corn borers fluctuates considerably year by year, making calculating the changes in practice brought about by *Bt* maize very difficult.

Field maize engineered with the Cry1F toxin protects plants from corn borers as well as root-feeding *Agrotis* spp. (black cutworm) and *Spodoptera* spp. (armyworm), although not corn rootworm. In areas where corn rootworm can be controlled by other means (e.g. rotation), Cry1F maize is expected to lead to more clear-cut reductions in insecticide use. However, these plants have not been available long enough for data to have been collected.

The Environmental Protection Agency also predicts that reduced levels of insecticides will be required for cultivating *Bt* sweetcorn. Average insecticide spray rates for sweetcorn are 5.5 times per year, and computer-simulated projections estimate that this could be reduced by 4.8 times with the use of *Bt* sweetcorn (Environmental Protection Agency, 2001). Field studies on the effect of *Bt* sweetcorn on beneficial insects revealed no differences in abundance of natural enemies between conventional and *Bt* crops, although there was some indication that populations of the coccinellid predator, *Coleomegila maculata*, were higher in *Bt* sweetcorn (Wold *et al.*, 2001).

Potatoes

The adoption of GMIR potatoes has been slow in the USA; only 50,000 ha or 4% of the national crop was planted with GM potatoes in 1999 (Environmental Protection Agency, 2001). Plants engineered with Cry3A toxin (derived from *B. thuringiensis*) for controlling the Colorado potato beetle, *Leptinotarsa decemlineata*, have been commercially available as 'NewLeaf' potatoes since 1996, and plants engineered with both *Cry3A* and resistance to potato leaf roll virus were licensed in 1998 ('NewLeafPlus' potatoes). Field studies in Ohio and Wisconsin have shown that use of *Bt* potatoes can lead to consistent increases in the within-field abundance of generalist predators and parasitoids relative to conventionally farmed potatoes (Hoy *et al.*, 1998). Systemic insecticides that are applied to the soil at planting also do not negatively affect the abundance of natural enemies (Reed *et al.*, 2001). In the absence of broad-spectrum pesticides, natural enemies can reduce aphid populations to below economic thresholds (Hoy *et al.*, 1998), and natural control of aphids on *Bt* crops can out-perform broad-spectrum pyrethroids (Reed *et al.*, 2001). Spraying for aphid outbreaks (which cause damage primarily as vectors of viruses) can lead to outbreaks of red spider mite in some growing areas, which can require additional chemical control (Hoy *et al.*, 1998). It is likely that the use of *Bt* potatoes, especially those engineered for resistance to leaf roll virus, would lead to the conservation of aphid and mite natural enemies and reduce the need for this control. Farmer survey data indicate that in the eastern USA, where the Colorado potato beetle is the major pest and aphids are a minor pest, 1.35 fewer insecticide applications are required per year in NewLeaf potatoes (Carpenter and Gianessi, 2001). There are several reasons for their slow adoption, including the recent emergence of systemic long-acting insecticides (e.g. imidacloprid) that are very effective against Colorado beetle, and the reluctance of some buyers currently to use GM potatoes (Carpenter and Gianessi, 2001; Environmental Protection Agency, 2001).

With the exception of *Bt* field maize, all survey data have supported the expectation that GMIR crops will lead to reduced applications of insecticides. GMIR crops have the potential to cause increased demand for pesticides to control secondary pests (Wolfenbarger and Phifer, 2001), and there is some evidence that the need to control secondary pests has moderated the impact of *Bt* cotton on pesticide use (Environmental Protection Agency, 2001). However, in many cases, reducing the need for broad-spectrum insecticides through GMIR crop adoption can lead to increased opportunities for integrated pest management and the conservation of natural enemies. For example, the use of pheromones for mating disruption in the pink bollworm, *P. gossypiella*, is only effective when this pest is at low density (Roush, 1997). In some instances, such as greenhouse horticulture in

Europe, an increasing reliance on natural enemies snowballs and produces a strong incentive for growers to avoid broad-spectrum insecticides altogether (van Lenteren, 2000).

Thus far, this chapter has ignored any potential impact of GMIR crops on other environmental factors such as water quality and fossil fuel consumption. Given the reduced pesticide input on these crops, GMIR crops are likely to have reduced environmental impact relative to conventional crops. Questionnaire data show that farmers are keen to plant *Bt* crops in their most distant or most difficult to spray fields; thus, benefits in fossil fuel use are also not just a simple function of how many fewer hectares need to be sprayed (ReJesus *et al.*, 1997).

Conclusions

Three areas of environmental impact for GMIR crops have been briefly reviewed: (i) potential impacts on non-target insects; (ii) the potential for insect toxin transgenes to cause natural enemy release in wild relatives; and (iii) the potential impact of GMIR crops on agricultural practice. Potential impacts of a GM crop on non-target species should always be considered in comparison with the crop it is intended to replace. However, given the strong evidence for biodiversity decline over the last few decades, it may be appropriate to set higher biodiversity standards for future crops and their associated practices (Hails, 2002). The issue is therefore one of comparative biodiversity arising from GMIR crop management compared with conventional crop management. However, most studies investigating the impact of GMIR crops on non-target species have focused on either identifying hazards (as in the early monarch butterfly work by Losey *et al.*, 1999), or on estimating potential exposure of the non-target species to the hazard (as in the resolution of the monarch butterfly story by Sears *et al.*, 2001). There are few studies comparing biodiversity at the field scale as a consequence of GMIR crop management, especially in Europe where the commercial growing of these crops is yet to be adopted (with the exception of *Bt* maize in certain regions of Spain). The data to date suggest that exposure of non-target arthropods is low – quite probably lower than to more conventional synthetic pesticides. Nevertheless, field-scale effects on farmland biodiversity remain an issue to be addressed – either by experiments akin to the farm-scale evaluations of herbicide-tolerant crops conducted in the UK (Firbank *et al.*, 1999), or in the early stages of commercialization.

The evidence that *Bt* transgenes may cause natural enemy release is as yet incomplete. Looking at the broad patterns illustrated by ecological studies, there is little experimental evidence that invertebrate herbivores have major impacts on the abundance of their host plants, and that therefore their exclusion is unlikely to lead to ecological release. However, there are plants with specific life history traits for which this is not the case (those with seed-limited recruitment and lacking in seed dormancy), and so the case by case scrutiny of the wild relatives of transgenic crops should continue.

A related consideration is the rate at which populations of target insects can evolve resistance to *Bt* crops, which indicates that co-evolution may also occur in natural populations. The rate at which co-evolution will occur is highly dependent upon the initial frequency of resistance alleles in field populations. In the few cases where this has been estimated in pest populations, it has been found to be high (Gould *et al.*, 1997), possibly because lepidopteran populations have been exposed to *B. thuringiensis* as part of their natural ecology. Consequently, any enhanced fitness due to possession of a *Bt* gene may well be transient.

Finally, the greatest impact associated with *Bt* crops may be the change in associated management practices, particularly the reduction in synthetic pesticide use. Given the dynamic relationship between agriculture and biodiversity, and the many different ways in which a crop and its associated management practices can impact upon the environment, the challenge will be how to integrate these different aspects of risk assessment to compare the environmental footprint of different agricultural options (Hails, 2002).

References

Agrawal, A.A., Strauss, S.Y. and Stout, M.J. (1999) Costs of induced responses and tolerance to herbivory in male and female fitness components of wild radish. *Evolution* 53, 1093–1104.

Begon, M., Harper, J.L. and Townsend, C.R. (1996) *Ecology.* Blackwell Science Limited, Oxford.

Benbrook, C. (2001) Do GM crops mean less pesticide use? *Pesticide Outlook* October, 204–207.

Benton, T.G., Bryant, D.M., Cole, L. and Crick, H.Q. (2002) Linking agricultural practices to insect and bird populations: a historical study over three decades. *Journal of Applied Ecology* 39, 673–687.

Brickle, N.W., Harper, D.G.C., Aebischer, N.J. and Cockayne, S.H. (2000) Effects of agricultural intensification on the breeding success of corn buntings, *Miliaria calandra. Journal of Applied Ecology* 37, 742–755.

Brown, M.W. and Schmitt, J.J. (2001) Seasonal and diurnal dynamics of beneficial insect populations in apple orchards under different management intensity. *Environmental Entomology* 30, 415–424.

Burgess, E.P.J., Malone, L.A. and Christeller, J.T. (1996) Effects of two proteinase inhibitors on the digestive enzymes and survival of honey bees (*Apis mellifera*). *Journal of Insect Physiology* 42, 823–828.

Carpenter, J.E. and Gianessi, L.P. (2001) Agricultural biotechnology: updated benefit estimates. National Center for Food and Agricultural Policy. Available at: http://www.ncfap.org/reports/biotech/updatedbenefits.pdf

Chamberlain, D.E., Fuller, R.J., Bunce, J.C., Duckworth, J.C. and Shrubb, M. (2000) Changes in the abundance of farmland birds in relation to the timing of agricultural intensification in England and Wales. *Journal of Applied Ecology* 37, 771–788.

Couty, A., Clark, S.J. and Poppy, G.M. (2001a) Are fecundity and longevity of female *Aphelinus abdominalis* affected by development in GNA-dosed *Macrosiphum euphorbiae*? *Physiological Entomology* 26, 287–293.

Couty, A., de la Vina, G., Clark, S.J., Kaiser, L., Pham-Delegue, M.-H. and Poppy, G.M. (2001b) Direct and indirect sublethal effects of *Galanthus nivalis* agglutinin (GNA) on the development of a potato-aphid parasitoid *Aphelinus abdominalis* (Hymenoptera: Aphelinidae). *Journal of Insect Physiology* 47, 553–561.

Crawley, M.J. (1983) *Herbivory: the Dynamics of Animal–Plant Interactions*. Blackwell Scientific Publications, Oxford.

Crawley, M.J. (1989) The relative importance of vertebrate and invertebrate herbivores in plant population dynamics. In: Bernays, E.A. (ed.) *Insect–Plant Interactions*, Vol. 1. CRC Press, Boca Raton, Florida, pp. 45–71.

Crawley, M.J. (1990) Rabbit grazing, plant competition and seedling recruitment in acid grassland. *Journal of Applied Ecology* 27, 803–820.

Crawley, M.J., Brown, S.L., Hails, R.S., Kohn, D. and Rees, M. (2001) The performance of transgenic crops in natural habitats: a 10-year perspective. *Nature* 409, 682–683.

Dixon, R.A. (2001) Natural products and plant disease resistance. *Nature* 411, 843–847.

Ellstrand, N.C., Prentice, H.C. and Hancock, J.F. (1999) Gene flow and introgression from domesticated plants into their wild relatives. *Annual Review of Ecology and Systematics* 30, 539–563.

Elton, C.S. (1958) *The Ecology of Invasions by Animals and Plants*. Methuen, London.

Environmental Protection Agency (2001) Bt-plant incorporated protectants. 15 October; Biopesticides Registration Action Document. Part E, Benefits Assessment. US Environmental Protection Agency, Washington, DC.

Ferré, J. and van Rie, J. (2002) Biochemistry and genetics of insect resistance to *Bacillus thuringiensis*. *Annual Review of Entomology* 47, 501–533.

Firbank, L.G., Dewar, A.M., Hill, M.O., May, M.J., Perry, J.N., Rothery, P., Squire, G.R. and Woiwod, I.P. (1999) Farm-scale evaluation of GM crops explained. *Nature* 399, 727–728.

Fitt, G.P. (2000) An Australian approach to IPM in cotton: integrating new technologies to minimise insecticide dependence. *Crop Protection* 19, 793–800.

Fitt, G.P., Mares, C.L. and Llewellyn, D.J. (1994) Field evaluation and potential ecological impact of transgenic cottons (*Gossypium hirsutum*) in Australia. *Biocontrol Science and Technology* 4, 535–548.

Flint, H.M., Henneberry, T.J., Wilson, F.D., Holguin, E., Parks, N. and Buehler, R.E. (1995) The effects of transgenic cotton, *Gossypium hirsutum* L., containing *Bacillus thuringiensis* toxin genes for the control of the pink bollworm, *Pectinophora gossypiella* (Saunders) and other arthropods. *Southwestern Entomologist* 20, 281–292.

Gould, F., Anderson, A., Jones, A., Sumerford, D., Heckel, D.G., Lopez, J., Micinski, S., Leonard, R. and Laster, M. (1997) Initial frequency of alleles for resistance to *Bacillus thuringiensis* toxins in field populations of *Heliothis virescens*. *Proceedings of the National Academy of Sciences USA* 94, 3519–3523.

Hails, R. (2002) Assessing the risks associated with new agricultural practices. *Nature* 418, 2–5.

Hald, A.B. (1999) The impact of changing the season in which cereals are sown on the diversity of the weed flora in rotational fields in Denmark. *Journal of Applied Ecology* 36, 24–32.

Head, G., Freeman, B., Moar, W., Ruberso, J. and Turnipseed, S. (2001) Natural enemy abundance in commercial Bollguard® and conventional cotton fields. *Proceedings of the Beltwide Cotton Conference* 2, 796–798.

Hellmich, R.L., Siegfried, B.D., Sears, M.K., Stanley-Horn, D.E., Daniels, M.J., Mattial, H.R., Soencer, T., Bidne, K.G. and Lewis, L.C. (2001) Monarch larvae sensitivity to *Bacillus thuringiensis* purified proteins and pollen. *Proceedings of the National Academy of Sciences USA* 98, 11925–11930.

Hilbeck, A., Baumgartner, M., Fried, P.M. and Bigler, F. (1998a) Effects of transgenic *Bacillus thuringiensis* corn fed prey on the mortality and development time of immature *Chrysoperia carnea* (Neuroptera, Chrysopidae). *Environmental Entomology* 17, 480–487.

Hilbeck, A., Moar, W.J., Pusztai-Carey, M., Filippini, A. and Bigler, F. (1998b) Toxicity of *Bacillus thuringiensis* CryIAb toxin to the predator *Chrysoperia carnea* (Neuroptera, Chrysopidae). *Environmental Entomology* 27, 1255–1263.

Hosking, J.R. (1995) The impact of seed- and pod-feeding insects on *Cytisus scoparius*. In: Delfosse, E.S. and Scott, R.R. (eds) *Proceedings of the Eighth International Symposium on Biological Control of Weeds*. CSIRO Publishing, Melbourne, pp. 45–51.

Hoy, C.W., Feldman, F., Gould, F., Kennedy, G.G., Reed, G. and Wyman, J.A. (1998) Naturally occurring biological controls in genetically engineered crops. In: Barbosa, P. (ed.) *Conservation Biological Control*. Academic Press, London, pp. 185–205.

Janzen, D.H. (1980) When is it coevolution? *Evolution* 34, 611–612.

Keane, R.M. and Crawley, M.J. (2002) Exotic plant invasions and the enemy release hypothesis. *Trends in Ecology and Evolution* 17, 164–170.

Kenmore, P. (1991) How rice farmers clean up the environment, conserve biodiversity, raise more food, make higher profits: Indonesian IPM – a model for Asia. FAO Inter-country programmes for IPV in rice in South and South East Asia. Manila, Philippines.

Kolar, C.S. and Lodge, D.M. (2001) Progress in invasion biology: predicting invaders. *Trends in Ecology and Evolution* 16, 199–204.

Losey, J.E., Rayor, L.S. and Carter, M.E. (1999) Transgenic pollen harms monarch larvae. *Nature* 399, 214.

Louda, S.M. and Potvin, S.A. (1995) Effect of inflorescence feeding insects on the demography and lifetime fitness of a native plant. *Ecology* 76, 229–245.

Mäder, P., Fliessbach, A., Dubois, D., Gunst, L., Fried, P. and Niggli, U. (2002) Soil fertility and biodiversity in organic farming. *Science* 296, 1694–1697.

Maron, J.L. and Vilà, M. (2001) When do herbivores affect plant invasion? Evidence for the natural enemies and biotic resistance hypotheses. *Oikos* 95, 361–373.

McEvoy, P., Cox, C. and Coombs, E. (1991) Successful biological control of ragwort, *Senecio jacobaea*, by introduced insects in Oregon. *Ecological Applications* 1, 430–442.

Oberhauser, K.S., Prysby, M.D., Mattila, H.R., Stanley-Horn, D.E., Sears, M.K., Dively, G., Olson, E., Pleasants, J.M., Lam, W.K.F. and Hellmich, R.L. (2001) Temporal and spatial overlap between monarch larvae and corn pollen. *Proceedings of the National Academy of Sciences USA* 98, 11913–11918.

Pleasants, J.M., Hellmich, R.L., Dively, G.P., Sears, M.K., Stanley-Horn, D.E., Mattila, H.R., Foster, J.E., Clark, T.L. and Jones, G.D. (2001) Corn pollen deposition on milkweeds in and near cornfields. *Proceedings of the National Academy of Sciences USA* 98, 11919–11924.

Pyke, B. and Fitt, G.P. (1998) Field performance of INGUARD cotton – the first two years. In: Zalucki, M.P., Drew, R.A.I. and White, G.G. (eds) *Pest Management – Future Challenges*. University of Queensland Press, Brisbane, Australia, pp. 230–238.

Rands, M.R.W. (1986) The survival of gamebird (Galliformes) chicks in relation to pesticide use in cereal fields. *Ibis* 128, 57–64.

Redman, A.M., Cipollini, D.F. and Schultz, J.C. (2001) Fitness costs of jasmonic acid-induced defense in tomato, *Lycopersicon esculentum*. *Oecologia* 126, 380–385.

Reed, G.L., Jensen, A.S., Riebe, J., Head, G. and Duan, J.J. (2001) Transgenic potato and conventional insecticides for Colorado potato beetle management: comparative efficacy and non-target impacts. *Entomologia Experimentalis et Applicata* 100, 89–100.

ReJesus, R.M., Greene, J.K., Hammig, M.D. and Curtis, C.E. (1997) Farmer's expectations in the production of transgenic *Bt* cotton: results from a preliminary survey in South Carolina. *Proceeding of the Beltwide Cotton Conferences* 1, 253–256.

Robinson, R.A. and Sutherland, W.J. (2002) Post-war changes in arable farming and biodiversity in Great Britain. *Journal of Applied Ecology* 39, 157–176.

Roush, R. (1997) Managing resistance to transgenic crops. In: Carozzi, N. and Koziel, M. (eds) *Advances in Insect Control: the Role of Transgenic Plants*. Taylor and Francis, London, pp. 271–292.

Sala, O.E., Chapin, F.S., Armesto, J.J., Berlow, E., Bloomfield, J., Dirzo, R., Huber-Sanwald, E., Huenneke, L.F., Jackson, R.B., Kinzig, A., Leemans, R., Lodge, D.M., Mooney, H.A., Oesterheld, M., Poff, N.L., Sykes, M.T., Walker, B.H., Walker, M. and Wall, D.H. (2000) Global biodiversity scenarios for the year 2100. *Science* 287, 1770–1774.

Saxena, D. and Stotzky, G. (2000) Insecticidal toxin from *Bacillus thuringiensis* is released from roots of transgenic *Bt* corn *in vitro* and *in situ*. *FEMS Microbiology Ecology* 33, 35–39.

Schuler, T.H., Poppy, G.M., Kerry, B.R. and Denholm, I. (1999a) Potential side effects of insect-resistant transgenic plants on arthropod natural enemies. *Trends in Biotechnology* 17, 210–216.

Schuler, T.H., Potting, R.P.J., Denholm, I. and Poppy, G.M. (1999b) Parasitoid behaviour and *Bt* plants. *Nature* 400, 825–826.

Scott, S.E. and Wilkinson, M.J. (1998) Transgene risk is low. *Nature* 393, 320.

Sears, M.K., Hellmich, R.L., Stanley-Horn, D.E., Oberhauser, K.S., Pleasants, J.M., Mattila, H.R., Siegfried, B.D. and Dively, G.P. (2001) Impact of *Bt* corn pollen on monarch butterfly populations: a risk assessment. *Proceedings of the National Academy of Sciences USA* 98, 11937–11942.

Shelton, A.M., Robertson, J.L., Tang, J.D., Perez, C., Eigenbrode, S.D., Preisler, H.K., Wilsey, W.T. and Cooley, R.J. (1993) Resistance of Diamondback moth (Lepidoptera: Plutellidae) to *Bacillus thuringiensis* in the field. *Journal of Economic Entomology* 86, 697–705.

Shelton, A.M., Zhao, J.-Z. and Roush, R.T. (2002) Economic, ecological, food safety, and social consequences of the deployment of *Bt* transgenic plants. *Annual Review of Entomology* 47, 845–881.

Snow, A.A., Pilson, D., Rieseberg, L.H., Paulsen, M.J., Pleskac, N., Reagon, M.R., Wolf, D.E. and Selbo, S.M. (2003) A *Bt* transgene reduces herbivory and enhances fecundity in wild sunflowers. *Ecological Applications* 13, 279–286.

Stanley-Horn, D.E., Dively, G.P., Hellmich, R.L., Mattila, H.R., Sears, M.K., Rose, R., Jesse, L.C.H., Losey, J.E., Obrycki, J.J. and Lewis, L. (2001) Assessing the impact of Cry1Ab-expressing corn pollen on monarch butterfly larvae in field studies. *Proceedings of the National Academy of Sciences USA* 98, 11931–11936.

Strauss, S.Y., Rudgers, J.A., Lau, J.A. and Irwin, R.E. (2002) Direct and ecological costs of resistance to herbivory. *Trends in Ecology and Evolution* 17, 278–285.

Turnbull, L.A., Crawley, M.J. and Rees, M. (2000) Are plant populations seed limited? A review of seed sowing experiments. *Oikos* 88, 225–238.

van Lenteren, J. (2000) A greenhouse without pesticides: fact or fantasy? *Crop Protection* 19, 375–384.

van Tol, N.B. and Lentz, G.L. (1998) Influence of *Bt* cotton on beneficial arthropod populations. *Proceedings of the Beltwide Cotton Conferences* 2, 1052–1054.

Wilby, A. and Thomas, M.B. (2002) Natural enemy diversity and pest control: patterns of pest emergence with agricultural intensification. *Ecology Letters* 5, 353–360.

Williams, M.R. (2002) *Cotton Insect Losses*. Mississipi University and the Cotton Foundation. Available at: http://www.msstate.edu/Entomology/CTNLOSS

Williamson, M. (1996) *Biological Invasions*. Chapman & Hall, London.

Wilson, F.D., Flint, H.M., Deaton, W.R., Fischhoff, D.A., Perlak, F.J., Armstrong, T.A., Fuchs, R.L., Berberich, S.A., Parks, N.J. and Stapp, B.N. (1992) Resistance of cotton lines containing a *Bacillus thuringiensis* toxin to pink bollworm (Lepidoptera: Gelechiidae) and other insects. *Journal of Economic Entomology* 85, 1516–1521.

Wold, S.J., Burkness, E.C., Hutchison, W.D. and Venetta, R.C. (2001) In-field monitoring of beneficial insect populations in transgenic corn expressing a *Bacillus thuringiensis* toxin. *Journal of Entomological Science* 36, 177–187.

Wolfenbarger, L.L. and Phifer, P.R. (2001) Response to Carpenter. *Science* 292, 638–639.

22 Risk Assessment of Genetically Modified Undomesticated Plants

ANDERS WENNSTRÖM

Department of Ecology and Environmental Science, Umeå University, SE-901 87 Umeå, Sweden, E-mail: anders.wennstrom@eg.umu.se

Abstract

There is a consensus that we need a comprehensive risk assessment of the ecological impacts of genetically modified (GM) plants. However, to date, ecological risk analysis is still in a developmental phase and we do not have the data available to allow conclusions to be drawn. The reason for this is that so far we have only studied parts of the process. Furthermore, we have to be aware that the bias towards agricultural systems will limit our ability to understand the impacts when GM undomesticated plants, such as forest trees, will be used in natural environments.

The only way to improve risk analysis is to gain as much detailed knowledge as possible of the various ecological processes and functional groups of organisms which might be influenced by the introduction of a GM plant. Here, I discuss that we need to focus more on interactions between GM plants, non-modified related species and natural enemies.

Introduction

The techniques for making GM plants are now well developed, and a wide range of transgenics now exist and are used commercially in agriculture. Modifications include increased resistance against natural enemies and specific herbicides, and growth and flowering modifications. For some crop plants, transgenic varieties dominate the market. For example, more than half (51%) of the 72 Mha of soybean grown worldwide were GM in 2002, and in the People's Republic of China more than half (51%) of the national cotton area of 4.1 Mha was planted with *Bt* cotton (James, 2002).

Before GM plants can approach commercialization and widespread deployment in Europe, it is of utmost importance to obtain a thorough understanding of the impact of their use. Risk analysis is the general process of identifying the elements that pose a risk, and analysing their likelihood, significance and management. It is also necessary to communicate any conclusions and to enter into dialogue with stakeholders and the broader community (Fig. 22.1). A risk analysis has to start as early as possible; ideally before the development of GM plants. The overall goal should be to determine if introduced traits are likely to result in any environmental, ecological or economic/human damage and how possible damage can be avoided.

Risk Analysis of Today

In contrast to the technology for making modified plants, ecological risk analysis is still in a developmental phase in Europe (Raybould and Gray, 1994; Hails, 2000). For example, we still need to establish baselines and we also need to some extent to identify what to study and how to study it most effectively.

The results from major research programmes in Europe were presented at two recent European meetings ('GMOs – Ecological Dimensions', Reading, UK, 2002; and 'Introgression from GMPs into wild relatives and its consequences', Amsterdam, The Netherlands, 2003, this volume) and clearly indicated that ecological risk assessments are difficult to make and that science has a long way to go before we reach an acceptable knowledge base. There seem to be a few trends in the research today. First, the focus is more or less entirely on agricultural crop plants, with few or no related wild plants. Secondly, there is an emphasis on questions regarding hybridization and, in most cases, it is limited to questions concerning whether or not hybridization occurs, over what distances and to some extent also the frequency of hybridization. A third focus seeks evidence of past introgression in wild relatives from non-modified crop plants.

There are a number of problems with these approaches. They are not comprehensive enough and only rarely allow conclusions to be drawn even in specific cases. The data collected seem to have done little to increase our ability to understand and quantify the possible impact that might result from the growing of GM plants. One reason may be that so far we have only studied parts of a multifaceted process (e.g. gene flow) and are not able to link what we have measured to variables of interest in risk assessment analysis.

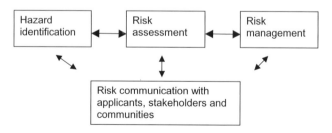

Fig. 22.1.   A basic risk analysis framework.

For example, it cannot come as a surprise to any plant ecologist that plants hybridize when they have the chance. Such information is not very useful when we do not ask: 'what are the consequences and will they matter?' Until these questions are asked, we will not see progress in our assessments of ecological risks. In addition, the limited number of studies and the fact that most risk studies concerning GM organisms have been conducted on agricultural crops (Raybould and Gray, 1994; Hilbeck *et al.*, 1999) make the generality of output data questionable.

Modification of Undomesticated Plants

One result of the focus on annual crop plants may be that the identified risks will be underestimated for other types of plants, such as long-lived trees (Bradshaw and Strauss, 2001). As we now are close to another phase in the development of GM plants, i.e. the development of undomesticated plants, such as forest trees, to be used in natural environments, we have to be aware that we do not have the necessary data in these systems.

There are many differences between agricultural systems and natural systems such as forests. For example, crop plants are mostly annual and short lived, and therefore changes in ecological and physiological behaviour will be relatively easy to detect. In contrast, trees in a natural system persist in the landscape for long periods of time. This time frame not only increases the probability that the trees will be subjected to a much wider range of stress conditions, including temperature extremes, enemy attacks and climate change, but also that the consequences of these changes potentially may be much harder to detect and manage. Furthermore, most of the agricultural crop production in the world takes place in highly modified and managed systems. In contrast, tree plantations will continue to be established in close proximity to stands of natural forest (see, for instance, Slavov *et al.*, Chapter 8, this volume). Such practices are likely to increase the interactions between the modified trees and their naturally occurring relatives.

Why Forestry?

Taking Sweden as an example of why there are interests in developing transgenic trees in forestry, it is possible to arrive at justifications. More than 60% of the Swedish land area is forest land and, as a consequence, it is an important driver of natural resource policy (Bernes, 1994). Forest products are in great demand all over the world, and forest industries in Sweden in 2000 generated a net export worth 100 million SEK. Despite this, up to 20% of the wood and fibre required by Sweden's industries is imported (Anonymous, 2001).

This suggests that the demand for wood and wood fibres, and pressure on forests, will increase substantially in the future. In particular, there will be a need for wood with specific properties that can be used for novel products, but also wood to be used as biofuels to replace nuclear power and oil-based petroleum. Furthermore, there are a number of other characteristics that are of interest to modify. For example, trees in general are slow growing, and growth characteristics may not always be optimal for the industry. Finally, forest trees are subjected to a vast number of natural enemies that may kill seedlings as well as adult trees. Recent advances in the 'new' biology, such as transgene technology, have made it possible to address these issues and to consider designing changes to trees.

Direct and Indirect Impacts

There is no doubt that transgene spread to natural environments is an undesirable possible nuisance. Unsurprisingly, today there is ample evidence that introgression will occur between GM plants (including trees) and wild species, when closely related plants adjoin one another (DiFazio *et al.*, 1999; Hails, 2000). New growth characteristics or pest resistance may allow GM plants to increase their invasive potential, which ultimately will lead to a replacement of native plants (Raybould and Gray, 1994; Hails, 2000). This has been highlighted for insect-resistant GM plants where toxins from the bacterium *Bacillus thuringensis* (*Bt* resistance) are used to reduce the damage by natural pests. However, the emphasis in current risk assessments is too strongly fixated on the probability that GM plants or transgenic genes (by means of introgression) will spread to natural environments. I think it is time to widen the base of knowledge to include the direct and indirect interactions between GM plants/genes and other organisms, areas where there are very few data.

One major concern is that GM plants could have unexpected effects on organisms and ecosystems. There are speculations that GM plants could affect non-target organisms and, as a result, change the ecosystem functions (Hails, 2000). For example, when a non-target species has an important function in the food web, i.e. a keystone function, the introduction of GM plants could cause perturbation throughout the ecosystem. In systems where host plants share a common pathogen, changes in one host may select for changes in characters of the pathogen that will result in effects on the alternative host. However, to a large degree, we lack information on how non-target organisms might be influenced, but there are suggestions that some of these might be negatively affected (Birch *et al.*, 1999; Hilbeck *et al.*, 1999; Losey *et al.*, 1999; see also Pimentel and Raven, 2000). The only way to minimize the risk for these unexpected events is to try to gain as much detailed knowledge as possible of the various ecological processes and functional groups of organisms which might be influenced by the introduction of a GM plant (Obrycki *et al.*, 2001). Here, I suggest that we need to focus more on interactions between GM plants, non-modified related species and natural enemies.

The Role of Natural Enemies

Enemies play an important role in natural ecosystems. Studies show that they may affect species abundance and diversity (Burdon, 1987). Natural enemies are strong selective agents that may change genetic compositions in plant populations (Alexander, 1992) and they may also affect succession (van der Putten *et al.*, 1993). Furthermore, Lambrinos (2002) recently suggested that the invasive ability of a plant is strongly dependent on its response to natural enemies. Many successful 'invaders' lack natural enemies that control the population growth (Simberloff, 1997; Prieur-Richard *et al.*, 2002). Therefore, it is fundamental to study natural enemies in a risk analysis of GM plants.

However, herbivores and pathogens cannot be treated as a homogeneous group with respect to their response to GM plants. Different functional groups of herbivores and pathogens will respond differently to changes in plant characters (Wennström, 1994; Fritz, 1999; Hjältén and Hallgren, 2002) and different mechanisms are involved in the plant resistance to herbivores and pathogens, respectively (Wennström, 1994; Hjältén, 1998; Felton and Korth, 2000).

The response of natural enemies (target and non-target species) to GM plants also provides some insight into how ecosystem processes will be influenced if GM plants or their genes (by introgression) invade natural systems. Attacks by herbivores and pathogens may induce various defensive responses (induced defences) in plants, and these responses can be very important for the outcome of plant–herbivore interactions (Karban and Baldwin, 1997). It is therefore important to assess if GM trees will differ in defensive responses compared with non-modified plants. Let us consider three scenarios dealing with consequences of modifications on: (i) selection pressures; (ii) competition between enemies; and (iii) susceptibilities in hybrids.

Selection pressure

In traditional breeding, the products of natural variation within a plant (or animal) are used to produce an offspring that has a novel and desirable combination of traits that enhance productivity and profitability. All changes in a plant, whether they have a basis in traditional breeding or in *in vitro* gene transfer technology, may affect the interaction between the plant and its neighbours. Up until now, we have not been concerned so much with the consequences of these shifts in traits on interactions with natural enemies. Nevertheless, this area of ignorance must be substantially diminished when modifications are made to undomesticated plants. History is full of examples of what may happen when the rules of the game change. One example comes from the southern corn leaf blight (1970–1971). It is the greatest plant disease epidemic recorded in the USA. A male sterility gene called *Tcms* (Texas cytoplasm male sterile) was discovered and used in breeding. There were several economic advantages of using *Tcms* in hybrid seed production that ensured the gene became widely distributed, and, by 1970, it was present in about 85% of the American maize crop. Unfortunately, the *Tcms* gene had the pleiotropic effect of conferring acute susceptibility to a variant of *Helminthosporium maydis*, a pathogen which up to that time had generally caused little damage. The lesions were first noticed in October 1969 on ears and stalks of samples from a seed field in Iowa. A fungus was isolated and identified as *H. maydis*. Inoculation tests showed that maize lines containing the cytoplasmic gene designated *Tcms* were acutely susceptible; those with normal cytoplasm only showed mild symptoms. The following year, the epidemic struck; the disease was most severe in the Mid-west and south of the USA, with some areas reporting 50–100% loss (Levings, 1990). It is clear that we need to have a better understanding of what the effects will be when selection pressure changes. From the *Tcms* example, we may learn that changes in a plant are not always predictable because they do not affect only the trait first targeted. In future production of new plant varieties, regardless of whether they are products of traditional selection breeding or modern technologies, it is appropriate to seek answers to questions such as 'will new races of pathogens be selected for or will initially uncommon races increase in their frequency?' Figure 22.2 shows a schematic illustration of what may happen when selection pressure changes. Races of the pathogen that earlier

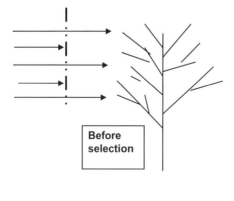

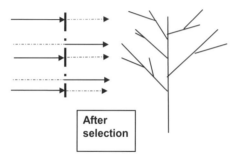

Fig. 22.2. Illustration showing that with changed selection pressure, previously avirulent pathotypes may become virulent. The vertical dotted lines symbolize resistance mechanisms. Horizontal lines symbolize different pathotypes. Previously virulent pathotypes have dotted horizontal lines.

have been non-virulent may, after alterations in the host plant, become virulent. Without testing, there is no way of knowing what effects, if any, this will have.

In natural situations, plants and their enemies interact in complex but very poorly understood ways. This further affects our ability to predict the effect of genetic modifications in plants. The rust *Melampsora pinitorqua* that needs two host species to complete its life cycle provides an example. The primary host is pine (*Pinus sylvestris*) and the alternative host is *Populus tremula*. Both hosts are needed for the pathogen to complete its sexual cycle (Gjaerum, 1974). *Populus* is one of the genera proposed to be modified in Sweden to produce, among many things, a higher yield. Therefore, changes that may occur in one host may also have secondary effects on another host plant (Fig. 22.3). If the selection pressure for the enemy changes on one host species, there may also be effects on the other host species.

Competition changes between enemies

Interspecific competition has long been regarded as important in natural communities (Fritz *et al.*, 1986). At all trophic levels, organisms interact with each other. Therefore, the dynamics of one species are often tightly linked to the dynamics of other species (Schmitz, 1998). As a result, changes to these interactions may have strong effects on the whole system. A common situation in natural systems is that one enemy is more abundant and suppresses other enemies (Fig. 22.4). When such dominance is broken up, other enemy

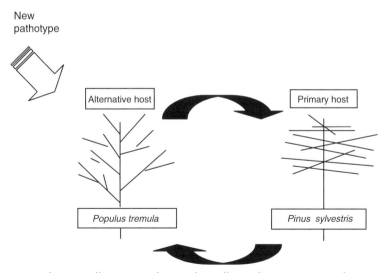

Fig. 22.3.   Schematic illustration of secondary effects after occurrence of a new pathotype of the rust *Melampsora pinitorqua*. The effects of the new pathotype may occur on both the alternative and primary hosts. Black arrows show how the spores are transmitted during the life cycle of the rust fungus.

species increase in abundance (Fig. 22.4). The net result on the host plant, when, for example, a gene for resistance is incorporated into the system, is therefore difficult to envisage. The effect on the host plant may be even stronger than before.

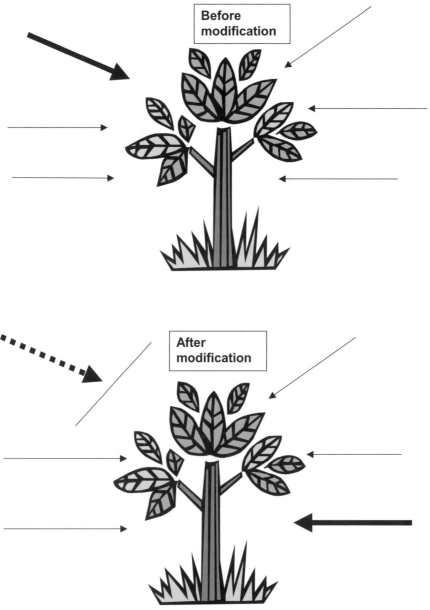

Fig. 22.4.   Competition changes between enemies. Each arrow symbolizes different enemies, of which one is dominant (thick arrow). After modification that made the plant resistant to the dominant enemy (now dotted line), another enemy is released from competition and takes the role as the dominant enemy (thick arrow).

Susceptibility of hybrids

Hybridization is a common process in nature which has been argued to have evolutionary significance (Rieseberg and Wendel, 1993; van Tienderen, Chapter 2, this volume). However, hybridization has also been considered a great hazard in conservation biology such as, for example, by interspecific hybridization with alien species (Ellstrand, 1992), and it may also affect interactions with other trophic levels such as naturally occurring enemies (Ericson *et al.*, 1992). For example, it has been suggested that hybrid zones may be important zones for the evolution of virulence in pathogens (Whitham *et al.*, 1994). Recent research has also found that hybrid plants may affect plant–enemy interactions in several ways, such as: (i) host shifts, where hybrids can be used as stepping stones to move from one host to another (Hjältén *et al.*, 2000); and (ii) disease sink, where hybrids may act as sinks on generalist enemies, which may affect other species (Floate and Whitham, 1993).

Earlier work has typically found that hybrids possess greater levels of resistance than their parents (Day, 1974; Maxwell and Jennings, 1980). However, more recent studies on hybrids within the genera *Populus* and *Salix* show that reduced resistance may also be the result. For example, Hjältén (1998) tested the susceptibility of the parent plants *Salix caprea* and *S. repens* and the hybrid between them to the rust *Melampsora* sp. He found that *S. repens* was totally resistant to the rust and that *S. caprea* had a low susceptibility. In contrast, the hybrid between the two species was highly susceptible to the rust. Thus, we have no way of knowing, without testing, what the effects will be after hybridization.

In conclusion, there are still a lot more questions to ask and a lot more data to gather before we can start drawing general conclusions, but we need to do this since, without a sound scientific basis, there will be no grounds for political decisions on the use of transgenic trees in agriculture and forestry. Furthermore, there will be no public acceptance without a comprehensive risk analysis.

Acknowledgement

This paper was financed by a grant from the Swedish Research Council for Environment, Agriculture Science and Spatial Planning.

References

Alexander, H.M. (1992) Fungal pathogens and structure of plant populations and communities. In: Carroll, G.C. and Wicklow, D.T. (eds) *The Fungal Community: Its Organization and Role in the Ecosystem*. Marcel Dekker, New York, pp. 481–497.
Anonymous (2001) Forestry and the forest products industry. Swedish Industry. *The Swedish Institute*, Fact sheet 129.

Bernes, C. (1994) *Biologisk Mångfald I Sverige: En landstudie. Monitor 14* (in Swedish). Naturvårdsverket, Solna.

Birch, N.A., Goeghegan, I.E., Majerus, M.E.N., MacNicol, J.W., Hackett, C.A., Gatehouse, A.M.R. and Gatehouse, J.A. (1999) Tri-trophic interactions involving pest aphids, predatory 2-spot ladybirds and transgenic potatoes expressing snowdrop lectin for aphid resistance. *Molecular Breeding* 5, 75–83.

Bradshaw, A.H. and Strauss, S.H. (2001) Plotting the course for GM forestry. *Nature Biotechnology* 19, 1103–1104.

Burdon, J.J. (1987) *Diseases and Plant Population Biology.* Cambridge Academic Press, Cambridge.

Day, P.R. (1974) *Genetics of Host–Parasite Interactions.* Freeman, San Fransisco, California.

DiFazio, S.P., Leonardi, S., Cheng, S. and Strauss, S.H. (1999) Assessing potential risks of transgene escape from fiber plantations. In: Lutman, P.W. (ed.) *Gene Flow and Agriculture: Relevance for Transgenic Crops.* Symposium Proceedings No. 72. British Crop Protection Council, Farnham, UK, pp 171–176.

Ellstrand, N.C. (1992) Gene flow by pollen – implication for plant conservation genetics. *Oikos* 63, 77–86.

Ericson, L., Burdon, J.J. and Wennström, A. (1992) Inter-specific host hybrids and Phalacrid beetles implicated in local survival of smut pathogens. *Oikos* 68, 393–400.

Felton, G.W. and Korth, K.L. (2000) Trade-offs between pathogen and herbivore resistance. *Current Opinions in Plant Biology* 3, 309–314.

Floate, K.G. and Whitham, T.G. (1993) The hybrid bridge hypothesis: host shifting via plant hybrid swarms. *American Naturalist* 141, 651–662.

Fritz, R.S. (1999) Resistance of hybrid plants to herbivores: genes, environment or both. *Ecology* 80, 382–391.

Fritz, R.S., Sacchi, C.F. and Price, P.W. (1986) Competition vs. host plant phenotype in species composition in willow sawflies. *Ecology* 67, 1608–1618.

Gjaerum, H.B. (1974) *Nordens Rustsopper* (in Norwegian). Fungiflora, Oslo, 321 pp.

Hails, R. (2000) Genetically modified plants – the debate continues. *Trends in Ecology and Evolution* 15, 14–15.

Hilbeck, A., Moar, W.J., Filippini, C.M. and Bigler, F. (1999) Prey-mediated effects of Cry1AB toxin and protoxin on the predator *Chrysoperla carnea. Entomologia Experimentalis et Applicata* 91, 305–316.

Hjältén, J. (1998) An experimental test of hybrid resistance to insects and pathogens using *Salix caprea, S. repens* and their F_1 hybrids. *Oecologia* 111, 127–132.

Hjältén, J. and Hallgren, P. (2002) The resistance of hybrid willows to specialist and generalist herbivores and pathogens: the potential role of secondary chemistry and parent host plant status. In: Wager, M.R., Clancy, K.M., Lieutier, F. and Paine, T.D. (eds) *Mechanisms and Deployment of Resistance in Trees to Insects.* Kluwer Academic Publishers, Dordrecht, The Netherlands, pp. 153–168.

Hjältén, J., Ericson, L. and Roininen, H. (2000) Resistance of *Salix caprea, S. phylicifolia* and their F_1 hybrids to herbivores and pathogens. *Ecoscience* 7, 51–56.

James, C. (2002) *Global Status of Commercialized Transgenic Crops.* ISAAA Briefs No. 27.

Karban, R. and Baldwin, I.T. (1997) *Induced Responses to Herbivores.* University of Chicago Press, Chicago, Illinois.

Lambrinos, J.G. (2002) The variable invasive success of *Cortaderia* species in a complex landscape. *Ecology* 83, 518–529.

Levings, C.S. III (1990) Texas cytoplasm of maize. *Science* 250, 942–947.

Losey, J.E., Raynor, L.S. and Carter, M.E. (1999) Transgenic pollen harms monarch larvae. *Nature* 399, 214.

Maxwell, F.G. and Jennings, P.R. (1980) *Breeding Plants Resistant to Insects.* Wiley, New York.

Obrycki, J.J., Losey, J.E., Taylor, O.R. and Jesse, L.C. (2001) Transgenic insecticidal corn: beyond insecticidal toxicity to ecological complexity. *BioScience* 51, 353–538.

Pimentel, D.S. and Raven, P.H. (2000) Bt corn pollen impacts on non-target lepidopterans: assessment of the effects in nature. *Proceedings of the National Academy of Sciences USA* 97, 8198–8199.

Prieur-Richard, A.H., Lavorel, S., Linhart, Y.B. and Dos-Santos, A. (2002) Plant diversity, herbivory and resistance of a plant community to invasion in Mediterranean annual communities. *Oecologia* 130, 96 –104.

Raybould, A.F. and Gray, A.J. (1994) Will hybrids of genetically modified crops invade natural communities? *Trends in Ecology and Evolution* 9, 85–89.

Rieseberg, L. and Wendel, J.F. (1993) Introgression and its consequences in plants. In: Harrison, R.G. (ed.) *Hybrid Zones and the Evolutionary Process.* Oxford University Press, New York, p. 374.

Schmitz, O. (1998) Direct and indirect effects of predation and predation risk in old-field interaction webs. *American Naturalist* 151, 327–342.

Simberlof, D. (1997) Eradications. In: Simberlof, D. (ed.) *Strangers in Paradise: Impacts and Managements of Nonindigenous Species in Florida.* Island Press, Washington, DC, pp. 221–228.

van der Putten, W.H., van Dijk, C. and Peters, B.A.M. (1993) Plant-specific soilborne diseases contribute to succession in foredune vegetation. *Nature* 362, 53–56.

Wennström, A. (1994) Systemic diseases on hosts with different growth patterns. *Oikos* 69, 535–538.

Whitham, T.G., Morrow, P.A. and Potts, M.B. (1994) Plant hybrid zones as centers of biodiversity: the herbivore community of two endemic Tasmanian eucalypts. *Oecologia* 97, 481–490.

A Tiered Approach to Risk Assessment of Virus Resistance Traits Based on Studies with Wild Brassicas in England

DENISE W. PALLETT[1], MILOU I. THURSTON[1], MARY-LOU EDWARDS[1], MARTIN NAYLOR[1], HUI WANG[1], MATTHEW ALEXANDER[2], ALAN J. GRAY[2], EMILY MITCHELL[2], ALAN F. RAYBOULD[3], JOHN A. WALSH[4] AND J. IAN COOPER[1]

[1]CEH Oxford, Mansfield Road, Oxford OX1 3SR, UK; [2]CEH Dorset, Winfrith Technology Centre, Winfrith Newburgh, Dorchester, Dorset DT2 8ZD, UK; [3]Syngenta, Jealott's Hill International Research Centre, Bracknell, Berks RG42 6E, UK; [4]Horticulture Research International, Wellesbourne, Warwick CV35 9EF, UK; E-mail: jic@ceh.ac.uk

Abstract

Concerns about the introduction of genetically modified crops frequently centre on the possibility of gene transfer to wild relatives, resulting either in the disruption of natural patterns of genetic diversity by introgressing into species gene pools or in the addition of traits which may cause wild plants to become more abundant or invasive. This chapter describes a phased approach to the assessment of possible harm in the context of a specific transgenic stress-tolerant trait, namely virus tolerance in inter-breeding species. We assessed the hazard of harm to the 'natural' environment as opposed to agricultural productivity. Our baseline was the distribution and relative abundance in field-grown wild or long-established 'naturalized' *Brassica* species (*B. oleracea*, *B. nigra* and *B. rapa*) of six viruses, and then glasshouse assessments of components of fitness. Because these observations suggested that generic risk assessment was unlikely to be possible, we focused on the economically significant turnip mosaic virus (TuMV), genus *Potyvirus*. TuMV is a target for transgenic (capsid-coding sequence-based) approaches to disease management in brassicas as an alternative to natural sources of resistance/tolerance to the virus and because insecticides do not kill the aphid vectors of the virus before they effect inoculation. TuMV was not found in *B. rapa* growing on the banks of the River Thames in Oxfordshire, UK. Glasshouse tests showed that *B. rapa* from these populations died within a few days of manual inoculation with some isolates of TuMV, but we found that the pathogenicity of three TuMV isolates from the UK was not uniform. We made crosses in which natural

B. rapa lines, genome designation *AA*, *n* = 10, were the female partners and *B. napus*, genome designation *AACC*, 2*n* = 38, were pollen donors. *B. napus* included untransformed lines and lines that contained a transgenic capsid-coding sequence from a potyvirus. As judged using polymerase chain amplification, 'C' genome transfer frequencies varied from 0 to 84% depending on pollen donor, but there was statistically significant within-population variation among *B. rapa*, *P* < 0.001 at Culham and *P* < 0.05 at Clifton Hampden, in the efficiency of transgene flow from one *B. napus* cultivar ('Drakkar'). When manually challenged with TuMV, the transformed *B. napus* was infectible but the virus was not lethal. In contrast, the untransformed counterparts of these plants were sensitive to the same challenge inocula although two cultivars ('Westar' and 'Drakkar') differed in their absolute infectibility by one of the three isolates of TuMV we assessed. Importantly, when F_1 hybrid progeny, identified on the basis of the presence of 'C'-specific sequence and capsid-coding sequence as judged by polymerase chain amplification, were manually challenged with TuMV, these plants tended to be more TuMV tolerant than their maternal parents. Thus, our contained glasshouse-based gene flow and pathogenicity tests, even though done in non-competitive conditions and with an incomplete knowledge of factors regulating the wild populations, provided prima facie evidence of a potential for ecological release from that natural virus constraint following introgression of a resistance trait.

The assessments of putative fitness impacts in hybrids between transformed (virus-tolerant) crop plants and wild (virus-sensitive) crop relatives were assembled during an ongoing EU-funded project (VRTP-IMPACT; QLK3-CT-2000-00361).

Introduction

Dozens of viruses that infect green vascular plants have been targets for transgenic tolerance/resistance because natural genes for that purpose are not always available. The first studies in the context of parasite-derived resistance showed that seedlings from the primary transformed lines expressing a virus-derived capsid protein were infectible but disease was delayed or indeed did not develop following challenge. In other words, the transgenic (i.e. 'obtained') plants were tolerant of infection but also somewhat resistant to virus invasion (*sensu* Cooper and Jones, 1983). Trials of crops with 'obtained' virus resistance/tolerance have been held in a number of countries including the USA, People's Republic of China, Australia and, less commonly, in the European Union. Most of the crops known to be under investigation express virus-derived sequences that code for capsid proteins, although sequences for non-structural virus genes have also been used in transgenic crops (Walsh, 2000), and some of these are also close to market.

Before crops expressing virus-derived sequences had been produced and shown to be promising for the management of specific viruses, there had been few studies on virus impacts other than in crops. As a consequence, there was essentially no knowledge about whether or how much ecological release might occur as an unintended consequence of the nullification of viruses that hitherto constrained genetic diversity or kept wild species in check. At the outset of our work, it was known that viruses have the potential to act in a density-dependent manner and can be lethal, although it is not known whether or not they regulate populations. There were general concerns that transgenic virus-derived tolerance/resistance genes might facilitate virus evolution as a result of recombination and/or ecological release, but we only addressed the latter. Driven by interest in possible biodiversity change following pollen–seed-mediated transgene flow, this chapter describes studies in the context of pathologically distinct isolates (= pathotypes *sensu* Jenner and Walsh, 1996) of one virus, turnip mosaic virus (genus *Potyvirus*; TuMV). TuMV is one of the world's most economically important crop pathogens (Shattuck, 1992) that naturally infects a diverse range of horticultural species including peas, rhubarb, watercress, *Abutilon*, wallflowers and also *Brassica* spp. in which it is associated with diminished quality or plant death (Walsh and Jenner, 2002).

The wild annual/biennial species *Brassica rapa* ssp. *sylvestris* L. (wild turnip) (previously named *Brassica campestris* L.) is locally common near the River Thames in Oxfordshire (Killick *et al.*, 1998; Wilkinson *et al.*, 2000), and is established at scattered sites elsewhere. *B. rapa* ssp. *sylvestris* is suspected to be non-native (introduced and naturalized before 1500) and a close relative of the cultivated turnip (*B. rapa* ssp. *rapa* L.) and also turnip rape (*B. rapa* ssp. *oleifera* DC.) that is grown predominantly in northern Britain. *B. rapa* is a diploid species ($n = 10$) with the genome designation '*AA*' and forms one of the apices of U's triangle (U, 1935). The other apices of U's triangle are *B. nigra* (L.) Koch ('*BB*', $n = 8$) and *B. oleracea* L. ('*CC*', $n = 9$).

B. rapa is a potential recipient of transgenes from the allotetraploid oil-seed rape (*B. napus* ssp. *oleifera* L. *AACC*; $2n = 38$) (e.g. Raybould and Gray, 1993; Chèvre *et al.*, 2000, Chapter 18, this volume), and the frequency of hybridization depends, among other things, on the spatial arrangement of populations of the two species (e.g. Jørgensen and Anderson, 1994; Scott and Wilkinson, 1998; Davenport *et al.*, 2000). Speculation about the consequences of the 'flow' of transgenes via pollen hitherto has largely centred on herbicide tolerance traits or, more recently, insect resistance (Halfhill *et al.*, 2002) in the explicit context of crop competition. With increasing numbers of genetic modifications to obtain stress tolerance in *Brassica* species, the need for data to enable informed environmental risk assessments has been highlighted.

There are large amounts of data on the occurrence and economic impacts of viruses in *Brassica* cultivars in the UK (e.g. Broadbent, 1957; Broadbent and Heathcote, 1958; Mowat, 1981; Walsh *et al.*, 1989; Hardwick *et al.*, 1994). To complement this knowledge, we initially focused on virus distribution and impacts in wild/naturalized *B. oleracea* (Raybould *et al.*, 1999a; Maskell *et al.*, 1999) and then on a congeneric species (*B. nigra*; Thurston *et al.*, 2001; Raybould *et al.*, 2003). Overall, our observations showed that these species had different virus loads and suggested that the viruses have different effects on the different species, even among populations of the same species. Subsequently, we investigated the prevalence of the six viruses previously studied in *B. oleracea* and *B. nigra*, in field populations of *B. rapa*. We elected to study the following six viruses because, except for a very few instances of

infection with cucumber mosaic virus (genus *Cucumovirus*), we detected
no others: TuMV, beet western yellows virus (genus *Polerovirus*; BWYV),
cauliflower mosaic virus (genus *Caulimovirus*; CaMV), turnip crinkle virus
(genus *Carmovirus*; TCV), turnip rosette virus (genus *Sobemovirus*; TRoV) and
turnip yellow mosaic virus (genus *Tymovirus*; TYMV).

Materials and Methods

Viruses and inoculation techniques

For challenge inoculations, the following virus isolates were used: BWYV
from oilseed rape (*B. napus* ssp. *oleifera*) as supplied by Dr H.G. Smith, IACR,
Broom's Barn, Bury St Edmunds, Suffolk; CaMV from *B. oleracea* at Winspit,
Dorset; TCV from *B. nigra* at Ringstead, Dorset; TuMV from *B. oleracea* at
Llandudno, Conwy (pathotype 3) or Chapman's Pool, Dorset (pathotype 1)
or from *B. napus* from a Canadian source (pathotype 4); TRoV was obtained
from *B. nigra* at Kimmeridge (TRoV-K) or Ringstead (TRoV-R), Dorset, and
TYMV from *B. oleracea* at Chapman's Pool, Dorset.

In insect-screened, temperature-controlled glasshouses, manual inoculation was performed by rubbing the carborundum-coated abaxial leaf
surfaces of plants grown from seed with sap from virus-infected plants.

Study sites

The field populations of *B. rapa* chosen for this study span approximately
3 km and are situated at Clifton Hampden, Abingdon and Culham on the
banks of the River Thames in Oxfordshire.

Virus detection in field samples

Leaf samples were tested for the presence of viruses as described by Maskell
et al. (1999), Thurston *et al.* (2001) and Pallett *et al.* (2002), and largely using
the enzyme-linked immunosorbent assay (ELISA) method of Edwards and
Cooper (1985).

Pathogenicity testing in *B. rapa*

To determine the effect of specific virus infection in plants of *B. rapa*, seed
collected from naturally established populations on the banks of the Thames
was grown in Fisons Universal compost. To facilitate synchronized germination, the seeds were chilled overnight at −20°C before sowing. At least one

representative isolate of each detected virus was tested for its effects on individuals from different half-sib families of *B. rapa* from both Oxfordshire sites, and this work is continuing to investigate the reactions of plants from seed of *B. rapa* growing in other places in England. Virus accumulation in glasshouse-grown *B. rapa* was measured and statistically analysed as described in Thurston *et al.* (2001).

Pathogenicity testing in *B. rapa* × *B. napus* hybrids

Observations on natural *B. rapa* genotypes were extended to investigate the impact of isolates of TuMV on hybrids made between natural *B. rapa* and, as pollen donors, transgenic *B. napus* containing a virus-derived capsid-coding sequence conferring virus tolerance/resistance. The lines of *B. napus* used as parents in the crossing programme included *B. napus* cv. 'Westar' containing a capsid-coding sequence from TuMV (Lehmann *et al.*, 2003; seeds of 'Westar' were supplied by J.A. Walsh, Wellesbourne, 2001) and *B. napus* cv. 'Drakkar' containing a capsid-coding sequence from lettuce mosaic virus (genus *Potyvirus*; LMV) that protects against heterologous potyviruses (Dinant *et al.*, 1993). The seeds of 'Drakkar' were supplied by M. Tepfer, Versailles, 2001. These two *B. napus* pollen donor cultivars differ in one important respect that needed to be considered when designing and interpreting data concerning virus challenge; 'Westar' has been reported to be intrinsically resistant to TuMV (pathotype 1) (Walsh *et al.*, 1999), whereas 'Drakkar' is not. Both cultivars are infectible by the other pathotypes of the virus known to occur in the UK. The pollen recipients were manually emasculated *B. rapa* that had been transplanted from the field into pots and repeatedly treated with a fumigant insecticide (nicotine). At the two-leaf stage, F_1 and F_2 progeny were assayed for the presence of the C genome and evidence of the virus-derived transgenic capsid-coding sequences using polymerase chain reaction (PCR) (Szewc-McFadden *et al.*, 1996; Dinant *et al.*, 1997; Raybould *et al.*, 1999b). The TuMV capsid-coding sequence was detected using PCR amplification and primers designed as described by J.A. Walsh and P. Hunter (Wellesbourne, 2001, personal communication).

Results and Discussion

Occurrence of viruses in field-grown wild *Brassica* species

As reported in detail by Pallett *et al.* (2002), no evidence of TuMV in natural *B. rapa* was obtained following tests on plants in communities on the banks of the River Thames. Our earlier studies of brassicas on the Dorset cliffs had revealed a low frequency (5/597) of infection by TuMV in *B. nigra* populations (Thurston *et al.*, 2001) and a much greater frequency (158/723) in *B. oleracea* (e.g. Raybould *et al.*, 1999a).

Explaining the differences in TuMV incidence

The first approach we used when seeking an explanation addressed infectibility and pathogenicity of the virus in plants grown from seed and manually inoculated at 4 weeks. Following challenge with TuMV pathotype 1, 100% (18/18) of seedlings became infected and all developed systemic necrosis within 10 days. Two to three weeks after inoculation, 56% (10/18) of the challenged plants had died. Eight-week-old plants of *B. rapa* were also challenged and, in them, necrosis took longer to develop. Nevertheless, at 8 weeks post-inoculation, 15/19 plants had died and the remaining four plants were stunted and had necrosis of the apical leaves. When pathotype 3 of TuMV was used as inoculum, *B. rapa* plants became infected systemically but this virus isolate was not uniformly lethal. Pathotype 4 of TuMV was similarly non-lethal, but nevertheless severely stunted the growth of *B. rapa* manually inoculated with the virus.

It is possible and we think it most likely that the absence/rarity of TuMV is due to the plants being susceptible and highly sensitive, but there is a need for more evidence to show exposure. When *B. nigra* and *B. oleracea* occur together (Thurston *et al.*, 2001), detection of TuMV in the latter, but not the former indicates exposure. Unfortunately, when *B. rapa* was found to be free of TuMV, we did not test adjoining species that might have indicated exposure. Thus the possibility exists that the virus had swept through the population, suffering 'burnout', and left a mixture of individuals tolerant, resistant or immune. We have little evidence for this scenario but, at one site where we have studied *B. rapa*, we have observed a very low incidence of TuMV and can thereby infer exposure. Investigation continues to characterize this population in terms of tolerant, immune and escaped residuals and to assess the possibility that resistance is so costly that there is a strong selection pressure against resistance between epidemics.

There might be alternative explanations that we did not investigate (e.g. whether vectors of TuMV avoid wild *B. rapa* and *B. nigra* or are otherwise unable to transmit the virus), but we were aware that variation in TuMV was a possibility; Jenner and Walsh (1996) found that the relative frequency of TuMV pathotypes varied with location. Furthermore, there is significant genetic variation among *B. nigra* populations growing only a few kilometres apart (Lowe *et al.*, 2002), so genetic differences in the response of *B. rapa* to TuMV are also possible.

Comparison of the proportions of *B. rapa* from two sites in Oxfordshire (Culham and Abingdon) that became infected following manual challenge with viruses other than TuMV showed that although most behaved uniformly in both *B. rapa* provenances, one virus, TRoV, did not. More plants from Culham than from Abingdon became infected with one isolate of TRoV ($\chi^2 = 13.9$, $P < 0.0001$). All the *B. rapa* (from both Oxfordshire sites) showed symptoms when infected with CaMV, TYMV or TuMV. With CaMV, plants became severely stunted and the leaves that formed post-inoculation were crinkled. TYMV infection caused the plants to produce a bright yellow mottling of the leaves, but was not lethal.

Interpreting the effects of TuMV resistance/tolerance traits in wild *Brassica* species

TuMV infected wild *B. oleracea* following manual inoculation and, in a field experiment conducted over 18 months, the virus decreased survival, growth rate and fecundity (Maskell *et al.*, 1999). Therefore, it seems reasonable to expect selection for uniformly high resistance or tolerance. However, we currently lack information about the heritability of response to TuMV through the whole life cycle of the plant, and the long life cycle of *B. oleracea* makes measurement particularly difficult. Appropriate demographic data are now being sought for *B. rapa*.

One explanation for the variation in *B. oleracea* is that there is a 'cost' to virus resistance and that while TuMV is distributed patchily in space and/or time, susceptible plants may regain a fitness advantage over resistant plants in the absence of the pathogen. We have no data on whether, for instance, unchallenged susceptible plants produce more seeds than unchallenged resistant plants. At present, we can only identify resistant and susceptible genotypes by challenging them with TuMV. With the imminent development of genomic tools and sequence-specific gene silencing, it should be possible directly to inform this point. In the interim, to address the question of 'cost' to transgenic virus resistance, we used a non-GM surrogate gene. *TuRB01* is a *Brassica A* genome dominant gene that confers resistance to certain isolates of TuMV, including pathotype 1 (Walsh *et al.*, 1999). During winter and spring 2000–2001, we used a pair of near-isogenic *B. napus* lines that differ only at and around the *TuRB01* locus. We tested for the presence of TuMV using ELISA, and the virus was detected in all inoculated plants but was not detected in any of the controls. As detailed in Raybould *et al.* (2003), after 26 weeks, we detected no cost of resistance in this trial.

When considering sources of a trait for virus tolerance/resistance, *B. napus* pollen presents a recognized hazard for *B. rapa* and, in CEH Oxford, we therefore synthesized hybrids with which to address the consequences of introgression in the context of TuMV.

Effects of TuMV in *B. napus* 'Drakkar' and hybrids with *B. rapa*

In 2001, a large number of reciprocal glasshouse crosses were made between wild *B. rapa* and untransformed *B. napus* cv. 'Drakkar' or 'Drakkar' containing a homozygous transgenic cassette including the capsid-coding sequence derived from LMV (Dinant *et al.*, 1993). When crosses with a common transgenic 'Drakkar' parent were analysed, we observed statistically significant differences in the efficiencies of 'C' transfer when the maternal parents were *B. rapa* from different Thames-side locations. 'C' genome transfer frequencies ranged from 0 to 36% in hybrids with *B. rapa* from Culham, compared with 0–66% in hybrids with *B. rapa* from Clifton Hampden. Furthermore, at these two sites, there was significant within-population variation in transgene transfer rates ($\chi^2 = 36.494$, df = 4, $P < 0.001$ for *B. rapa* from Culham and $\chi^2 = 23.831$, df = 12, $P < 0.05$ for *B. rapa* from Clifton Hampden). This indicates that these *B. rapa* populations were not uniform recipients of transgenic pollen from *B. napus* cv. 'Drakkar'. A reciprocal observation was reported in North America when individual *B. rapa* were used to cross with different *B. napus* lines containing insecticidal *Bacillus thuringiensis* (*Bt*) transgenes in a field experiment (Halfhill *et al.*, 2002).

A test of TuMV impact in hybrid progeny between *B. rapa* and 'Drakkar'

Given the relative ease and frequency with which hybridization occurred, a short-term experiment was designed to examine whether or not the transgenic gene conferring virus tolerance had a fitness impact. Seedlings from glasshouse crosses made between wild *B. rapa* and 'Drakkar' were raised and, once the plants had produced two true leaves, one leaf was tested for the presence of the LMV-derived sequence and the 'C' genome using PCR. Sixty-four hybrid plants (containing the 'C' genome) were identified, and half of this number (16 in which we obtained evidence of LMV capsid-coding sequence and 16 without) were inoculated with an isolate of TuMV (pathotype 3). Pathotype 3 of TuMV was chosen for this challenge because it was less pathogenic (and not uniformly lethal) in *B. rapa* seedlings assessed under similar glasshouse conditions. The remaining plants (16 containing the LMV-derived sequence and 16 containing only evidence of the 'C' genome) were 'mock' inoculated with water. All seedlings were maintained in a fully randomized layout, and components of fitness (plant height, survival and time to flower) were recorded.

Plants confirmed as containing the LMV-derived transgene were manually infectible but were not visibly changed by TuMV (pathotype 3). Water-inoculated plants that contained the virus-derived transgene were smaller than those without the gene. There was no difference in height between TuMV-inoculated individuals containing the LMV-derived transgene and those in which the virus-derived sequence was not detected. Over a 16-week period, plants infected with TuMV flowered earlier, and fewer plants flowered compared with the other treatments. These results suggest that the presence of the LMV-derived transgene prevents TuMV affecting time of flowering.

Effects of TuMV in *B. napus* 'Westar' and hybrids with *B. rapa*

In 2002, four crosses were made in which *B. rapa* was pollen recipient and transgenic *B. napus* 'Westar' was pollen donor. In these instances, few seeds were obtained but the 'C' genome was transferred to 84% of progeny and PCR revealed the TuMV-derived CP sequence in all (77) of those plants so far tested.

None out of five untransformed *B. napus* 'Westar' was invaded by pathotype 1 of TuMV but, in unmodified *B. napus* 'Westar', TuMV (pathotypes 3 and 4) caused crinkling and severely diminished growth. In further tests using pathotype 3 of TuMV, 4/5 plants of *B. napus* 'Westar' containing the TuMV-derived capsid-coding sequence were not invaded systemically following manual challenge, yet 5/5 *B. rapa* were invaded systemically. Importantly, none (out of five) of the progeny from the cross in which *B. rapa* from Culham was the pollen recipient and 'Westar' containing the TuMV-derived sequence was the pollen donor was invaded by the virus isolate that was invasive and pathogenic in the *B. rapa* parent.

Conclusion

Following publication of the article on insect-resistant transgenic oilseed rape hybridization with *B. rapa* (Halfhill *et al.*, 2002), the *New Scientist* (30 November 2002) used the headline 'Modified crop breeds toxic hybrid'. Our work shows that the following headline might be appropriate: 'Gene flow may give hybrids an edge'. Undoubtedly, we have observed that *B. napus* containing, as a transgenic gene, a capsid-coding sequence from a potyvirus breeds virus-tolerant hybrids with *B. rapa*. However, this is a trend, not an all or nothing effect. The crucial 'ecological' question is; does virus tolerance give the hybrids any 'edge' in the struggle to perpetuate their genes?

TuMV is notable in lessening seed output in *B. oleracea* (Maskell *et al.*, 1999). We currently lack crucial information about what really determines the fitness of any *Brassica* species, but there is prima facie potential for a novel introgressing trait conferring TuMV resistance/tolerance to change the composition of communities (assuming that population growth rate is seed limited). We do not know if an introgressed virus-derived TuMV resistance factor would spread and be maintained, but small-scale and short-term experiments did not indicate an ecological cost to a 'traditional' gene (*TuRB01*) conferring resistance to some TuMV pathotypes. On that limited basis, such a gene under selection might indeed stabilize in a *Brassica* population.

The near complete absence of particular viruses from field-grown populations of particular wild species (in this instance, TuMV from *B. nigra* and *B. rapa*) is most easily explained in terms of our glasshouse challenge data; both species were manually infectible and fatally sensitive. If a plant's sensitivity to manual challenge inoculation explains the absence of TuMV in *B. rapa*, then it may be reasonable to infer a risk of consequential ecological release from this natural constraint following introgression of a resistance/tolerance trait active against that particular virus. Published tests of efficacy of different transgenic lines usually have been restricted to viruses homologous to the source of the transgenic sequence. Now that the possibility of unintended environmental consequences cannot be excluded, the breadth of heterologous activity of a transgenic sequence conferring virus tolerance must be considered in an 'environmental' risk assessment.

When attempting an environmental hazard assessment, non-uniformity in the receptiveness of individuals in *B. rapa* communities to *B. napus* pollen must be taken into account with knowledge of heterogeneity in responses to virus challenge. Since we found that multiple infection by different viruses is not uncommon (Thurston *et al.*, 2001; Pallett *et al.*, 2002), knowledge of synergy between different viruses is also desirable. Nevertheless, the effect of a target virus in pathogenicity tests in the glasshouse seems to be a correlate/predictor of the prevalence of the virus in wild populations. In our observations, TuMV provided the clearest example, but CaMV showed the same pattern in being common in the tolerant species (*B. oleracea*) and rare or absent in populations of the more sensitive species (*B. nigra* and *B. rapa*). Thus, it is tempting to conclude that the rarity of certain viruses in *B. rapa* may be due to the uncompetitiveness of infected plants whether the viruses are directly lethal or merely debilitating. To confirm this, detailed long-term monitoring of infection will be required.

Tantalizingly, the scientifically and ecologically interesting information on the extent to which hybrids between transgenic *B. napus* and wild *B. rapa* are virus resistant/tolerant is not yet complete. Many of the essential background tests have been done, but as a series of small-scale experiences rather than in one replicated and fully controlled experiment. Nevertheless, the reaction to challenge by specific TuMV pathotypes of both unmodified *B. napus* cultivars was known to be severe (e.g. Walsh and Tomlinson, 1985), and our observations support the prior art. Furthermore, it was known that the LMV-derived capsid-coding sequence (which in our tests was a transgenic element in *B. napus* 'Drakkar') had been linked to resistance to heterologous potyviruses (Dinant *et al.*, 1993). We confirmed and extended this finding in showing that the virus tolerance trait was perpetuated in hybrid progeny following the crossing with *B. rapa*. Although we have detected the transgenic virus-derived sequence in the F_2 generation, it is now crucial to backcross the F_1 progeny with parental *B. rapa* and to assay fecundity as well as transgene transfer rates and reaction to virus challenge in the F_2 generation.

Commercial oilseed rape lines are not routinely tested for their infectibility/resistance by TuMV, but the fact that *B. rapa* and *B. napus* cultivars have been used to distinguish TuMV pathotypes (e.g. Hughes *et al.*, 2002; Walsh *et al.*, 2002) indicates that there is variation in this property. We are now embarked on a detailed experimental assessment of consequences of transgene transfer, but have not yet obtained answers to the key questions that depend on life stage and demographic data. Until proved otherwise, it seems reasonable to assume that different pathotypes of a virus will differ in their potential fitness impacts.

When hybridization between *B. napus* and *B. rapa* is successful, and the transgenic parent is homozygous for the transgene (as in these instances with 'Drakkar' and 'Westar'), the predicted rate of transfer should approximate to 100%. We did not always observe this and do not know why. Unknown ploidy levels and allocations of the '*A*' and the '*C*' genetic components among hybrid progeny introduce uncertainties, and the extent, if any, to which positional or other lethal effects from the transgene have an impact on the outcome when hybridization is successful cannot yet be factored into the interpretation.

We observed significant within-population variation in transgene transfer rates from *B. napus* to *B. rapa*, and this further complicates assessment of harm potentially associated with gene flow. Our experience with TuMV is in accord with the earlier findings of Pallett *et al.* (2002); TCV in *B. rapa* at one site showed that individuals of *B. rapa* differed significantly ($P < 0.001$) in the concentration of viral antigen that accumulated in plants with or without symptoms. Furthermore, following manual challenge, two clear patterns were observed in the relationship between TCV concentration (capsid protein) in the *B. rapa* and the presence of visible change in the foliage. These observations support experience in which *B. nigra* from different provenances reacted very differently to manual challenge by TRoV (Thurston *et al.*, 2001; Thurston, 2002). Therefore, we suggest that when assessing the potential hazard from an introduced resistance trait, an early step should be to assess the diversity of the wild crop relative to specific virus challenge in the contained glasshouse or laboratory (having regard to information about the diversity of viruses in the region).

When creating a commercial justification for the development of a transgenic approach for crop protection, it is very likely that some knowledge of pathotype variation (as manifested in crop productivity) will be available. However, it is by no means certain that an appropriate knowledge of pathotype occurrence/prevalence in natural communities will be available. We suggest that this issue must be addressed, as should the intrinsic virus resistance properties of pollen donors to the virus under assessment. Such knowledge is crucial when designing appropriate challenge inocula, but also when interpreting data. If the wild relatives suspected to be at risk are immune (or are tolerant) to virus isolates in the proposed release zone, the 'obtained' resistance gene may be assumed to have no environmental impact. When a virus does infect the wild crop relative in controlled (glasshouse) conditions (Fig. 23.1), the next step should be to assess whether infection occurs in the field. If no such infection is detected, again the gene (trait) may reasonably be assumed to have no effect on the fitness of the relative and further experiments are unnecessary. However, if infection does occur, then further experimental work, perhaps involving the exclusion of vectors, is necessary to determine if virus infection controls the population dynamics of the host.

Deregulated (= commercial) use in the USA of squash (*Cucurbita pepo* subsp. *ovifera*) containing virus-derived capsid-coding sequences was first authorized in 1994 (USDA/USDA Petition 92–204–01 for determination of non-regulated status of ZW-20 squash). When deregulating this transgenic crop, the competent authority in the USA explicitly stated that its decision was based on the fact that the target viruses were not known to occur in natural populations of the related wild species (*Cucurbita texana*; Kling, 1996). This presumption of safety by the US Department of Agriculture may have been fundamentally flawed.

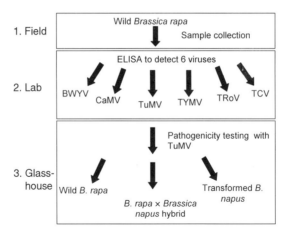

Fig. 23.1. Tiers of risk assessment. (1) Field. Samples of wild *B. rapa* were collected from different field sites. (2) Lab. Samples were tested by ELISA for the presence of plant pathogenic viruses (BWYV, beet western yellows virus; CaMV, cauliflower mosaic virus; TuMV, turnip mosaic virus; TYMV, turnip yellow mosaic virus; TRoV, turnip rosette virus; TCV, turnip crinkle virus). (3) Glasshouse. TuMV was isolated from wild *B. rapa* and was used to challenge glasshouse-grown *B. rapa*, virus-resistant transgenic *Brassica napus* and *B. rapa* × transgenic *B. napus* hybrids.

Acknowledgements

The authors thank Mrs Delia McCall for maintenance of glasshouse plants. The work that was done under DEFRA Licence No. PHL 147A/4344 (12/2002) benefited from a diverse range of funding: the NERC Ecological Dynamics and Genes (EDGE) Thematic Programme (GST/02/1848), the European Union (VRTP-IMPACT; QLK3-CT-2000-00361) and DETR (EPG1/5/132). M.T. was funded by a MAFF/DEFRA studentship (HH1762).

References

Broadbent, L. (1957) Investigation of virus diseases of *Brassica* crops. *Agricultural Research Council Report* series 14. Cambridge University Press, Cambridge.

Broadbent, L. and Heathcote, G.D. (1958) Properties and host range of turnip crinkle, rosette and yellow mosaic viruses. *Annals of Applied Biology* 46, 585–592.

Chèvre, A.M., Eber, F., Darmency, H., Fleury, A., Picault, H., Letanneur, J. and Renard, M. (2000) Assessment of interspecific hybridization between transgenic oilseed rape and wild radish under normal agronomic conditions. *Theoretical and Applied Genetics* 100, 1233–1239.

Cooper, J.I. and Jones, A.T. (1983) Responses of plants to viruses: proposals for the use of terms. *Phytopathology* 73, 127–128.

Davenport, I.J., Wilkinson, M.J., Mason, D.C., Charters, Y.M., Jones, A.E., Allainguillaume, J., Butler, H.T. and Raybould, A.F. (2000) Quantifying gene movement from oilseed rape to its wild relatives using remote sensing. *International Journal of Remote Sensing* 21, 3567–3573.

Dinant, S., Blaise, F., Kusiak, C., Astiermanifacier, S. and Albouy, J. (1993) Heterologous resistance to potato virus-Y in transgenic tobacco plants expressing the coat protein gene of lettuce mosaic potyvirus. *Phytopathology* 83, 818–824.

Dinant, S., Maisonneuve, B., Albouy, J., Chupeau, Y., Chupeau, M.C., Bellec, Y., Gaudefroy, F., Kusiak, C., Souche, S., Robaglia, C. and Lot, H. (1997) Coat protein gene-mediated protection in *Lactuca sativa* against lettuce mosaic potyvirus strains. *Molecular Breeding* 3, 75–86.

Edwards, M.-L. and Cooper, J.I. (1985) Virus detection using a new form of indirect ELISA. *Journal of Virological Methods* 11, 309–319.

Halfhill, M.D., Millwood, R.J., Raymer, P.L. and Stewart, C.N. (2002) Bt-transgenic oilseed rape hybridization with its weedy relative, *Brassica rapa*. *Environmental Biosafety Research* 1, 19–28.

Hardwick, N.V., Davies, J.M.L. and Wright, D.M. (1994) The incidence of three virus diseases of winter oilseed rape in England and Wales in the 1991/92 and 1992/93 growing seasons. *Plant Pathology* 43, 1045–1049.

Hughes, L., Green, S.K., Lydiate, D.J. and Walsh, J.A. (2002) Resistance to turnip mosaic virus in *Brassica rapa* and *B. napus* and the analysis of genetic inheritance in selected lines. *Plant Pathology* 51, 567–573.

Jenner, C.E. and Walsh, J.A. (1996) Pathotype variation in turnip mosaic virus with special reference to European isolates. *Plant Pathology* 45, 848–856.

Jørgensen, R.B. and Anderson, B. (1994) Spontaneous hybridization between oilseed rape (*Brassica napus*) and weedy *B. campestris* (Brassicaceae): a risk of growing genetically modified oilseed rape. *American Journal of Botany* 81, 1620–1626.

Killick, J., Perry, R. and Woodell, S. (1998) *The Flora of Oxfordshire*. Pisces Publications, Information Press, Oxford.

Kling, J. (1996) Could transgenic supercrops one day breed superweeds? *Science* 274, 180–181.

Lehmann, P., Jenner, C.E., Kozubek, E., Greenland, A.J. and Walsh, J.A. (2003) Coat protein-mediated resistance to turnip mosaic virus in oilseed rape (*Brassica napus*). *Molecular Breeding* 11, 83–94.

Lowe, A.J., Jones, A.M.E., Raybould, A.F., Trick, M., Moule, C.J. and Edwards, K.J. (2002) Transferability and genome specificity of a new set of microsatellite primers among *Brassica* species of the U triangle. *Molecular Ecology Notes* 2, 7–11.

Maskell, L.C., Raybould, A.F., Cooper, J.I., Edwards, M.-L. and Gray, A.J. (1999) Effects of turnip mosaic virus and turnip yellow mosaic virus on the survival, growth and reproduction of wild cabbage (*Brassica oleracea*). *Annals of Applied Biology* 135, 401–407.

Mowat, W.P. (1981) Turnip crinkle virus in swede and turnip. *Scottish Crop Research Institute First Annual Report* 117.

Pallett, D.W., Thurston, M.I., Cortina-Borja, M., Edwards, M.L., Alexander, M., Mitchell, E., Raybould, A.F. and Cooper, J.I. (2002) The incidence of viruses in wild *Brassica rapa* ssp. *sylvestris* in Southern England. *Annals of Applied Biology* 141, 163–170.

Raybould, A.F. and Gray, A.J. (1993) Genetically modified crops and hybridisation with wild relatives: a UK perspective. *Journal of Ecology* 30, 199–219.

Raybould, A.F., Maskell, L.C., Edwards, M.-L., Cooper, J.I. and Gray, A.J. (1999a) The prevalence and spatial distribution of viruses in natural populations of *Brassica oleracea*. *New Phytologist* 141, 265–275.

Raybould, A.F., Mogg, R.J., Clarke, R.T., Gliddon, C.J. and Gray, A.J. (1999b) Variation and population structure at microsatellite and isozyme loci in wild cabbage (*Brassica oleracea* L.) in Dorset (UK). *Genetic Resources and Crop Evolution* 46, 351–360.

Raybould, A.F., Alexander, M.J., Mitchell, E., Thurston, M.I., Pallett, D.W., Hunter, P., Walsh, J.A., Edwards, M.-L., Jones, A.M.E., Moyes, C.L., Gray, A.J. and Cooper, J.I. (2003) The ecology of turnip mosaic virus in populations of wild *Brassica* species. In: Beringer, J., Godfray, C.H.J. and Hails, R.A. (eds) *Genes in the Environment*. Blackwell Scientific Press, Oxford, pp. 226–244.

Scott, S. and Wilkinson, M. (1998) Transgene risk is low. *Nature* 393, 320.

Shattuck, V.I. (1992) The biology, epidemiology and control of turnip mosaic virus. *Offprints from Plant Breeding Reviews* 14, 199–238.

Szewc-McFadden, A.K., Kresovich, S., Bliek, S.M., Mitchell, S.E. and McFerson, J.R. (1996) Identification of polymorphic, conserved simple sequence repeats (SSRs) in cultivated *Brassica* species. *Theoretical and Applied Genetics* 93, 534–538.

Thurston, M. (2002) Virus accumulation dynamics, pathogenesis and fitness implications for *Brassica* spp. PhD Thesis, University of Oxford.

Thurston, M.I., Pallett, D.W., Cortina-Borja, M., Edwards, M.-L., Raybould, A.F. and Cooper, J.I. (2001) The incidence of viruses in wild *Brassica nigra* in Dorset (UK). *Annals of Applied Biology* 139, 277–284.

U, N. (1935) Genome analysis in *Brassica* with special reference to the experimental formation of *B. napus* and the peculiar mode of fertilization. *Japanese Journal of Botany* 7, 389–452.

Walsh, J.A. (2000) Transgenic approaches to disease resistant plants as exemplified by viruses. In: Dickinson, M. and Beynon, J. (eds) *Molecular Plant Pathology*. Sheffield Academic Press, Sheffield, UK, pp. 218–252.

Walsh, J.A. and Jenner, C.E. (2002) *Turnip mosaic virus* and the quest for durable resistance. *Molecular Plant Pathology* 3, 289–300.

Walsh, J.A. and Tomlinson, J.A. (1985) Viruses infecting winter oilseed rape (*Brassica napus* ssp. *oleifera*). *Annals of Applied Biology* 107, 485–495.

Walsh, J.A., Perrin, R.M., Miller, A. and Laycock, D.S. (1989) Studies on beet western yellows virus in winter oilseed rape (*Brassica napus* ssp. *oleifera*) and the effect of insecticidal treatment upon its spread. *Crop Protection* 8, 137–143.

Walsh, J.A., Sharpe, A.G., Jenner, C.E. and Lydiate, D.J. (1999) Characterisation of resistance to turnip mosiac virus in oilseed rape (*Brassica napus*) and genetic mapping of *TuRB01*. *Theoretical and Applied Genetics* 99, 1149–1154.

Walsh, J.A., Rusholme, R.L., Hughes, S.L., Jenner, C.E., Bambridge, J.M., Lydiate, D.J. and Green, S.K. (2002) Different classes of resistance to turnip mosaic virus in *Brassica rapa*. *European Journal of Plant Pathology* 108, 15–20.

Wilkinson, M.J., Davenport, I.J., Charters, Y.M., Jones, A.E., Allainguillaume, J., Butler, H.T., Mason, D.C. and Raybould, A.F. (2000) A direct regional scale estimate of transgene movement from genetically modified oilseed rape to its wild progenitors. *Molecular Ecology* 9, 983–991.

24 Environmental and Agronomic Consequences of Herbicide-resistant (HR) Canola in Canada

Suzanne I. Warwick[1], Hugh J. Beckie[2], Marie-Josée Simard[3], Anne Légère[3], Hari Nair[2] and Ginette Séguin-Swartz[2]

[1]Agriculture and Agri-Food Canada (AAFC), Ottawa, ON, Canada K1A 0C6; [2]AAFC, Saskatoon, SK, Canada S7N 0X2; [3]AAFC, Ste-Foy, QC, Canada G1V 2J3; E-mail: warwicks@agr.gc.ca

Abstract

Herbicide-resistant (HR) canola (oilseed rape, *Brassica napus*) was first introduced in Canada in 1995. By 2001, HR cultivars accounted for 80% of the 4.0 Mha of *B. napus* grown in Canada: 47% glyphosate resistant, 13% glufosinate resistant, 20% imidazolinone resistant (non-transgenic) and less than 1% bromoxynil resistant. The environmental and agronomic consequences of HR *B. napus* canola in Canada will be reviewed in this chapter. Both *B. napus* and *B. rapa* are grown as canola in Canada. The introduction and preference for HR *B. napus* lines have resulted in a decline in *B. rapa* production from 50% in 1995 to approximately 5% in 2002. From an initial bottle-neck in HR *B. napus* cultivar diversity, HR cultivar number has increased rapidly from two in 1996 to 33 in 2002. The large-scale use of HR *B. napus* canola has provided an opportunity to test current models of intraspecific and interspecific pollen flow on a realistic field scale. The HR trait is easy to monitor, provides accurate assessments and is highly suited for large-scale screening programmes. Recent Canadian studies utilizing the HR trait to evaluate pollen-mediated gene flow in *B. napus* on a commercial field scale, and the effect of gene flow on HR volunteers (feral canola) in subsequent years, are described. These studies have shown that both pollen and seed movement are important means of escape for the HR trait in canola. Unintentional gene stacking of HR traits in *B. napus* canola volunteers resulting from intraspecific pollen flow is common. Data from two recent studies indicate that the adventitious presence of off-types (contaminants) in certified seed lots often exceeds stipulated thresholds. Recommended control measures for multiple HR volunteer canola in commercial fields are reviewed. Interspecific hybridization, on the other hand, is a less likely consequence of gene flow. Results from a 3-year gene flow study between *B. napus* and four related weedy species (*B. rapa*, *Raphanus raphanistrum*, *Erucastrum gallicum* and *Sinapis arvensis*) in Canada are summarized. These results include data from experimental field trials and commercial HR *B. napus* canola fields.

Hybridization between HR *B. napus* and natural wild populations of *B. rapa* was confirmed in two commercial HR *B. napus* canola fields in Québec, thus representing the first documented occurrence of transgene escape into a natural weed population (studies to confirm introgression of the trait into the *B. rapa* genome are in progress). The most obvious consequence of large-scale HR crop introduction is the selection of HR biotypes in both related and unrelated weed species. There is, to date, no evidence of selection of HR biotypes in unrelated weed species as a result of HR *B. napus* canola production in Canada. There is also no evidence for changes in weed diversity, i.e. towards more tolerant species, due to herbicide use patterns associated with HR *B. napus* canola. Although stewardship plans are in place to mitigate agronomic problems associated with HR *B. napus* canola, widespread implementation depends on greater awareness among growers of best management practices.

Introduction

During the period from 1995 to 2001, herbicide resistance has consistently been the dominant trait of global commercial transgenic crop production. In 2001, herbicide-resistance (HR) canola or oilseed rape (*Brassica napus* L.) ranked as the third most abundant transgenic crop, accounting for approximately 5% of the global transgenic crop area. A total of 2.7 out of the 25 Mha of canola grown was transgenic (James, 2001).

B. napus (*AACC*, $2n = 38$) and *B. rapa* L. (*AA*, $2n = 20$) are both grown as canola in Canada. Commercial production of glyphosate- and glufosinate-resistant *B. rapa* cultivars was limited in western Canada in 1998; however, these cultivars are no longer registered. Since its introduction in 1995, HR *B. napus* canola has been rapidly and widely adopted by Canadian growers for several reasons, including easier and better weed control, higher seed yields and higher financial net returns based primarily on the higher yield, reduced dockage (i.e. percentage of weeds seeds as a weight basis in harvested *B. napus* that reduces crop value) and lower herbicide costs (Devine and Buth, 2001). HR *B. napus* canola has allowed growers to reduce annual herbicide usage by 6000 t, reduce tillage previously required for weed control and incorporation of soil-applied herbicides, and consequently reduce fuel consumption by 3.2 Ml (Canola Council of Canada, 2001). In 2001, of the 4.0 Mha of *B. napus* grown in Canada, 20% were non-HR, 47% were glyphosate resistant, 13% were glufosinate resistant, less than 1% were bromoxynil resistant and 20% were imidazolinone resistant (Fig. 24.1). Imidazolinone-resistant cultivars are non-transgenic, derived by chemically induced genetic mutation.

Regulations that permit the production of HR *B. napus* vary greatly from country to country. In Canada, all four HR types of *B. napus* are regulated identically as 'plants with novel traits' (PNTs). The Canadian Food Inspection Agency defines PNTs as

> plant varieties/genotypes that are not considered 'substantially equivalent', in terms of their specific use and safety both for environment and for human health, to plants of the same species in Canada, having regard to weediness

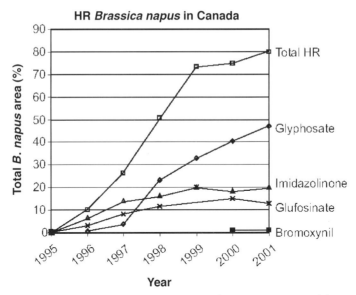

Fig. 24.1. Herbicide-resistant (HR) *B. napus* canola as a percentage of the total area grown in western Canada (crop area of 4.0 Mha in 2001).

potential, gene flow, plant pest potential, impact on non-target organisms, impact on biodiversity, anti-nutritional factors and nutritional composition.

PNTs may be produced by conventional breeding, mutagenesis or, more commonly, by recombinant DNA techniques (Canadian Food Inspection Agency, 2002). This procedure has scientific merit, because it is the impact of the trait that requires regulation rather than the process by which the trait is introduced into the crop.

Changes in Crop Diversity

One of the main consequences of the adoption of HR *B. napus* canola cultivars has been a reduction in crop diversity due to a reduction of *B. rapa* production from 50% in 1995 to approximately 5% of canola production in 2002. The decline in *B. rapa* production was due to low yields and was accelerated by the advent of HR *B. napus* cultivars. From an initial bottle-neck in *B. napus* HR cultivar diversity, the number of HR cultivars has increased rapidly from two in 1996 to 33 in 2002 (similar to non-HR cultivar number).

Environmental Impact

The introduction of HR crops potentially can have both direct and indirect effects on other species. These effects would be restricted primarily to the agroecosystem where herbicides are regularly applied. The HR trait (or transgene) may escape via either seed or pollen flow. The HR trait may be confined to the crop, but the crop itself may become a weedy volunteer through seed escape. Gene flow can be problematic to control in a crop such as canola that forms volunteer (feral) populations. The persistence of the HR trait in the seed bank and subsequent pollen flow from the volunteer populations can result in escape of the HR trait to adjacent non-HR canola fields or to other HR cultivars, resulting in gene stacking, i.e. the combining of two or more independent genes in a single plant. Seed escape via farming operations (e.g. seeding and harvesting equipment, trucks and storage facilities) may contribute to seed admixtures. Alternatively, the HR trait may escape to wild relatives via pollen flow and interspecific hybridization, potentially increasing the weediness of the recipient species. Thus, the potential environmental impact of HR *B. napus* canola is of particular concern, as the crop is partially outcrossing, forms a persistent seed bank and volunteer weed populations in subsequent crops, and has several wild relatives present in cultivated areas in Canada. Even when the HR trait is confined to the crop species, non-related weed species may be affected indirectly as a result of herbicides used in the production of such HR crops, namely evolution of HR weed biotypes or shifts in weed communities to more tolerant species.

Intraspecific Pollen Flow

Canola is self-fertile, with pollen movement by both wind and insects. Several Canadian studies have indicated inter-plant outcrossing rates averaging 30% (e.g. Rakow and Woods, 1987; reviewed in Beckie *et al.*, 2003). The degree of outcrossing between populations of canola is sharply reduced as the distance between the pollen source and the recipient population increases. Although the vast majority of pollen produced by canola plants falls within a few metres, pollen flow from fields of canola by wind and insects can be substantial and extend over long distances. Studies have found evidence of pollen flow in Canada up to 366 m (Stringam and Downey, 1982), in the UK at 400 m (Scheffler *et al.*, 1995), 2.5 km (Timmons *et al.*, 1995) and 4 km (Thompson *et al.*, 1999), and in Australia up to 3 km away from the pollen source (Rieger *et al.*, 2002). Pollen flow in canola is affected by many variables, including canola variety (male-fertile versus male-sterile lines), the relative size of pollen donor and recipient populations, presence of a pollen trap or border row, environmental conditions (temperature, wind speed and direction, relative humidity, etc.), and presence of insect vectors (Ingram, 2000; Staniland *et al.*, 2000).

   To test pollen flow models on a realistic field scale, gene flow between adjacent commercial fields of glyphosate- and glufosinate-resistant canola was examined at 11 locations in Saskatchewan, Canada in 1999 (Beckie *et al.*, 2001, 2003). Seed was collected from 0 to 800 m along a transect perpendicular to the field border and screened for HR in the greenhouse. Resistance arising from pollen flow was confirmed with laboratory tests (protein test strips and/or molecular polymerase chain reaction (PCR) analyses). In 1999, *ex situ* estimation of gene flow between the paired fields ranged from 1.4% at the common border to 0.4% at 400 m, with no gene flow detected at 600 or 800 m (Fig. 24.2). *In situ* estimates of gene flow were assessed in three of the 11 paired fields by mapping double HR volunteers that survived sequential herbicide applications the following spring. Gene flow as a result of pollen flow in 1999 was orders of magnitude higher (2.5–10%) and detectable to the limits of the study areas (800 m). Large variations in gene flow levels and patterns were evident among the three sites. Such variability will make modelling and the accurate prediction of gene flow in canola grown on a commercial scale very difficult.

Weedy Canola Crop Volunteers

Volunteer canola is often a common weed in subsequent crops as a result of shattering and seed loss during harvesting operations. In a study in Saskatchewan in 1999 and 2000, average *B. napus* yield losses of 5.9% (3000 viable seeds/m^2) were measured in 35 growers' fields (Gulden *et al.*, 2003) and ranged from 3.3 to 9.9% yield loss or 9–56 times the normal seeding rate of canola. In western Canada, volunteer canola occurs in 11% of cropped fields and is ranked 18th in relative abundance of all weeds (Beckie *et al.*,

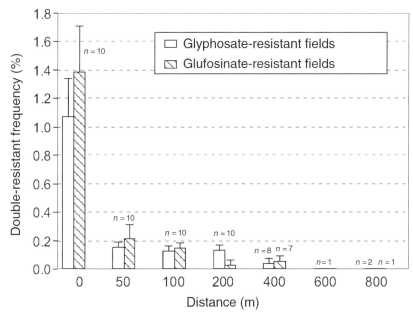

Fig. 24.2. Outcrossing (bars = SE) between adjacent glyphosate- and glufosinate-resistant *B. napus* canola fields in Saskatchewan in 1999, based on frequency of occurrence of confirmed double-resistant plants (glyphosate and glufosinate) as a function of distance from the common border (*n* = number of fields; when one value is shown per distance, it denotes the number of glyphosate- and glufosinate-resistant fields) (from Beckie *et al.*, 2003).

2001; Légère *et al.*, 2001). Comparative weed survey data (1997 and 2002) from Manitoba and Alberta (Leeson *et al.*, 2002a, b) indicated an increase in both the frequency of occurrence and relative abundance of volunteer canola since the introduction of HR *B. napus* canola. Volunteer canola densities in the following crop can be high prior to herbicide control, often exceeding 100 plants/m². Volunteer canola can persist for a minimum of 4–5 years after production in Canada (Légère *et al.*, 2001; Simard *et al.*, 2002), compared with up to 10 years in Europe (Lutman and López-Granados, 1998). Volunteers can serve as a potential pollen source for wild relatives and canola crops that follow in rotation or are located in nearby fields, and thereby extend the potential for gene flow spatially and temporally (Warwick *et al.*, 1999, 2003; Beckie *et al.*, 2003).

The potential weediness of HR volunteer canola is obviously affected by any associated fitness cost to HR. In contrast to triazine-resistant canola, which is less fit than susceptible plants (Beversdorf *et al.*, 1988), no associated fitness cost has been observed in the four HR *B. napus* canola types grown commercially in Canada (Kumar *et al.*, 1998; Cuthbert *et al.*, 2001; Simard *et al.*, 2003).

Following the introduction of different HR *B. napus* canola types, the possibility of gene stacking increases over time. Hall *et al.* (2000) reported that pollen flow between cultivars with different HR traits had resulted in canola volunteers with multiple resistance at a field site in western Canada. In this instance, a field of glyphosate-resistant canola was grown adjacent to a field containing both glufosinate-resistant and imidazolinone-resistant canola. Volunteers from the latter field were selected with glyphosate the following year. The surviving volunteers flowered and produced seeds that contained individuals resistant to glyphosate and glufosinate, glyphosate and imazethapyr, and glyphosate, imazethapyr and glufosinate. Two triple-resistant individuals were detected, with one of these plants located 550 m from the glyphosate-resistant pollen source. The results of the above study and that of Beckie *et al.* (2003) suggest that gene stacking in *B. napus* canola volunteers in western Canada may be common.

Multiple HR volunteers have, to date, not caused problems to the average Canadian grower, with the exception of those who elect not to grow HR *B. napus* canola or organic growers. The management of HR volunteers does require a specific stewardship plan, particularly when broadleaf crops with few in-crop herbicide options are grown in a rotation. Herbicides with alternative modes of action, such as phenoxy herbicides (e.g. 2,4-D, MCPA) or photosystem II inhibitors applied alone or in a mixture, provide effective control of canola volunteers with single or stacked HR traits, such that yield of subsequent crops is not affected. There is no evidence for altered herbicide sensitivity of single or multiple resistant plants to herbicides of alternative modes of action due to the genetic transformation (i.e. pleiotrophic effects) (Senior *et al.*, 2002; Beckie *et al.*, 2004). However, effective control of target weeds by a herbicide is defined in Canada as more than 80% efficacy and, by that definition, HR *B. napus* canola volunteers are not eliminated. Where initial volunteer density is high (the year following *B. napus* cultivation), a substantial number of individuals can survive to maturity, resulting in replenishment of the seed bank.

Seed Contamination

The presence of off-types in certified seed lots of canola must be expected as a result of pollen and seed movement. In Canada, a maximum of 0.25% off-types or adventitious presence is permitted in commercial certified canola seed (Canadian Seed Growers Association, 2002). Prior to the introduction of the HR trait, there were no definitive genetic markers to quantify precisely levels of genetic purity in canola cultivar seed lots. By definition, off-types in non-HR *B. napus* canola seed lots may include individual seeds that contain HR genes, singly or stacked; in the case of HR *B. napus* canola cultivars, non-HR seed or another HR type would be considered an off-type. In order to reduce pollen flow, an isolation distance of 800 m is required for growers of certified seed of hybrid *B. napus* canola in Canada, whereas a 100 m isolation distance is stipulated currently for certified seed growers of pedigree-derived cultivars (Canadian Seed Growers Association, 2002). However, these distances do not preclude gene flow (as discussed above). and it is not surprising that data from two recent studies in Canada have provided evidence that off-type levels (in particular stacked HR) in certified seed frequently exceed the stipulated threshold. Downey and Beckie (2002) found that 35 of 70 certified *B. napus* seed lots tested from 14 herbicide-susceptible, open-pollinated cultivars produced in 2000 contained the gene conferring glyphosate resistance, and 41 seed lots (59%) contained the glyphosate or glufosinate resistance gene. Only two cultivars were free of both genes. Friesen *et al.* (2003) also found that 14 of 27 commercial certified seed lots tested had contamination levels above 0.25%.

Unexpected contamination, even at 0.25%, can cause problems for growers that practice direct seeding and depend on glyphosate for non-selective broad-spectrum weed control. The consequence of the presence of a 0.25% contamination level in certified seed in two commercial fields was documented by Beckie *et al.* (2003). Data from two adjacent glyphosate- and glufosinate-resistant commercial fields of *B. napus* canola are shown in Fig. 24.3. In the subsequent spring, putative double HR volunteers that survived sequential herbicide applications were mapped in the two fields using GPS (global positioning system; shown in Fig. 24.3, left) and resistance in sampled plants was characterized. Double HR adventitious seed was known to be present in the glyphosate-resistant seedlot at 0.30%. The adventitious seed had a different gene (*pat* gene) for glufosinate resistance compared with the adjacent glufosinate-resistant field (*bar* gene). It was therefore possible to separate double HR plants that grew as volunteers in these two fields in the subsequent year into two categories: those with the *bar* gene and therefore produced as a result of pollen flow between the two fields (Fig. 24.3, middle), and those with the *pat* gene and therefore derived from the adventitious seed (Fig. 24.3, right). Results from this study indicated a

Glufosinate-resistant

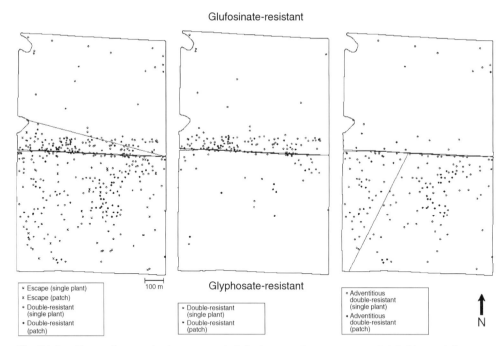

Fig. 24.3. Two adjacent glyphosate- and glufosinate-resistant commercial fields, and the location of the hybrids formed (growing season 2000 for one of the 11 paired fields sites shown in Fig. 24.2). Left: distribution of confirmed double-resistant (glyphosate and glufosinate) *B. napus* canola volunteers (single plant or patch) and volunteers that escaped herbicide control. Middle: occurrence of confirmed double-resistant volunteers as a result of gene flow. Right: occurrence of confirmed double-resistant volunteers due to the presence of adventitious double herbicide-resistant seeds in the planted seed lot (from Beckie *et al.*, 2003).

rapid build-up of the double HR adventitious seed in the volunteer seed bank in the subsequent year. They also showed that the HR adventitious plants in the glyphosate-resistant field served as a pollen source in gene flow events to the adjacent glufosinate field.

Gene Flow or Hybridization to Related Species

A major concern about the agricultural release of genetically modified organisms (GMOs) is the escape of transgenes in the environment through hybridization with their wild relatives (Warwick *et al.*, 1999; Snow, 2002). Canola has numerous wild relatives present in cultivated areas in Canada and worldwide (reviewed in Chèvre *et al.*, Chapter 18, this volume). Several studies have indicated the possibility for genetic exchange between *Brassica* species and related weedy species found in Canada, including crossing of *B. napus* with *B. rapa* (Jørgensen and Andersen, 1994; Bing *et al.*, 1996; Landbo *et al.*, 1996; Hansen *et al.*, 2001; Halfhill *et al.*, 2002) and *Raphanus raphanistrum*

L. (Darmency *et al.*, 1998; Chèvre *et al.*, 2000; Rieger *et al.*, 2001) under field conditions.

Results from a 3-year Canadian gene flow study between *B. napus* and four related weedy species (*B. rapa*, *R. raphanistrum*, *Erucastrum gallicum* (Willd.) O.E. Schulz and *Sinapis arvensis* L.) are summarized below (Warwick *et al.*, 2003). These results include data from experimental field trials and commercial HR *B. napus* canola fields. *S. arvensis* is a major weed in canola-growing areas in western North America, whereas the other three weed species have more limited distributions. Gene flow was inferred by screening seed collected from the wild populations for the presence of the HR trait found in adjacent HR *B. napus* canola fields.

Hybridization between *B. rapa* and *B. napus* was expected and, indeed, occurred at a frequency of approximately 7% in two field experiments where plants of *B. rapa* were grown at a density of 1 plant/m² with HR *B. napus*. *B. rapa* × *B. napus* F_1 hybrids were also detected in two *B. rapa* populations growing in or near commercial HR *B. napus* canola fields in Québec. This represents the first case of transgene escape into a natural weed population (studies to confirm introgression of *B. napus* traits into the *B. rapa* genome are in progress). An even higher frequency of hybridization (13.6%) was observed in one of the wild *B. rapa* populations and was probably due to greater isolation distance between *B. rapa* plants. All F_1 hybrids were morphologically similar to *B. rapa*, but hybrids were confirmed by the presence of the HR trait, triploid ploidy level (*AAC*, $2n = 29$ chromosomes) and presence of *B. napus*- and *B. rapa*-specific amplified fragment length polymorphism (AFLP) molecular markers. The hybrids had reduced pollen viability (~55%) and segregated for both self-incompatible and self-compatible individuals (the latter being a *B. napus* trait). In contrast, gene flow between *R. raphanistrum* and *B. napus* was rare. A single *R. raphanistrum* × *B. napus* F_1 hybrid was obtained in an HR *B. napus* field plot experiment where *R. raphanistrum* plants were grown at a density of 1 plant/m² with HR *B. napus*, and no hybrids were detected in HR commercial fields in Québec and Alberta (22,114 seedlings). Except for the presence of a *B. napus* trait 'opening of the seed pod by valves' and distortion of the seed pods, the hybrid was morphologically similar to *R. raphanistrum*. This hybrid had a genomic structure consistent with the fusion of an unreduced gamete of *R. raphanistrum* and a reduced gamete of *B. napus* (*RrRrAC*, $2n = 37$), both *B. napus*- and *R. raphanistrum*-specific AFLP markers, and less than 1% pollen viability. No *S. arvensis* × or *E. gallicum* × *B. napus* hybrids were detected (42,828 and 21,841 seedlings, respectively) from commercial HR *B. napus* canola fields in Saskatchewan. These findings suggest that the probability of gene flow from *B. napus* to *R. raphanistrum*, *S. arvensis* or *E. gallicum* is very low (< 2–5×10^{-5}). Our results are in accordance with gene flow predictions based on previous studies.

The two natural populations of *B. rapa*, where *B. rapa* × *B. napus* F_1 hybrids were found, will be monitored for persistence of the glyphosate resistance trait and for evidence of introgression of the HR transgene into the *B. rapa* genome (Warwick *et al.*, studies in progress). Previous studies have shown that an HR transgene can be passed between the two species and be active in successive generations (Mikkelsen *et al.*, 1996; Metz *et al.*, 1997). Such HR transgenic hybrids would have an obvious selective advantage in an agroecosystem where the herbicide is applied. Similar to HR volunteer canola, these hybrids may require altered or additional control measures but can generally be controlled in subsequent crops with herbicides with different modes of action. Also, we may assume no fitness cost (as discussed above) in the acquisition of HR to glyphosate, glufosinate and bromoxynil in a hybrid, and therefore persistence of these genes in the weedy populations in the absence of herbicide selection pressure. Snow *et al.* (1999) showed that fitness costs associated with transgenic glufosinate resistance introgressed from *B. napus* into weedy *B. rapa* were negligible.

Evolution of HR Biotypes

The most obvious consequence of the release of HR crops is a shift in the weed spectrum towards more tolerant weed species or the selection of biotypes in both related and unrelated weed species resistant to herbicides used in such crops (Warwick *et al.*, 1999). Worldwide, the number of HR weeds has expanded exponentially during the past two decades from the first documented occurrence of triazine resistance in 1968 to 272 HR weed biotypes (162 species) in 2003 (Heap, 2003).

To date, no bromoxynil-resistant weed biotypes have evolved as a result of herbicide use in bromoxynil-resistant canola, nor is it expected because such cultivars have not been grown in Canada since 2001. There is only one report of a bromoxynil-resistant weed biotype, *Senecio vulgaris* L., from the USA (Oregon) in 1995 (Mallory-Smith, 1998; Heap, 2003). Glyphosate- or glufosinate-resistant weed biotypes have not evolved as a result of the use of these herbicides in commercial HR *B. napus* canola fields in Canada. Glufosinate is used only in glufosinate-resistant canola. Glufosinate resistance in weeds in Canada is unlikely because of the low frequency of canola in typical rotations (25–30%) due to disease pressure, the relatively low acreage of glufosinate-resistant canola, and low risk status of glufosinate to select for resistance in weeds. Although glyphosate also is a low-risk herbicide, the rapidly increasing use of this herbicide in Canadian farming systems will increase selection pressure for weed resistance. It would be difficult to attribute glyphosate-resistant weeds solely to glyphosate use in glyphosate-resistant canola because of its use pattern (e.g. chemical fallow; pre-seeding, preharvest, postharvest) in Canadian cropping systems.

Worldwide, only four glyphosate-resistant species have been reported (Heap, 2003). However, the recent reports (Heap, 2003) of large areas of glyphosate resistance in the dicot weed, *Conyza canadensis* (L.) Cronq., in areas of extensive glyphosate-resistant crop (cotton, soybean and maize) production in eastern North America (Delaware (2000) (Van Gessel, 2001), Tennessee (2001), Indiana, Maryland, New Jersey and Ohio (2002)) are of concern. Suspect Ontario populations of *C. canadensis* are currently under investigation (F. Tardiff, University of Guelph, Ontario, 2002, personal communication). The number of glyphosate-resistant weeds is expected to increase gradually given the increasing number of glyphosate-resistant crops. Estimates of mutation rates for resistance clearly vary with herbicide mode of action and may be higher or lower than the likelihood of a wild relative of canola acquiring the HR trait through interspecific hybridization, depending on the wild species involved. For example, the probability of evolved glyphosate resistance in *B. rapa* would be much less than resistance acquired via hybridization through gene flow with canola.

Similar to glyphosate resistance in weeds, quantifying the contribution of acetolactate synthase (ALS) inhibitor herbicide use in imidazolinone-resistant canola to selection of ALS-resistant weed biotypes is difficult because ALS herbicides (includes sulphonylureas and imidazolinones) are frequently applied in cereal and legume crops in Canada. For example, ALS herbicides are currently applied to 35% of the agricultural land in Alberta based on recent weed survey data from Leeson *et al.* (2002b). ALS weed biotypes are the most readily growing class of HR weeds in Canada (Warwick, 1999), with several new biotypes described in 2001 and 2002. It is also the most rapidly growing class of HR weed biotypes worldwide, with 73 reported ALS-resistant weed biotypes (Heap, 2003). The approval of several ALS-resistant crops, including canola, soybean (*Glycine max* (L.) Merr.) and wheat (*Triticum aestivum* L.), and the subsequent increased use of group 2 herbicides are likely to contribute to the high selection pressure for ALS-resistant biotypes in cropping systems in Canada.

Sound management strategies, such as crop and herbicide rotations, must be implemented to prevent or delay the evolution of HR weed biotypes. Inclusion of HR crops in the rotation may complicate herbicide rotation options, particularly when crops in a rotation permit repeated use of herbicides of the same herbicide mode of action. On the positive side in Canada, inclusion of glufosinate- and glyphosate-resistant canola in rotations has facilitated opportunities for rotation of herbicide modes of action and proactive or reactive management of acetyl-CoA carboxylase (ACCase)- and ALS-resistant wild oat (*Avena fatua* L.) that are pervasive in the Canadian prairies.

Changes in Weed Species Dominance and Biodiversity

As a result of large-scale adoption of HR crops and associated changes in management practices, changes more subtle than weed resistance are expected in the diversity and structure of weed communities. For example, differential levels of tolerance to glyphosate have led to changes in weed succession and a shift from annual to perennial weed species (reviewed in Baylis, 2000; Shaner, 2000). In contrast, Stoltenberg and Jeschke (2003) concluded that weed community changes were no greater in glyphosate-resistant maize and soybean systems than those based on conventional pre-emergence soil-residual herbicide programmes. Orson (2002) indicated that gene stacking of HR traits in volunteer canola has implications for biodiversity in cropped and non-cropped areas. Attempts to minimize the number of volunteers may result in intensification in the use of cultural or chemical control methods in subsequent crops within the rotation or in non-cropped areas, such as roadsides or waste ground. Changes in weed biodiversity in Canada due to HR *B. napus* canola in rotations have not yet been detected (Derksen *et al.*, 1999). Given the relatively low frequency of canola in cropping systems (i.e. one in four in the rotation), it is unlikely that there will be a significant impact on weed biodiversity within Canadian agroecosystems. Weed species diversity (and dominance) may, in the future, result from the growing of a greater range of crops resistant to herbicides with the same mode of action or to herbicides which provide exceptional weed control.

Acknowledgements

We wish to thank Brian Miki and Connie Sauder, AAFC-Ottawa for providing reviews of earlier drafts of this chapter.

References

Baylis, A.D. (2000) Why glyphosate is a global herbicide: strengths, weaknesses and prospects. *Pest Management Science* 56, 299–308.

Beckie, H.J., Hall, L.M. and Warwick, S.I. (2001) Impact of herbicide-resistant crops as weeds in Canada. In: *Proceedings Brighton Crop Protection Conference – Weeds*. British Crop Protection Council, Farnham, UK, pp. 135–142.

Beckie, H.J., Warwick, S.I., Nair, H. and Séguin-Swartz, G. (2003) Gene flow in commercial fields of herbicide-resistant canola. *Ecological Applications* 13, 1276–1294.

Beckie, H.J., Séguin-Swartz, G., Nair, H., Warwick, S.I. and Johnson, E. (2004) Control of canola (*Brassica napus*) with multiple herbicide resistance traits by alternative herbicides. *Weed Science* 52, 152–157.

Beversdorf, W.D., Hume, D.J. and Donnelly-Vanderloo, M.J. (1988) Agronomic performance of triazine-resistant and susceptible reciprocal spring canola hybrids. *Crop Science* 28, 932–934.

Bing, D.J., Downey, R.K. and Rakow, G.F.W. (1996) Hybridizations among *Brassica napus*, *B. rapa* and *B. juncea* and their two weedy relatives *B. nigra* and *Sinapis arvensis* under open pollination conditions in the field. *Plant Breeding* 115, 470–473.

Canadian Food Inspection Agency (2002) *The Regulation of Plants with Novel Traits in Canada*. Plant Health and Production Division, Plant Biosafety Office. Available at: http://www.inspection.gc.ca/english/plaveg/pbo/pbobbve.shtml

Canadian Seed Growers Association (2002) Regulations and procedures for pedigreed seed crop inspection. Circular 6–94 (revised). Available at: http://www.seedgrowers.ca

Canola Council of Canada (2001) *An Agronomic and Economic Assessment of Transgenic Canola*. Report. Winnipeg, Manitoba, January.

Chèvre, A.M., Eber, F., Darmency, H., Fleury, A., Picault, H., Letanneur, J. and Renard, M. (2000) Assessment of interspecific hybridization between transgenic oilseed rape and wild radish under normal agronomic conditions. *Theoretical and Applied Genetics* 100, 1233–1239.

Cuthbert, J.L., McVetty, P.B.E., Freyssinet, G. and Freyssinet, M. (2001) Comparison of the performance of bromoxynil-resistant and susceptible near-isogenic populations of oilseed rape. *Canadian Journal of Plant Science* 81, 367–372.

Darmency, H., Lefol, E. and Fleury, A. (1998) Spontaneous hybridizations between oilseed rape and wild radish. *Molecular Ecology* 7, 1467–1473.

Derksen, D.A., Harker, K.N. and Blackshaw, R.E. (1999) Herbicide-tolerant crops and weed population dynamics in western Canada. In: *Proceedings Brighton Crop Protection Conference – Weeds.* British Crop Protection Council, Farnham, UK, pp. 417–422.

Devine, M.D. and Buth, J.L. (2001) Advantages of genetically modified canola: a Canadian perspective. In: *Proceedings Brighton Crop Protection Conference – Weeds.* British Crop Protection Council, Farnham, UK, pp. 367–372.

Downey, R.K. and Beckie, H. (2002) *Isolation Effectiveness in Canola Pedigree Seed Production.* Report to Canadian Seed Growers' Association. Agriculture and Agri-Food Saskatoon, Canada.

Friesen, L.F., Nelson, A.G. and Van Acker, R.C. (2003) Evidence of contamination of pedigreed canola (*Brassica napus*) seedlots in Western Canada with genetically engineered herbicide resistance traits. *Weed Science Society of America Abstracts* 43, 83.

Gulden, R.H., Shirtliffe, S.J. and Thomas, A.G. (2003) Harvest losses of canola (*Brassica napus*) cause large seedbank inputs. *Weed Science* 51, 83–86.

Halfhill, M.D., Millwood, R.J., Raymer, P.L. and Stewart, C.N. Jr (2002) Bt-transgenic oilseed rape hybridization with its weedy relative, *Brassica rapa. Environmental Biosafety Research* 1, 19–28.

Hall, L., Topinka, K., Huffman, J., Davis, L. and Good, A. (2000) Pollen flow between herbicide-resistant *Brassica napus* is the cause of multiple-resistant *B. napus* volunteers. *Weed Science* 48, 688–694.

Hansen, L.B., Siegismund, H.R. and Jørgenson, R.B. (2001) Introgression between oilseed rape (*Brassica napus* L.) and its weedy relative *B. rapa* L. in a natural population. *Genetic Resources and Crop Evolution* 48, 621–627.

Heap, I. (2003) International survey of herbicide-resistant weeds. Online. Available at: http://www.weedscience.org/in.asp

Ingram, J. (2000) The separation distance required to ensure cross-pollination is below specified limits in non-seed crops of sugar beet, maize and oilseed rape. *Plant Varieties Seeds* 13, 181–199.

James, C. (2001) *Global Review of Commercialized Transgenic Crops: 2001.* ISAAA Briefs No. 24–2001 ISAAA (International Service for the Acquisition of Agri-Biotech Applications), Ithaca, New York.

Jørgensen, R.B. and Andersen, B. (1994) Spontaneous hybridization between oilseed rape (*Brassica napus*) and weedy *B. campestris* (Brassicaceae): a risk of growing genetically modified oilseed rape. *American Journal of Botany* 81, 1620–1626.

Kumar, A., Rakow, G. and Downey, R.K. (1998) Isogenic analysis of glufosinate-ammonium tolerant and susceptible summer rape lines. *Canadian Journal of Plant Science* 78, 401–408.

Landbo, L., Andersen, B. and Jørgensen, R.B. (1996) Natural hybridisation between oilseed rape and a wild relative: hybrids among seeds from weedy *B. campestris. Hereditas (Lund)* 125, 89–91.

Leeson, J.Y., Thomas, A.G., Andrews, T., Brown, K.R. and Van Acker, R.C. (2002a) *Manitoba Weed Survey of Cereal and Oilseed Crops in 2002.* Weed Survey Series Publication 02–2. Agriculture and Agri-Food Canada, Saskatoon Research Centre, Saskatoon, Saskatchewan.

Leeson, J.Y., Thomas, A.G. and Hall, L.M. (2002b) *Alberta Weed Survey of Cereal, Oilseed and Pulse Crops in 2001.* Weed Survey Series Publication 02–1. Agriculture and Agri-Food Canada, Saskatoon Research Centre, Saskatoon, Saskatchewan.

Légère, A., Simard, M.-J., Thomas, A.G., Pageau, D., Lajeunesse, J., Warwick, S.I. and Derksen, D.A. (2001) Presence and persistence of volunteer canola in Canadian cropping systems. In: *Proceedings Brighton Crop Protection Conference – Weeds.* British Crop Protection Council, Farnham, UK, pp. 143–148.

Lutman, P.J.W. and López-Granados, F. (1998) The persistence of seeds of oilseed rape (*Brassica napus*). *Aspects of Applied Biology* 51, 147–152.

Mallory-Smith, C. (1998) Bromoxynil-resistant common groundsel (*Senecio vulgaris*). *Weed Technology* 12, 322–324.

Metz, P.L.J., Jacobsen, E., Nap, J.-P., Pereira, A. and Stiekema, W.J. (1997) The impact of biosafety of the phosphinothricin-tolerance transgene in inter-specific *B. rapa* × *B. napus* hybrids and their successive backcrosses. *Theoretical and Applied Genetics* 95, 442–450.

Mikkelsen, T.R., Andersen, B. and Jørgensen, R.B. (1996) The risk of crop transgene spread. *Nature* 380, 31.

Orson, J. (2002) Gene stacking in herbicide tolerant oilseed rape: lessons from the North American experience. *English Nature Research Reports* No. 443.

Rakow, G. and Woods, D.L. (1987) Outcrossing in rape and mustard under Saskatchewan prairie conditions. *Canadian Journal of Plant Science* 67, 147–151.

Rieger, M.A., Potter, T., Preston, C. and Powles, S.B. (2001) Hybridization between *Brassica napus* L. and *Raphanus raphanistrum* L. under agronomic conditions. *Theoretical and Applied Genetics* 103, 555–560.

Rieger, M.A., Lamond, M., Preston, C., Powles, S.B. and Roush, R.T. (2002) Pollen mediated movement of herbicide resistance between commercial canola fields. *Science* 296, 2386–2388.

Scheffler, J.A., Parkinson, R. and Dale, P.J. (1995) Evaluating the effectiveness of isolation distances for field plots of oilseed rape (*Brassica napus*) using a herbicide-resistance transgene as a selectable marker. *Plant Breeding* 114, 317–321.

Senior, I.J., Moyes, C. and Dale, P.J. (2002) Herbicide sensitivity of transgenic multiple herbicide-tolerant oilseed rape. *Pest Management Science* 58, 405–412.

Shaner, D.L. (2000) The impact of glyphosate-tolerant crops on the use of other herbicides and on resistance management. *Pest Management Science* 56, 320–326.

Simard, M.J., Légère, A., Pageau, D., Lajeunnesse, J. and Warwick, S.I. (2002) The frequency and persistence of canola (*Brassica napus*) volunteers in Québec cropping systems. *Weed Technology* 16, 433–439.

Simard, M.J., Légère, A., Séguin-Swartz, G., Nair, H. and Warwick, S.I. (2003) Fitness of double (imidazolinone + glufosinate) vs single herbicide resistant canola (*Brassica napus*). *Weed Science Society of America Abstracts* 43, 75–76.

Snow, A.A. (2002) Transgenic crops – why gene flow matters. *Nature Biotechnology* 20, 542.

Snow, A.A., Andersen, B. and Jørgensen, R.B. (1999) Costs of transgenic herbicide resistance introgressed from *Brassica napus* into weedy *Brassica rapa*. *Molecular Ecology* 8, 605–615.

Staniland, B.K., McVetty, P.B.E., Friesen, L.F., Yarrow, S., Freyssinet, G. and Freyssinet, M. (2000) Effectiveness of border areas in confining the spread of transgenic *Brassica napus* pollen. *Canadian Journal of Plant Science* 80, 521–526.

Stoltenberg, D.E. and Jeschke, M.R. (2003) Weed management and agronomic risks associated with glyphosate-resistant corn and soybean cropping systems. *Weed Science Society of America Abstracts* 43, 37.

Stringam, G.R. and Downey, R.K. (1982) Effectiveness of isolation distances in seed production of rapeseed (*Brassica napus*). *Agronomy Abstracts* 136–137.

Thompson, C.E., Squire, G., Mackay, G.R., Bradshaw, J.E., Crawford, J. and Ramsay, G. (1999) Regional patterns of gene flow and its consequences for GM oilseed rape. In: Lutman, P.J.W. (ed.) *Gene Flow and Agriculture: Relevance for Transgenic Crops*. Proceedings of the British Crop Protection Council Symposium No. 72, Farnham, UK, pp. 95–100.

Timmons, A.M., O'Brien, E.T., Charters, Y.M., Dubbels, S.J. and Wilkinson, M.J. (1995) Assessing the risks of wind pollination from fields of genetically modified *Brassica napus* ssp. *oleifera*. *Euphytica* 85, 417–423.

Van Gessel, M.J. (2001) Glyphosate-resistant horseweed from Delaware. *Weed Science* 49, 703–705.

Warwick, S.I., Beckie, H. and Small, E. (1999) Transgenic crops: new weed problems for Canada? *Phytoprotection* 80, 71–84.

Warwick, S.I., Simard, M.J., Légère, A., Beckie, H.J., Braun, L., Zhu, B., Mason, P., Séguin-Swartz, G. and Stewart, C.N. Jr (2003) Hybridization between transgenic *Brassica napus* L. and its wild relatives: *B. rapa* L., *Raphanus raphanistrum* L., *Sinapis arvensis* L., and *Erucastrum gallicum* (Willd.) O.E. Schulz. *Theoretical and Applied Genetics* 170, 528–539.

25 Prospects of a Hybrid Distribution Map Between GM *Brassica napus* and Wild *B. rapa* Across the UK

MIKE WILKINSON[1], LUISA ELLIOTT[2],
JOEL ALLAINGUILLAUME[1], CAROL NORRIS[3], RUTH WELTERS[4],
MATTHEW ALEXANDER[4], GIULIA CUCCATO[1], JEREMY SWEET[3],
MIKE SHAW[1] AND DAVID MASON[2]

[1]*School of Plant Sciences, The University of Reading, Reading RG6 6AS, UK;*
[2]*Environmental Systems Science Centre, NERC Environmental Systems Science Centre, The University of Reading, Reading RG6 6AL, UK;* [3]*NIAB, Huntingdon Road, Cambridge CB3 0LE, UK;* [4]*Centre for Ecology and Hydrology, Windfrith, Dorset, UK; E-mail: m.j.wilkinson@reading.ac.uk*

Abstract

The formation of F_1 hybrids between a genetically modified (GM) crop and a wild or weedy relative represents the first of a series of exposure events that could ultimately result in an undesirable change to the environment (hazard). The abundance of hybrids directly influences the dynamics of subsequent exposure events that lead to the hazard and so must be quantified if an overall estimate of exposure is to be compiled for a hazard over a legislative region (i.e. a nation). Furthermore, the feasibility of risk management through the prevention of hybrid formation also relies on a reliable estimate of hybrid frequency within the region under consideration. In this study, we describe an approach to compile a nationwide, spatially explicit estimate of hybrid numbers using gene flow from oilseed rape into natural waterside populations of *Brassica rapa* as a model. A two-phase strategy is adopted: (i) local hybrid numbers are calculated from the frequency of hybridization in sites of co-occurrence (between crop and relative) and the abundance of such sites; and (ii) long-range hybrid numbers are modelled from the pollen dispersal characteristics of the crop and the inferred distribution of donor and recipient populations. The practical difficulties in assembling these data sets are explained, and their importance for quantitative risk assessment explored.

Introduction

Several workers have argued that widespread cultivation of genetically modified (GM) crops will radically increase the adaptability of farming and will provide benefits to farmers and, in some cases, to the environment (Trewavas and Leaver, 2001; Phipps and Park, 2002). Conversely, there are legitimate concerns over possible environmental consequences arising from the use of some GM cultivars (Rogers and Parkes, 1995; Peterson *et al.*, 2000; Dale *et al.*, 2002). The manner in which cultivation of GM crops could lead to unwanted changes to the environment can be broadly divided into changes relating to the crop itself (e.g. altered herbicide/pesticide application leading to changed abundance of weeds, pests or associated on-farm fauna and flora) and those resulting from the movement of the transgene from the crop and into a wild or weedy recipient (e.g. the production of more persistent weeds or wild relatives with enhanced fitness leading to secondary unwanted ecological changes). Whilst both routes can generate profound changes to the environment, it is the latter that has the greater evolutionary and ecological significance, and is the focus of this chapter.

The Importance of Quantifying Hybridization

Most modern crops have wild or weedy relatives with which they can exchange genetic material by hybridization (see Ellstrand *et al.*, 1999). However, the formation of F_1 hybrids represents only the first stage in the process by which gene flow of any transgene could lead to unwanted environmental change. The presence of a transgenic hybrid does not constitute an unwanted change in itself but rather provides the opportunity for several different categories of unwanted change (hazards) to occur. There is variation in the likelihood and number of steps needed for hazard realization, although almost all require F_1 hybrid formation followed by introgression and transgene spread. Events subsequent to transgene spread are likely to be specific to both the transgene and the hazard. For example, in order for a transgene conferring resistance to a herbivore that has become widespread in a recipient plant to cause serious decline in herbivore numbers, the recipient plant must commonly feature in the herbivore's diet (e.g. a resistant rare plant would have negligible impact on rabbit numbers). A hypothetical example of a sequence of prerequisite events leading to the local extinction of a specialist predator is illustrated in Fig. 25.1. The commonality in the early events of all pathways involving gene flow provides an opportunity to produce a quantitative measure of exposure that applies to the vast majority of this type of hazard. If the cumulative likelihood of a transgene becoming widespread after hybrid formation, introgression and subsequent spread by infraspecific transgene movement between populations is deemed negligible (approximately zero) for a crop–location combination, then efforts should be directed to other combinations with a higher probability of occurrence. It should also be remembered that

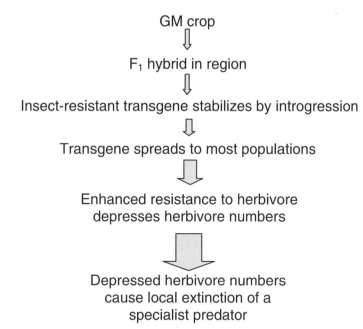

Fig. 25.1. Hypothetical scheme representing the sequence of events where transgene movement into a wild relative leads to the extinction of a specialist predator.

introgression and spread are not an inevitable consequence of a single hybridization event and so it is important to first estimate the number of hybrids in a legislative region to generate a truly quantitative measure of the probability of transgene spread.

There is also a second, more direct, reason for wishing to estimate hybrid numbers in a legislative region. Numerous workers have proposed measures that can be employed to reduce and possibly eliminate interspecific hybrid formation between a GM crop and its wild relatives. These include the enforcement of isolation distances, the use of male-sterile GM lines (Rosellini *et al.*, 2001), selection of 'safe' integration sites for a transgene (Metz *et al.*, 1997; Tomiuk *et al.*, 2000), transformation of the chloroplast (Daniell *et al.*, 1998) and the use of inducible promotor systems, popularly termed terminator technology (Oliver *et al.*, 1999). Such strategies will vary widely in their ability to repress hybrid frequency, and so it is important to have a broad estimate of hybrid numbers expected in a legislative region over a given time period in order to define how effective these measures must be to prevent hybrid formation or repress it sufficiently that transgene spread is deemed highly improbable.

Quantifying F₁ Hybrid Formation in a Legislative Region

The abundance of interspecific hybrids between a GM crop and its wild relatives is influenced largely by the strength of interspecific breeding barriers, the proximity and context of contact between the crop and recipient populations, and the capacity for long-range dispersal of viable pollen or, less frequently, seed. These factors are predominantly independent of the transgene but will vary between crop–recipient combinations and also according to location and geographical scale. Regulation generally applies at the national level, and so this is the most appropriate scale at which to describe hybrid distribution and frequency for legislative purposes. For a given GM crop–nation combination, the first task is to identify and then crudely rank the possible candidates for hybridization on the basis of ease of hybrid formation and introgression. The highest ranking candidate (i.e. most likely to form hybrids) should be examined first. In order to estimate hybrid numbers, it is important to consider context. This may necessitate separate examination of different recipient population types. Most hybrids form at relatively low frequencies and so empirical measures of hybrid numbers may only be possible when crop and recipient populations are sympatric (i.e. co-occur). It is probable that long-range hybridization will need to be estimated separately using a modelling approach. Thus, for most cases, the process of hybrid quantification will comprise four steps: (i) identify and rank recipients; (ii) quantify local hybrid numbers across a nation; (iii) estimate long-range hybrid numbers using a modelling approach; and (iv) combine estimates to describe number and distribution of hybrids.

Case Study: GM *Brassica napus* in the UK

In this section, we seek to illustrate the problems associated with generating even crude estimates of hybrid frequency and outline possible strategies to overcome these problems using GM *B. napus* in the UK as an example.

Identification and ranking of candidate recipients

Scheffler and Dale (1994) reviewed the possibility of hybridization between cultivated *B. napus* and its wild relatives in the UK. In all, there were four species identified as capable of spontaneous interspecific hybridization with the crop (*B. rapa*, *B. juncea*, *B. adpressa* and *Raphanus raphanistrum*) and a further 13 species were able to form hybrids when pollination is carried out manually (see also Chèvre *et al.*, Chapter 18, this volume). These were ranked according to the ease of hybrid formation, with the progenitor species *B. rapa* being ranked as most likely to generate hybrids. This assertion, coupled with the fact that *B. rapa* is the most widespread of the British wild brassicas (Preston *et al.*, 2002), means that effort should first centre on characterizing the likelihood of spontaneous hybridization, introgression and spread into *B. rapa* within the UK before progressing to species ranked as less likely to yield significant numbers of F_1 hybrids.

Compilation of national-scale estimates of *B. napus–B. rapa* hybridization

A measure of the annual number of hybrids can be compiled by combining estimates of hybrid frequency when crop and recipient co-occur (local hybridization) with the number of hybrids formed over long distances (remote hybridization). Local hybrid abundance can be estimated from empirical measures of hybrid frequency at sites of co-occurrence (sympatric hybridization rates) and the calculated number of sympatric sites. Long-range pollination cannot be measured empirically and so requires a modelling component. Each element will be addressed separately.

Local hybridization rates

B. napus is a dibasic allotetraploid with 38 chromosomes ($2n = 4x = 38$) and *B. rapa* is a diploid with 20 chromosomes, so that hybrids between them are almost always triploid ($2n = 3x = 29$). This feature means that it is possible to screen through large stands of predominantly *B. napus* or *B. rapa* plants or through seed samples for the presence of triploid hybrid candidates. Hybrid status can be confirmed by molecular analysis (e.g. Wilkinson *et al.*, 2000). When estimating the number of local hybrids formed over a geographical region, it is important to consider the different contexts in which the crop and wild relative come into close proximity. This requires an adequate knowledge of the ecology and distribution of the wild relative. In the case of *B. rapa*, natural populations occur in three types of locality: as casuals of disturbed land (typically waste ground and construction sites); as an arable weed (usually of *B. napus*); and as stable wild populations along riverbanks and other water courses (see also Norris *et al.*, Chapter 9, this volume). These settings each present contrasting opportunities for hybrid production and so warrant separate examination. Here, we consider only local hybrid formation into waterside populations. Local hybridization frequency into riverside *B. rapa* can be measured empirically using two approaches. The first estimates hybrid abundance among germinated seed samples collected from *B. rapa* plants in the year of co-occurrence with the crop. Scott and Wilkinson (1998) used this approach to describe hybrid frequency in two riverside populations growing beside commercial fields of *B. napus*. There were 15,341 seeds sown and 8647 of these germinated. The resultant offspring were screened by flow cytometry and inter-simple sequence repeat (ISSR) analysis, and found to contain 46 hybrids. This approximates to 0.4% of the germinated seed in a population that was separated from the crop by 5 m and 1.5% in a second population positioned only 1 m from another field. There are problems with using these values to estimate the abundance of hybrid plants. First, these seed-based estimates do not take into consideration the apparent difference in dormancy between the hybrids (which lack genetic dormancy) and the *B. rapa* (which shows strong dormancy) (Linder, 1998). Indeed, if the viability of *B. rapa* and hybrid seeds matches those reported elsewhere (> 90%; see Linder, 1998), then most of the ungerminated (*B. rapa*) seeds must be included in the percentage estimate of hybrid formation. This reduces the above values from 0.4 and 1.5% given above to less than 0.3% and 0.9%, respectively. A second important limitation of seed-based estimates of gene flow is that the number of hybrid seeds produced by a plant need not necessarily relate to the number of hybrid plants that subsequently appear in the population. This depends on the relative fitness of the hybrids relative to the surrounding *B. rapa* plants. Finally, seed-based estimates collected from the recipient population fail to include hybrids formed on *B. napus* but subsequently dispersed by seed into the sympatric *B. rapa* population. It can therefore be reasoned that direct measures of hybrid plant abundance represent the most appropriate measure of sponta-neous hybrid frequency. Data of this kind are largely absent from the literature at the time of writing but are the subject of concerted effort. To date, the only direct estimate of hybrid plants amongst riverside *B. rapa* comes from the study of Wilkinson *et al.* (2000), where one hybrid was recov-ered from 505 plants screened from two populations in the year following sympatry. Clearly, far larger samples are required to generate meaningful estimates. We currently are in the process of assembling such an estimate.

Frequency of co-occurrence

Once estimates have been produced for the number of hybrids that typically result from an incidence of co-occurrence, there is the need to estimate the frequency of co-occurrence to produce an overall estimate of local hybridization. This process inevitably requires a structured surveying approach coupled with a modelling component to allow extrapolation over the entire target area. In the *Brassica* example, the modelling component is made simpler by the ecological restriction of this ecotype of *B. rapa* to the banks of rivers and canals. This means that co-occurrence will be limited to those sites where *B. napus* fields are adjacent to a water system that contains *B. rapa*. The location of fields of the crop can be determined directly using remote sensing technology (Davenport *et al.*, 2000). Fields sited next to rivers can be identified by overlaying masked images indicating the positions of *B. napus* fields with the national rivers data set for the UK, as was employed by Wilkinson *et al.* (2000). Generation of a detailed distribution of riverside *B. rapa* presents more problems. In essence, it is first necessary to discover which river systems contain *B. rapa* and then to describe the pattern and variance of populations along rivers where *B. rapa* is present. The first has been achieved by reference to over 600 herbarium specimens, 82 local floras for vice counties across the UK, the Centre for Ecology and Hydrology countryside survey and by direct surveying. These data revealed that *B. rapa* is present on rivers throughout most of England (except the extreme South West and North West and parts of Lincolnshire and East Anglia) and parts of southern Wales, but is absent from Scotland. Direct surveys of over 300 km of eight rivers and four canals have been performed in which all populations were size-estimated by direct counts of plants at flowering and positioned by GPS (global positioning system). This work has revealed that *B. rapa* is far more abundant on rivers than on canals and has helped to define the pattern of population size and position distribution across the 12 river systems studied. Other factors limiting the appearance of *B. rapa* include altitude and the presence of brackish water. Combining these factors, it is possible to model the distribution of *B. rapa* populations along all UK rivers and canals known to contain the species. When this distribution is overlaid onto the positions of riverside *B. napus* fields, co-occurrence is found to be limited almost entirely to central and eastern England (Fig. 25.2). Once reliable estimates have been generated for hybrid abundance within sympatric sites, these data will then allow estimates to be made of the total number of 'local' hybrids generated in sites of co-occurrence.

Long-range hybridization

The number of hybrids arising from long-range pollination is determined by the size and abundance of non-sympatric populations, separation

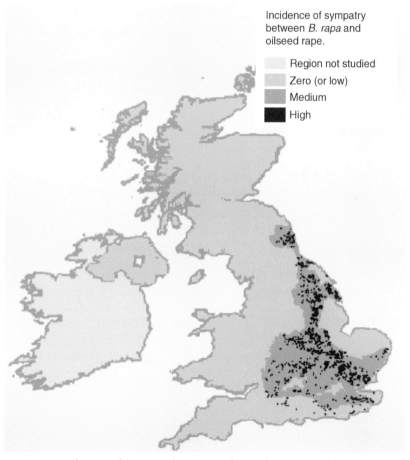

Fig. 25.2. Distribution of sympatry between cultivated *B. napus* and riverside populations of *B. rapa*.

profiles between the crop and recipient populations, pollen dispersal characteristics from B. *napus* fields and the efficacy of B. *napus* pollen in the presence of B. *rapa* pollen. Each of these elements will be addressed in turn.

Population size distributions were obtained as part of the riverbank survey outlined above. In all, over 1000 B. *rapa* populations were assessed for population size. A summary of population size profiles is presented in Table 25.1. Calculation of the total number of B. *rapa* plants in the UK is slightly more problematic, but can be estimated based on extrapolation from the distribution pattern observed along the surveyed rivers. Two strategies can be adopted to profile the separation between B. *napus* and riverbank B. *rapa* populations. The simplest approach is to make direct use of B. *rapa* location information generated by the survey. Isolation of all identified B. *rapa* populations from B. *napus* can be measured by reference to a remote sensing mask that reveals the locations of B. *napus* fields. Whilst this strategy generates accurate data for the survey area, it does not accommodate variation in the amount of B. *napus* cultivation between regions. An attractive alternative is therefore to use remote sensing over a large area to determine the locations of B. *napus* fields and to model the positions of B. *rapa* on all rivers on the basis of the distribution pattern observed during the survey.

Inference of the pollen dispersal characteristics from B. *napus* fields to B. *rapa* populations deserves careful consideration. Commercial fields of B. *napus* can disperse pollen by wind or insects (Timmons *et al.*, 1996), although the relative importance of each is still open to debate. These modes of pollen delivery should probably be addressed separately. There have been numerous works that describe the dispersal of pollen from B. *napus* fields (e.g. McCartney and Lacey, 1991; Timmons *et al.*, 1995). These works typically use empirical observations of airborne B. *napus* pollen densities as a basis on which to model the decay of pollen density with distance. We have used data from a work published by Timmons *et al.* (1995) in which pollen traps were placed at 0, 100, 360 m and 2 km from an isolated source field of B. *napus*. Decline in airborne pollen density was observed over 2 years

Table 25.1. Population size distribution along 98 km of the river Thames, UK.

Population size (plants)	No. of populations	Frequency (%)
1–5	480	42.6
10–25	374	33.2
50–70	153	13.6
75–150	72	6.4
200–500	33	2.9
750–1,500	14	1.2
10,000	1	0.1
Total	1,127	100

and could be described crudely by the inverse power law relationship $p = 299.5 \times [1/(20 + d)^{0.6}]$, where p is pollen density and d is distance (m), although the accuracy of this model currently is being refined. Combination of isolation profile data with the airborne pollen decay relationship allows for long-range pollen delivery to *B. rapa* populations to be estimated. Conversion of this to hybrid seed set relies on the relative competitiveness of *B. napus* pollen in the presence of *B. rapa* pollen. To some extent, this can be estimated from the hybrid frequency observed over short distances since pollen densities at this distance are known from empirical data. However, this estimate ignores the possibility of insect-mediated pollination. Alternatively, therefore, use should be made of pollen competition experiments such as those described by Pertl *et al.* (2002).

Modelling gene flow attributable to insect dispersal is more difficult. The model by Cresswell *et al.* (2002) provides a useful start, although more work is probably required to establish the effect of increasing distance on the likelihood of long-range delivery.

Conclusions

Estimation of total hybrid numbers between GM *B. napus* and riverbank *B. rapa* requires detailed information on the distribution of both parents, empirical estimates of hybrid formation at a local scale and realistic models to describe the rates of hybridization over large distances. Data currently are in the process of being assembled to estimate the frequency of hybrids when *B. napus* and riverside *B. rapa* co-occur and on the abundance of such sites. These data will allow for a UK-wide estimate of hybrids formed by 'local' gene flow. Early estimates indicate that annual numbers will be in the order of tens of thousands of hybrids. Long-range gene flow estimates are also in progress and are based on the separation profiles of both parental populations and on the relationship describing decay in pollen delivery from source with distance. To date, this relationship is far better modelled for wind pollination than for pollen delivery by insects. In consequence, the initial estimates of global hybrid numbers will have value primarily for the rather crude needs for setting efficacy limits for measures to prevent hybridization. Moreover, it is already clear that measures that reduce hybrid frequency by less than 10,000-fold would be insufficient to prevent hybrid formation altogether, although such measures would repress hybrid abundance. This finding has direct implications for assessing the feasibility of strategies proposed to contain constructs that are considered likely to enhance recipient fitness. Further refinement will be needed before the estimates will have quantitative utility to predict the likelihood of transgene recruitment, spread and, ultimately, to the realization of a defined hazard. For this, particular emphasis will need to be placed on producing realistic error estimates for all terms used.

Note Added at Proof Stage

Since writing this chapter, Wilkinson *et al.* (2003) succeeded in producing a provisional estimate of hybrids between these species across the UK using the approaches outlined above.

Acknowledgements

We thank the BBSRC and NERC of the UK government and the Perry Foundation for funding this work.

References

Cresswell, J.E., Osborne, J.L. and Bell, S.A. (2002) A model of pollinator-mediated gene flow between plant populations with numerical solutions for bumblebees pollinating oilseed rape. *Oikos* 98, 375–384.

Dale, P.J., Clarke, B. and Fontes, E.M.G. (2002) Potential for environmental impact of transgenic crops. *Nature Biotechnology* 20, 567–574.

Daniell, H., Datta, R., Varma, S., Gray, S. and Lee, S.-B. (1998) Containment of herbicide resistance through genetic engineering of the chloroplast genome. *Nature Biotechnology* 16, 345–348.

Davenport, I.J., Wilkinson, M.J., Mason, D.C., Jones, A.E., Allainguillaume, J., Butler, H.T. and Raybould, A.F. (2000) Quantifying gene movement from oilseed rape to its wild relatives using remote sensing. *International Journal of Remote Sensing* 21, 3567–3573.

Ellstrand, N.C., Prentice, H.C. and Hancock, J.F. (1999) Gene flow and introgression from domesticated plants into their wild relatives. *Annual Review of Ecology and Systematics* 30, 539–563.

Linder, C.R. (1998) Potential persistence of transgenes: seed performance of transgenic canola and wild × canola hybrids. *Ecological Applications* 8, 1180–1195.

McCartney, H.A. and Lacey, M.E. (1991) Wind dispersal of pollen from crops of oilseed rape (*Brassica napus* L.). *Journal of Aerosol Science* 22, 467–477.

Metz, P.L.J., Jacobsen, E., Nap, J.P., Pereira, A. and Stiekema, W.J. (1997) The impact on biosafety of the phosphinothricin-tolerance transgene in inter-specific *B. rapa* × *B. napus* hybrids and their successive backcrosses. *Theoretical and Applied Genetics* 95, 442–450.

Oliver, M.J., Quisenberry, J.E., Trolinder, N.L.G. and Keim, D.L. (1999) Control of plant gene expression. US Patent Number 5,925,808.

Pertl, M., Hauser, T.P., Damgaard, C. and Jørgensen, R.B. (2002) Male fitness of oilseed rape (*Brassica napus*), weedy *B. rapa* and their F-1 hybrids when pollinating *B. rapa* seeds. *Heredity* 89, 212–218.

Peterson, G., Cunningham, S., Deutsch, L., Erickson, J., Quinlan, A., Raez-Luna, E., Tinch, R., Troell, M., Woodbury, P. and Zens, S. (2000) The risks and benefits of genetically modified crops: a multidisciplinary perspective. *Conservation Ecology* 4, article no. 13.

Phipps, R.H. and Park, J.R. (2002) Environmental benefits of genetically modified crops; global and European perspectives on their ability to reduce pesticide use. *Journal of Animal and Food Sciences* 11, 1–18.

Preston, C.D., Pearman, D.A. and Dines, T.D. (2002) *New Atlas of the British and Irish Flora*. Oxford University Press, Oxford.

Rogers, H.J. and Parkes, H.C. (1995) Transgenic plants and the environment. *Journal of Experimental Botany* 46, 467–488.

Rosellini, D., Pezzotti, M. and Veronesi, F. (2001) Characterization of transgenic male sterility in alfalfa. *Euphytica* 118, 313–319.

Scheffler, J.A. and Dale, P.J. (1994) Opportunities for gene transfer from transgenic oilseed rape (*Brassica napus*) to related species. *Transgenic Research* 3, 263–278.

Scott, S.E. and Wilkinson, M. (1998) Transgene risk is low. *Nature* 393, 320.

Timmons, A.M., O'Brien, E.T., Charters, Y.M., Dubbels, S.J. and Wilkinson, M.J. (1995) Assessing the risks of wind pollination from fields of genetically modified *Brassica napus* ssp. *oleifera*. *Euphytica* 85, 417–423.

Timmons, A.M., Charters, Y.M., Crawford, J.W., Burn, D., Scott, S.E., Dubbels, S.J., Wilson, N.J., Robertson, A., O'Brien, E.T., Squire, G.R. and Wilkinson, M.J. (1996) Risks from transgenic crops. *Nature* 380, 487.

Tomiuk, J., Hauser, T.P. and Jørgensen, R.B. (2000) A- or C-chromosomes, does it matter for the transfer of transgenes from *Brassica napus*? *Theoretical and Applied Genetics* 100, 750–754.

Trewavas, A.J. and Leaver, C.J. (2001) Is opposition to GM crops science or politics? An investigation into the arguments that GM poses a particular threat to the environment. *EMBO Reports* 2, 455–459.

Wilkinson, M.J., Davenport, I.J., Charters, Y.M., Jones, A.E., Allainguillaume, J., Butler, H.T., Mason, D.C. and Raybould, A.F. (2000) A direct regional scale estimate of transgene movement from genetically modified oilseed rape to its wild progenitors. *Molecular Ecology* 9, 983–991.

Wilkinson, M.J., Elliott, L.J., Allainguillaume, J., Shaw, M.W., Norris, C., Welters, R., Alexander, M., Sweet, J. and Mason, D.C. (2003) Hybridization between *Brassica napus* and *B. rapa* on a national scale in the United Kingdom. *Science* 302, 457–459.

26 Potential and Limits of Modelling to Predict the Impact of Transgenic Crops in Wild Species

Claire Lavigne[1], Céline Devaux[1], Alexandra Deville[1],
Aurélie Garnier[1], Étienne K. Klein[1,2], Jane Lecomte[1],
Sandrine Pivard[1] and Pierre-Henri Gouyon[1]

[1]Laboratoire Ecologie, Systématique et Evolution, UMR 8079,
CNRS/Université Paris-Sud, Centre d'Orsay, Bâtiment 360, F-91405 Orsay
Cedex, France; [2]UMR BIA 518, INRA/INA-PG, 16 rue Claude Bernard,
F-75005 Paris, France; E-mail: claire.lavigne@ese.u-psud.fr

Abstract

Models have become a necessary tool for researchers addressing questions about complex long-term or large-scale processes, such as the impact of GM crops on wild species. They may be used for different purposes such as: (i) accurate quantitative predictions; (ii) ranking of scenarios; (iii) predictions, even inaccurate, of events difficult to observe; and (iv) understanding of relationships between processes. These purposes require different levels of precision, generality or explicative power. We argue that statistical models are the most precise, theoretical models are general, and mechanistic models offer the best explicative power. Intermediate types of models exist, but there is a trade-off between precision, generality and explicative power, and no model can achieve all three. We illustrate this trade-off by detailing a few examples of models for pollen dispersal and we present a non-exhaustive review of the literature on models addressing questions relative to gene flow from crops to related wild species.

Introduction

In general, questions from society deal with complex long-term or large-scale processes, and classical experiments cannot answer them. Models, together with field surveys, have become necessary tools for scientists addressing them (Legay, 1996).

Assessing the impact of transgenic crops in wild species is such a question. It involves either large numbers, or time or space scales that are inaccessible to researchers. As an example, in terms of pollen dispersal, one can easily calculate that a 1 ha field of maize produces about 10^{12} pollen grains (~500 kg) and that only 10^6 fertilize ovules within the field. There is therefore a need to predict the fate of the approximately 10^{12} remaining pollen grains. Of course, this prediction is difficult to achieve by experimentation. This difficulty is even more striking for very long-distance dispersal events. Conversely, at a short distances but a long time scale, it is difficult to estimate, by experiments, the length of time we will find a transgene in wild species after stopping the growth of the corresponding transgenic oilseed rape variety. One way to answer such questions would be to build models that would enable us to make reliable predictions.

Our aim in this chapter is to discuss the different existing types of models, as regards their ability to answer such questions, as well as their limitations, based on our experience with pollen dispersal modelling. Finally, we carry out a short (and non-exhaustive) review on models that attempt to understand or predict gene flow from crops into related species.

What are Models and What Do We Expect from Them?

In the current language, there are numerous definitions of models, from something we would like to copy (e.g. if copying a painting) to the copy itself (e.g. the model of a plane or building) (Legay, 1996). As they are used in science, models can generally be considered as tools that help us make a link between reality and concepts, i.e. models are built as an attempt to understand and/or predict observations from 'true life', and the conceptualization necessary for the design of models, in return, helps us improve our observation of reality.

There are different types of models depending on the tools used to build them. *Verbal models* are a verbal expression of variables of interest and processes that link them. Darwin's model of evolution by natural selection is probably one of the verbal models that most influenced recent biology. *Experimental models* are experiments that mimic reality while controlling many factors to make interpretation easier. Good examples are experimental evolution studies with bacteria (Lenski and Travisano, 1994) or colonization of empty islands (Simberloff and Wilson, 1970). *Numerical models* are based on calculations, and *analytical models* on resolution of equations.

By construction, a model is not an exact reflection of reality. It is both an abstraction and a simplification of reality. This simplification is based both on justified choices (e.g. the Mendelian inheritance of a gene) and on assumptions (e.g. it is often assumed that the fitness cost of a transgene is constant and does not evolve under the influence of modifier genes). A model therefore always reflects its designer's point of view. As models are simplifications, it can be stated that all models are false. However, as they help us to either understand or predict observations, some of them are useful ('all models are false, but some are useful', G.E.P. Box cited by Chatfield (1988)).

Depending on the question asked, we can have different expectations of models. We choose to distinguish four categories of expectations: (i) accurate quantitative predictions; (ii) ranking of scenarios; (iii) predictions, even inaccurate, of events difficult to observe; and (iv) understanding of relationships between processes. The first category is the most immediate. Accurate quantitative predictions are needed, for example, if we need to predict the exact rate of contamination of a field due to pollen flow (i.e. the proportion of seeds sired by pollen of an undesirable origin) depending on the crop grown and the shape and the size of the surrounding fields. Subsequently, such predictions are needed if we want to know how to isolate a field to ensure a given threshold of seed purity. Here we are not interested in the order of magnitude of the contamination, but in its exact value. The goal of the second category of models is not to give a precise output value of contamination but to rank scenarios depending on their contamination rate. This corresponds to models such as GENESYS (Colbach *et al.*, 2001). Questions asked are, for example, what cropping system and harvesting procedures will minimize the risk of GM presence in non-GM crops? Or, what is the best organization of refuge zones (in terms of distance, shape and size of plots) to minimize the speed of appearance of *Bt*-resistant insects (Onstad and Guse, 1999)? The third category concerns events that are difficult to observe, such as very long-distance dispersal events or rare hybridization events. Models can use indirect information related to these events in order to roughly estimate an order of magnitude of their occurrence. As an example, we attempted to predict gene flow from cultivated to wild beet by using the inverse gene flow model detailed below (Lavigne *et al.*, 2002). Finally, models can help us to understand the roles of the different processes acting to determine, for example, the fate of a transgene in a wild population. A well-studied interaction in that context is that of migration that increases the frequency of the transgene and selection that decreases it.

Potential and Limits of Models

Models allow the exploration of a wide range of input values rather easily. Changing input values allows the modeller to test the sensitivity of the output to their variation via sensitivity or elasticity analyses (Caswell, 2001), testing the impact of possible innovations and extrapolation in time and space. Models also produce a large number of replicates of a given situation, allowing the integration of stochasticity in the input variables or the modelled processes. As a consequence, it becomes possible to determine the probability of an observation under different scenarios. These are definite advantages over experiments that are often difficult to replicate.

Ideally, a model would make accurate predictions in very different situations (general) and at the same time would help us to understand the processes that lead to the predictions (explicative). However, we are not able to use a single type of model for both purposes, because, for example, the human mind has difficulty grasping multidimensional interactions and/or because different types of models make different assumptions. Different models are therefore needed depending on what we expect from them. This trade-off is particularly well illustrated by the triangle (verbal model) commonly used in ecology. This triangle describes the different types of models and their respective qualities (Fig. 26.1). It shows that any given model is a compromise between accuracy, generality and explanation, but none can focus on all three at the same time.

There are different ways to classify models (Legay, 1996). In this chapter, we will follow the classification associated with the preceding triangle, as it is most relevant to the questions of interest (estimation of efficient intra- and interspecific gene flow and its consequences). There are basically three different types of models. The most data-based models are statistical models. They consist of fitting functions on existing data or selecting variables most explicative of the data set. The most illustrative example is probably a multiple regression on data with a stepwise selection of explicative variables. Concerning gene flow, mathematically simple functions such as the negative exponential or inverse power law are often used to describe the decay of pollen with distance, and other effects (e.g. impact of source size) are often considered as linear. If the data set on which these models are based is good, these models can make accurate predictions within the range of observed input values. However, their explicative power and their generality are low because they are mainly black boxes, poorly adapted to cope with events outside of the range of observations from the original data set. Less dependent on the quality of data but still based on observations of biological processes are the mechanistic models. These models attempt to depict all these biological processes as accurately as possible. Building such models enables the modeller to think seriously about the critical input variables, the processes to model, as well as their interactions. Running the models improves our understanding by identifying inconsistencies in the

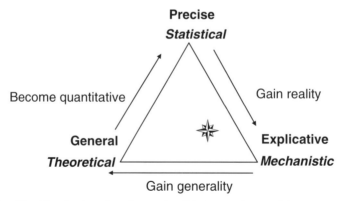

Fig. 26.1. The triangle describes the trade-off between the precision, generality and explicative power of statistical, mechanistic and theoretical models. The symbol in the triangle represents the localization of a given model.

predictions and thus compelling the modeller to think again about input variables and processes. Such models therefore have a high explicative value. Their drawbacks are their low generality, since they cannot take into account all processes in all situations, and their imprecision. Indeed, small errors are accumulated at each step and they can result in barely precise predictions. Finally, almost disconnected from data are theoretical models. These models aim at understanding interactions between very few processes (e.g. the interplay of migration and natural selection in population genetics, or that of competition and predation in population dynamics). Being very general, these models are necessarily far from any real specific situation and are therefore inadequate for predictions or selecting critical mechanisms.

Choice of a Type of Model

When trying to address a problem with a model, the first question deals with the type of model needed. In some situations, the choice is easy. Coming back to the four expectations that one can have regarding models, quantitative predictions can be achieved via statistical models (Fig. 26.2). If the goal is to predict in known situations, purely statistical models are sufficient. Moving along the side of the triangle to slightly more mechanistic models can help predictions in new situations, although some precision is lost. Ranking of scenarios is usually done by at least partially mechanistic simulation models which allow the exploration of different situations by changing combinations of parameter values. Sensitivity analyses can also be conducted on this type of model to help understand which interactions between processes mostly determine the output. Relationships between processes can also be explored, sometimes analytically, via theoretical models, in particular if few processes are modelled. Finally, there are different ways to make rough predictions, depending on the available data (Fig. 26.3A). In the following, we illustrate how these different types of models can be used to model pollen-mediated gene flow between fields, a topic with which we are most familiar.

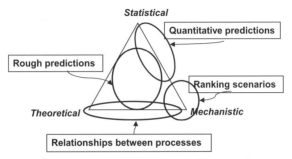

Fig. 26.2. Choice of a type of model depending on the type of question asked. The area of the triangle corresponding to each type of question is indicated by a circle.

From statistical to mechanistic models of pollen dispersal

A purely statistical model can be illustrated by the treatment of data published by Rieger *et al.* (2002) that describe the rate of contamination of fields as estimated by the presence of a herbicide resistance marker situated at different distances from a herbicide-resistant field. This treatment consists of fitting a function to assess how the rate of contamination declines with distance and how it depends on the crop variety or the sampling site within the field. This function cannot be used in new situations.

Moving towards a more mechanistic model can consist of describing the decrease of contamination with distance as a combination of the pollen dispersed by all source plants and recipient plants. The function that describes this decrease can then be considered as the convolution product of the dispersal function of each plant ('individual dispersal function' or 'dispersal kernel') and the function describing the localization and type (source or recipient) of each plant (Lavigne *et al.*, 1996, 1998; Nurminiemi *et al.*, 1998; Klein *et al.*, 2003). The easiest assumption is to consider that individual dispersal functions are probability density functions chosen for their simplicity, such as negative exponential, inverse power law, Weibull, etc. Such a modelling of dispersal can allow the impact of the individual function to be distinguished from that of the shapes and sizes of the source and receptor plots on the contamination rate. This modelling also allows modellers to make predictions for new sizes and shapes of plots. A shift towards mechanistic models can be made by understanding the processes that result in the individual dispersal function, as was achieved for maize pollen dispersal (Klein *et al.*, 2003). The individual dispersal function is reconstructed from the movement of individual pollen grains, modelled, for example, as the movement of particles following a Brownian movement with parameters describing the settling velocity of the pollen, the wind speed and turbulences, and the difference in height between male and female flowers. This further allows the assessment of the impact of wind speed or plant morphology on the pattern of pollen flow. However, the values of a few parameters are now necessary to make predictions, but these are not always available. Finally, fully mechanistic models of dispersal are available that actually describe pollen concentration in the different layers of the atmosphere (McCartney and Lacey, 1991; Foudhil, 2002; Foudhil *et al.*, 2002). Such models, however, contain many parameters and, while they help us to understand the impact of discontinuities such as roads or edges in the landscape, they cannot be used easily for quantitative predictions.

Half-way between statistical and theoretical models: rough predictions of the pollen cloud composition above wild populations

The basic motivation for this model (for details, see Lavigne *et al.* (2002)) is that for some crop species, populations of wild relatives are difficult to find and it will be difficult to make direct estimates of gene flow from the crop into these wild populations. In such a case, rough estimations are still better than nothing, and our idea was to use our knowledge of contamination from wild plants into crop plants to estimate the inverse contamination, i.e. from crop plants into populations of wild plants. Indeed, contamination of crop plants is usually easier to assess and to quantify because crops are rather homogeneous (at least for some traits) and fields are easily localized compared with wild populations. This model, which we named the 'inverse gene flow model' (Fig. 26.3A), is somewhat theoretical because it is based on two major assumptions: (i) all plants have the same individual dispersal; and (ii) this dispersal function is isotropic (or the distribution of wild populations around fields is isotropic). The model is somewhat statistical because it determines a function that links gene flow from crops to wild populations to the inverse, observable, gene flow. It is not mechanistic because it does not try to describe processes that lead to gene flow from crops to wild populations. Such a model is not very precise but quite general, since it can be adapted to any crop for which there are data on crop seed contamination by wild relatives.

Models Addressing Gene Flow from Crops into Related Species

In this last section, we try to provide a short (and probably non-exhaustive) review of published models that address the question of gene flow from crops to related species. Three steps can be considered before a transgene from a crop is established in a wild population: pollen/seed dispersal; hybridization and introgression; and establishment of the transgene in the wild population. These three steps have usually been modelled independently, but some models attempt to consider them at the same time (e.g. STEVE, see Slavov *et al.*, Chapter 8, this volume) or are now developed to do so (e.g. by the teams of Breckling (Germany) or of Wilkinson (UK), Chapter 25, this volume). We will focus on models that address each step.

Composition of the pollen/seed cloud

Different existing models are presented in Fig. 26.3A together with the names of authors who contributed to their development. On the right side of the triangle are models that we described previously for pollen dispersal, as well as their counterparts for seed dispersal. At the opposite left angle, are the very general genetic population models that describe pollen movement among populations following three distinct modalities. The island model assumes that the migration rate between populations does not depend on the distance that separates them, the stepping-stone model assumes that migration only occurs between adjacent populations, and the propagule pool model assumes that all migrants arriving in a population originate from one single population. On the theoretical/mechanistic side of the triangle are models of pollen dispersal by insects developed by Cresswell *et al.*

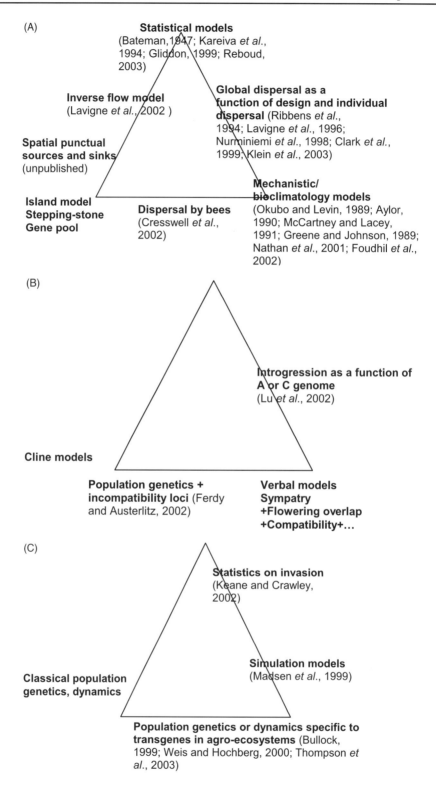

(A)

Statistical models
(Bateman, 1947; Kareiva *et al.*, 1994; Gliddon, 1999; Reboud, 2003)

Inverse flow model
(Lavigne *et al.*, 2002)

Spatial punctual sources and sinks
(unpublished)

**Island model
Stepping-stone
Gene pool**

Dispersal by bees
(Cresswell *et al.*, 2002)

Global dispersal as a function of design and individual dispersal (Ribbens *et al.*, 1994; Lavigne *et al.*, 1996; Nurminiemi *et al.*, 1998; Clark *et al.*, 1999; Klein *et al.*, 2003)

**Mechanistic/
bioclimatology models**
(Okubo and Levin, 1989; Aylor, 1990; McCartney and Lacey, 1991; Greene and Johnson, 1989; Nathan *et al.*, 2001; Foudhil *et al.*, 2002)

(B)

Introgression as a function of A or C genome
(Lu *et al.*, 2002)

Cline models

Population genetics + incompatibility loci (Ferdy and Austerlitz, 2002)

**Verbal models
Sympatry
+Flowering overlap
+Compatibility+…**

(C)

Statistics on invasion
(Keane and Crawley, 2002)

Simulation models
(Madsen *et al.*, 1999)

Classical population genetics, dynamics

Population genetics or dynamics specific to transgenes in agro-ecosystems (Bullock, 1999; Weis and Hochberg, 2000; Thompson *et al.*, 2003)

(2002). Finally, the inverse gene flow model described above stands on the statistical/theoretical side of the triangle.

As shown, numerous models have addressed the question of pollen/ seed dispersal, and they are very different depending on whether they were designed, for example, to make predictions, or to explore theoretically the evolution of genetic diversity in populations or the invasiveness of a gene/ species. Some general conclusions can be drawn from all these models that describe the composition of the seed/pollen cloud outside the fields. First, together with data, they teach us that there is more long-distance pollen dispersal than first suspected. Secondly, they help us to understand the effects of shape and size of the plots and those of discontinuities in the landscape (such as edges). Finally, they indicate that a total confinement of transgenic crops will not be manageable only through spatial isolation.

Hybridization/introgression

Fewer types of models exist for this step (Fig. 26.3B). In particular, we did not find any statistical model addressing this question. The most frequent models are verbal models that describe the different steps that have to be achieved for introgression to happen. These models are mechanistic as they detail most processes: sympatry of plants, overlapping of flowering periods, compatibility, etc. More formal models also exist. The most classical are theoretical population genetics models, known as cline models, that describe the diffusion of genes between two adjacent more or less inter-fertile populations. More recently, a population genetics model has been published that addresses the question of hybridization by specifically modelling compatibility loci (Ferdy and Austerlitz, 2002). These authors suggest that introgression is easier if incompatibility is determined by a large number of loci. Even more recently, Thompson *et al.* (2003) published a general model addressing the question of introgression of transgenes that they parameterized to fit broadly both *Brassica* and *Gossypium* species. This general but explicit model indicates that introgression is almost certain if the time period considered is long enough. Finally, Lu *et al.* (2002) recently published a model specifically addressing the question of the hybridization between *B. napus* and *B. rapa*. This model is partly statistical because, on the one hand, it is fitted on relative fitnesses and frequencies of chromosome numbers measured in specific experiments. On the other hand, it considers different steps of introgression one after the other, allowing some understanding of their respective roles.

Fig. 26.3. *(Opposite)* A non-exhaustive list of types of published models and their main contributors for each step leading to the introgression of transgenes in wild populations. (A) Pollen and seed flow; (B) hybridization; (C) fate after introgression.

Fate after introgression

As for the previous step, there exist verbal models for invasion. The most statistical is what has been termed the 10s rule (Williamson, 1996). This model states that on average, 10% of introduced species succeed, and 10% of these will have a significant ecological impact. It is thus a purely statistical model based on existing data sets. The Baker's list is another verbal model that describes the characteristics that a plant species must exhibit to become a weed. This model has been widely used in safety assessment of transgenic crops by seed companies (Parker and Kareiva, 1996). More formal are classical population genetics and population dynamics models that investigate the general conditions of invasion of genes or categories of individuals. Some of these models have been modified to address the question of invasion in the specific context of the release of GM plants. Examples are population dynamics models that consider the fate of feral populations of oilseed rape by taking into account demographic parameters specific to different life stages (Bullock, 1999). Also in that field is the model of Weis and Hochberg (2000) describing the asymmetry of competition between conventional and *Bt*-producing crops and its impact on the evolution of the frequency of *Bt* plants. Finally, simulation models exist that aim to predict the fate of a transgene under various cropping systems (Madsen *et al.*, 1999) (Fig. 26.3C).

Some conclusions can be drawn from all these models. Of course, a transgene that confers a selective advantage outside the field can invade. However, being at a selective disadvantage (s) outside the field is not sufficient to prevent a transgene from invading feral populations, provided the migration rate (m) is sufficient. There is, in particular, a threshold value of m/s above which a transgene will increase in frequency until it becomes fixed (Nagylaki, 1975). The initial increase is faster if the expression of the transgene is recessive. For example, transgenes that enhance growth in fishes could confer an increased reproductive success but a decreased viability. Such transgenes could invade in wild populations and even cause their extinction (Muir and Howard, 1999). Models that aim to optimize the introduction of transgenes in wild populations (for control of pests) also teach us that introgression will be faster if many individuals carrying few transgenes are released rather than few individuals carrying many transgenes (Davis and Fulford, 1999).

General Conclusions

In general, we would like to underscore once more that models are essentially a tool to formalize (verbally or mathematically) observations and to link these observations and concepts. We think that this formalization is necessary as soon as questions asked are complex or when questions are asked regarding the long term, such as when investigating the possible introgression of transgenes in wild related populations. They are also necessary when a hazard is suspected, and reliable predictions must precede any action.

We hope to have shown that there are multiple ways to build a model in order to investigate a single question. In general, we think that the choice of the model depends on the expected answer to two essential questions: (i) what is the aim of the model? and (ii) what is the precision needed of the description of the processes and of the output?

If the aim of the model is to make quantitative predictions in well-known situations, statistical models should be chosen. However, these models only make sense if they can be fitted and validated on good quality data sets. If, on the contrary, the model is used to mediate discussion between different players by indicating the possible outcome of different scenarios, then more mechanistic models are to be considered. Answering the second question is more difficult because, as biologists, we often feel that the model would be totally wrong if we do not include every detail of the biological processes we would like to investigate. However, if the aim of the model is the understanding of mechanisms that determine the output of the model, then too many details are detrimental because important processes are fragmented and the analysis of the models becomes inefficient. The difficult task then is to remove the unnecessary details so as not to become lost in an overly precise level of description. To sum up, 'A model should be as simple as possible but no simpler' (Einstein cited in Crawley (1999)). If the aim is to fit a simulation model on large data sets from monitoring, more details may be included in the model, by constantly going to and fro between models and data in order to calibrate the model. Models of weather forecast, for example, constantly adjust the parameter values by comparing the model output with new observations.

As a conclusion, we think that in the case of the introgression of transgenes in wild populations, we will need both models that help us to understand the biological processes, and models adjusted on data sets. We therefore think that the emergence of different models on this single question is to be favoured and that it will be interesting to see in the future if these different types of models lead to similar conclusions. Of course, this necessitates that the first model published does not serve as a reference for all subsequent models. Finally, we wish to point out that models will only make reliable predictions if they are based on good data sets.

References

Aylor, D.E. (1990) The role of intermittent wind in the dispersal of fungal pathogens. *Annual Review of Phytopathology* 28, 73–92.

Bateman, A.J. (1947) Contamination of seed crops. III Relation with isolation distance. *Heredity* 1, 303–336.

Bullock, J.M. (1999) Using population matrix models to target GMO risk assessment. *Aspects of Applied Biology* 53, 205–212.

Caswell, H. (2001) *Matrix Population Models. Construction, Analysis, and Interpretation.* Sinauer Associates Inc. Publishers, Sunderland, Massachusetts.

Chatfield, C. (1988) *Problem Solving: a Statistician's Guide.* Chapman and Hall, London.

Clark, J.S., Silman, M., Kern, R., Macklin, E. and HilleRisLambers, J. (1999) Seed dispersal near and far: patterns across temperate and tropical forests. *Ecology* 80, 1475–1494.

Colbach, N., Clermont Dauphin, C. and Meynard, J.M. (2001) GENESYS: a model of the influence of cropping system on gene escape from herbicide tolerant rapeseed crops to rape volunteers. I. Temporal evolution of a population of rapeseed volunteers in a field. *Agriculture Ecosystems and Environment* 83, 235–253.

Crawley, M.J. (1999) Bullworms, genes and ecologists. *Nature* 400, 501–502.

Cresswell, J.E., Osborne, J.L. and Bell, S.A. (2002) A model of pollinator-mediated gene flow between plant populations with numerical solutions for bumblebees pollinating oilseed rape. *Oikos* 98, 375–384.

Davis, S. and Fulford, G. (1999) Modelling the integration of a transgene by stocking. *Theoretical Population Biology* 55, 53–60.

Ferdy, J.B. and Austerlitz, F. (2002) Extinction and introgression in a community of partially cross-fertile plant species. *American Naturalist* 160, 74–86.

Foudhil, H. (2002) Développement d'un modèle numérique de dispersion atmosphérique de particules à l'échelle d'un paysage hétérogène. PhD Thesis, Université Paris-Sud, Paris, France.

Foudhil, H., Brunet, Y. and Caltagirone, J.P. (2002) Modélisation physique de la dispersion atmosphérique à l'échelle d'un paysage hétérogène. In: INRA (ed.) *AIP 'OGM et Environnement'. Restitution des Résultats 1998–2001.* INRA, Paris, pp. 27–30.

Gliddon, C.J. (1999) Gene flow and risk assessment. In: Lutman, P.J.W. (ed.) *Gene Flow and Agriculture. Relevance for Transgenic Crops.* British Crop Protection Council, University of Keele, UK, pp. 49–56.

Greene, D.F. and Johnson, E.A. (1989) A model of wind dispersal of winged or plumed seeds. *Ecology* 70, 339–347.

Kareiva, P., Morris, W. and Jacobi, C.M. (1994) Studying and managing the risk of cross-fertilization between transgenic crops and wild relatives. *Molecular Ecology* 3, 15–21.

Keane, R.M. and Crawley, M.J. (2002) Exotic plant invasions and the enemy release hypothesis. *Trends in Ecology and Evolution* 17, 164–170.

Klein, E.K., Lavigne, C., Foueillassar, X., Gouyon, P.H. and Larédo, C. (2003) Corn pollen dispersal: mechanistic models and field experiments. *Ecological Monographs* 73, 131–150.

Lavigne, C., Godelle, B., Reboud, X. and Gouyon, P.H. (1996) A method to determine the mean pollen dispersal of individual plants growing within a large pollen source. *Theoretical and Applied Genetics* 93, 1319–1326.

Lavigne, C., Klein, E.K., Vallée, P., Pierre, J., Godelle, B. and Renard, M. (1998) A pollen-dispersal experiment with transgenic oilseed rape. Estimation of the average pollen dispersal of an individual plant within a field. *Theoretical and Applied Genetics* 96, 886–896.

Lavigne, C., Klein, E.K. and Couvet, D. (2002) A method to estimate gene flow from crops to wild plants: application to sugar beet and consequences for the release of transgenic crops. *Theoretical and Applied Genetics* 104, 139–145.

Legay, J.M. (1996) *L'expérience et le Modèle. Un Discours sur la Méthode.* INRA éditions, Paris.

Lenski, R.E. and Travisano, M. (1994) Dynamics of adaptation and diversification: a 10,000-generation experiment with bacterial populations. *Proceedings of the National Academy of Sciences USA* 91, 6808–6814.

Lu, C.M., Kato, M. and Kakihara, F. (2002) Destiny of a transgene escape from *Brassica napus* into *Brassica rapa*. *Theoretical and Applied Genetics* 105, 78–84.

Madsen, K.H., Blacklow, W.M., Jensen, J.E. and Streibig, J.C. (1999) Simulation of herbicide use in a crop rotation with transgenic herbicide-tolerant oilseed rape. *Weed Research* 39, 95–106.

McCartney, H.A. and Lacey, M.E. (1991) Wind dispersal of pollen from crops of oilseed rape (*Brassica napus* L.). *Journal of Aerosol Sciences* 22, 467–477.

Muir, W.M. and Howard, R.D. (1999) Possible ecological risks of transgenic organism release when transgenes affect mating success: sexual selection and the Trojan gene hypothesis. *Proceedings of the National Academy of Sciences USA* 96, 13853–13856.

Nagylaki, T. (1975) Conditions for the existence of clines. *Genetics* 80, 595–615.

Nathan, R., Safriel, U.N. and Noy Meir, I. (2001) Field validation and sensitivity analysis of a mechanistic model for tree seed dispersal by wind. *Ecology* 82, 374–388.

Nurminiemi, M., Tufto, J., Nilsson, N.O. and Rognli, O.A. (1998) Spatial models of pollen dispersal in the forage grass meadow fescue. *Evolutionary Ecology* 12, 487–502.

Okubo, A. and Levin, S.A. (1989) A theoretical framework for data analysis of wind dispersal of seeds and pollen. *Ecology* 70, 329–338.

Onstad, D.W. and Guse, C.A. (1999) Economic analysis of transgenic maize and nontransgenic refuges for managing European corn borer (Lepidoptera: Pyralidae). *Journal of Economic Entomology* 92, 1256–1265.

Parker, I.M. and Kareiva, P. (1996) Assessing the risks of invasion for genetically engineered plants: acceptable evidence and reasonable doubt. *Biological Conservation* 78, 193–203.

Reboud, X. (2003) Effect of a gap on gene flow between otherwise adjacent transgenic *Brassica napus* crops. *Theoretical and Applied Genetics* 106, 1048–1058.

Ribbens, E., Silander, J.A. and Pacala, S.W. (1994) Seedling recruitment in forests: calibrating models to predict patterns of tree seedling dispersion. *Ecology* 75, 1794–1806.

Rieger, M., Lamond, M., Preston, C., Powles, S.B. and Roush, R.T. (2002) Pollen mediated movement of herbicide resistance between commercial canola fields. *Science* 296, 2386–2388.

Simberloff, D. and Wilson, E.O. (1970) Experimental zoogeography of islands: a two-year record of colonization. *Ecology* 51, 934–937.

Thompson, C.J., Thompson, B.J.P., Ades, P.K., Cousens, R., Garnier-Gere, P., Landman, K., Newbigin, E. and Burgman, M.A. (2003) Model-based analysis of the likelihood of gene introgression from genetically modified crops into wild relatives. *Ecological Modelling* 162, 199–209.

Weis, A.E. and Hochberg, M.E. (2000) The diverse effects of intraspecific competition on the selective advantage to resistance: a model and its predictions. *American Naturalist* 156, 276–292.

Williamson, M. (1996) *Biological Invasions*. Chapman and Hall, London.

27 Introgression of GM Plants and the EU Guidance Note for Monitoring

HANS C.M. DEN NIJS[1] AND DETLEF BARTSCH[2]

[1]Institute for Biodiversity and Ecosystem Dynamics, University of Amsterdam, Kruislaan 318, 1098 SM Amsterdam, The Netherlands; [2]Robert Koch Institute, Center for Gene Technology, 13187 Wollankstraße 15–17, Berlin, Germany; E-mail: nijs@science.uva.nl

Abstract

According to EU Directive 2001/18/EC and its Appendix VII, there will be mandatory monitoring of genetically modified organisms (GMOs) placed on the market to: (i) trace and identify eventual effects of the placing on the market of GMOs; and (ii) give feedback to the risk assessment procedure. A short overview is given of the accompanying Guidance Note for monitoring that has been implemented recently. This contribution will be restricted to aspects related to genetically modified higher plants (GMPs), and to their interactions, and to those aspects that directly relate to the escape and consecutive introgression of transgenes. Introgression of transgenes in a wild relative can be through GMP pollen siring hybrids directly, and indirectly through initial dispersal of the whole GMP (as seed or vegetative diaspore), and subsequent hybridization. According to the EU, the monitoring will have two focuses: (i) the possible effects of the GM crop, identified in the formal risk assessment (RA) procedure; and (ii) unforeseen effects. Where there is scientifically valid evidence of a potential adverse effect linked to the genetic modification, then, in the first part, 'case-specific monitoring' should be carried out after commercialization in order to confirm the assumptions of RA. Any evaluation of the potential consequence of that effect should be science based and compared with baseline information. In the second part, unanticipated effects of the environmental release are subject to a general surveillance programme that should be implemented independently of whether the RA found an indication of a harmful effect or not. The Guidance Note explicitly suggests that this latter part may well be long-term monitoring, given the fact that unexpected effects can be of delayed and long-term type. It is emphasized that, basic to the monitoring and the consecutive evaluation, is the availability of baseline data which give reference values for weighing of the GMP effects. It is acknowledged that there is great demand for such data sets. In any case, in order to be able to follow the consequences of transgene introgression, their identification is a prerequisite. Therefore, cost-effective and unambiguous markers may be placed in the GMPs, such as PCR-able eventually non-coding

©CAB International 2004. *Introgression from Genetically Modified Plants into Wild Relatives* (eds H.C.M. den Nijs, D. Bartsch and J. Sweet)

365

sequences in the constructs. In general, gene flow is mostly unavoidable. It is therefore the quality of the GMP traits that counts. A large amount of gene flow from a GMP with environmentally neutral traits is of less concern than minimal gene flow of plant fitness-enhancing traits. The Directive places the full responsibility for the establishment of the entire monitoring plan and the data report (to the competent state authorities) with the consent holder, in many cases private companies. It is questioned whether this is practical and feasible; therefore, it is suggested that part of the general surveillance, as well as additional monitoring elements outside the regulatory requirements, should be integrated in existing (state-owned) monitoring networks.

Introduction

In 2001, the EU Directive 2001/18/EC was implemented. This Directive deals with the deliberate release and introduction into the environment of genetically modified organisms (GMOs) in the EU. Part C of the Directive regulates the placing on the market of GMOs. The Directive also prescribes the monitoring that is foreseen to accompany the introduction of a GMO. To this end, a Guidance Note (GN; 2002/811/EG) for the design of the monitoring programme has been attached to the Directive. In this contribution, we will give an overview of the main framework of this GN, and we will especially elaborate those aspects that have a particular impact for the issue of gene flow and introgression from genetically modified higher plants (GMPs) to wild relatives.

The GN explicitly aims at regulating the monitoring of the market phase of the GMO introduction and not the pre-market experimental releases. It is designed to observe and evaluate any potential negative effects of the GMO on the environment or human health, either direct or indirect, immediate or delayed. In this contribution, only environmental aspects will be addressed.

We will start with very briefly introducing the relationship between risk assessment (RA) and monitoring, followed by an overview of the general outline of the GN. Following this, we will pay particular attention to those items that are connected with the events of gene flow and transgene introgression. Then, a series of monitoring strategy aspects, as prescribed in the GN, are presented, focusing on two main categories, the baseline data and background data, respectively. It will be evident that the baseline data in particular form an essential package for any monitoring plan, but also that there are large knowledge lacunas with respect to many species/wild relatives. The section on methods and tools for monitoring identifies the need for intensive recording designs and modelling tools to be developed, especially with respect to the predictability of changes in invasiveness.

Risk Assessment and Monitoring

What is risk assessment?

RA is the scientific standard used to assess the risk of any new technology that might affect human and animal health or environmental safety. At the outset of RA work, it is necessary to harmonize definitions and clearly link end-points to adverse effects in order to characterize results as hazardous. It is this RA procedure that will provide the necessary verification points and parameters for the flanking recording of the practical use of the crop in agriculture. In the debate on GMPs, there is much confusion as to how to use the term 'risk'. Many biosafety research studies (see Hoffman, 1990; Rissler and Mellon, 1996) tended to regard 'risk' as equivalent to 'exposure' or assume 'gene flow' equals 'hazard'. If this was the case, evolution would be classified as a harmful process, since hybridization, introgression and gene flow are essential to speciation, especially in plants. The simple definition of risk is: risk is a function of the effect of an event and the likelihood of the event occurring.

$$\text{Risk} = f \text{ effect } (= \text{hazard}) \times \text{likelihood (of the event)}$$

In this formula, effect is synonymous with hazard and, given the common meaning of the word, hazard is negative. However, it is useful to realize that (in this formula) not all effects are by definition negative. The outcome of the RA procedure must be the identification of any negative effect that as such leads to the quantification of the risk connected to the commercial exploitation of the GMPs.

Generally, five categories of possible hazards are identified, which may arise from the following: (i) direct and/or indirect effects of GMPs as a whole. Much concern is expressed with respect to GMPs running wild and developing increased weediness. Another possible consequence would be any adverse impact of changes in agricultural practices (specifically due to the use of the GMP) on the agricultural landscape and its (semi-)natural surrounding ecosystems. (ii) Effects of transgene movements and subsequent expression in other organisms. (iii) Non-target effects, due to GM products being active on other than pest organisms. (iv) Resistance evolution in pests (with adverse effects on non-GMO plant protection methods). (v) Direct or indirect effects on human health. Given the scope of this chapter, we will not consider the last effect further. Of course, these items are of utmost importance, but they are fully covered by other disciplines.

All risks related to the effects listed, of course, fully depend on the chance that the events actually occur, so estimating the likelihood is crucial to the RA process. The full RA procedure then is designed to predict expected undesirable consequences of introduction in both the long and short term.

What is environmental monitoring?

As stated in the Introduction, all the following basically involves the last phase of the development of a GM crop variety, its placing on the market after its approval and the consent for commercial growing. In general, this will only take place after the lengthy procedure of initial small-scale experiments and subsequent scaling-up. Thus, when discussing monitoring plans or guidelines for the commercial growing phase, it is done on the basis of previous RAs and experimental experience. The earlier phases of introduction will thus largely inform the RA that is applied for the commercial introduction of GM crops.

The concept of monitoring has been much debated, and it is widely interpreted as registering any sort of effect and changes taking place (in nature, in agricultural systems and in environmental parameters). Differentiation can be made due to the intensity and area of monitoring.

1. 'Case-specific monitoring' (CSM) is a specific, intensive, scientific measurement and data collection that often has experimental character and may actively use specific indicators or establish new networks.
2. 'General surveillance' (GSur) is a broad and less specific observation that may use passively indicators or existing monitoring networks.

However, in the framework of the commercialization of GM plants, monitoring is particularly defined in the case of CSM as the activities to test the hypotheses of identified adverse effects drawn from the RA procedure. In practice, this means that the monitoring programme should record the expected hazards that follow from the relevant introduction, to identify and measure their impact and frequency and to compare results with those predicted in the RA. In contrast to CSM, it is advocated in GSur that unexpected effects are to be addressed and surveyed as efficiently as possible. The ultimate goal of the monitoring should be to determine whether the data collected during CSM and GSur identify different risks due to commercialization by weighing the negative consequences of the introduction against its advantages in the managed and natural environments, and whether there is a net environmental cost. Consequently, environmental monitoring should be informative, giving feedback to the RA procedure, which ultimately may lead to reassessment and changes to the original introduction permit and its conditions of release (see Fig. 27.1). Depending on the outcome of the monitoring procedure, these changes may lead to a more liberal approach, or to extra restrictions, for continuation of the introduction.

How to achieve the monitoring goal: some basic considerations

The National Research Council (of the USA) authored a detailed survey on the possible environmental effects of transgenic plants, which focused on the general requirements and options for monitoring (National Research Council, 2002). The principle of a case by case approach is important in

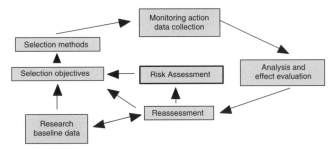

Fig. 27.1. The relationships and mutual interactions between components of risk assessment and monitoring. Adapted from Kapuscinski (2002) and Kjellsson and Strandberg (2001).

designing monitoring programmes, in the same way that risks are also assessed per taxon (organism), per trait and per environment. In the National Research Council study, it is advocated that there will be a two-phased monitoring: one for direct, short-term effects, and another design for long-term and unforeseen developments. Due to the wide array of possible effects over place and time, there are bottle-necks foreseen in the available versus required resources and in the personnel available for carrying out what could be a very time-consuming survey.

Even for monitoring and surveillance of short-term effects, there is still much debate with respect to the best and most practical investment of personnel and other resources. With respect to the direct effects, it frequently is suggested that trained observers are the best and most economic people to use. For instance, farmers, having experience in estimating their crops and the surroundings for variance and deviation in performance, could easily report abnormal events. The National Research Council report further suggests that nature reserve managers, bird watchers and networks of volunteers could be recruited as monitoring observers. These people would be able to observe sudden and obvious changes. However, this view has been criticized by Pool and Esnayra (2001). They did not believe that data from various ecosystems, taxa, trophic levels, from different climatic regions and from different states could be collected reliably by such a diversity of volunteer groups, who were probably not well equipped to collect and analyse data using tools such as geographical information system (GIS), population modelling and simulation, and statistics in a sufficiently reliable way.

The second level of monitoring, according to the US National Research Council, calls for looking at eventual long-term effects. This level is similar to the EU GSur and has a time axis for surveillance that may be impracticably, but necessarily, long in order to cover events such as a change in the invasive behaviour of GMP species which may be postponed for many generations depending on the nature of the relevant taxon. This aspect is complicated further by the fact that spatial and temporal aspects of the impact must be integrated, including the geographical location of the commercialization of the transgene, and the associated distribution patterns of wild and weedy relatives in the region (Fig. 27.2). We have learned from basic ecosystem

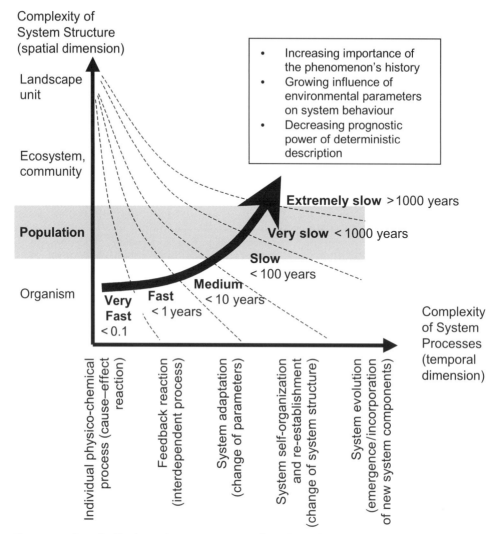

Fig. 27.2.   Speed of biological response to gene flow dependent on time and space. The
concept is based on the hierarchical concept of ecosystems (in relation to O'Neil *et al.*, 1986).
Environmental monitoring plans concerning GMP gene flow and introgression effects on the
population level are, more or less realistically, placed within the shaded area.

research how limited the available detection window is, since the speed
of environmental changes is slower the higher the level of biological
organization. Detecting environmental changes – due to gene flow – is very
difficult even if it is mediated by a single biological process. When we look at
the pattern and rate of biological processes in general, only a tiny fraction
can realistically be addressed by monitoring in time and space. Monitoring
natural plant populations on a global scale is difficult since most crops
hybridize with wild relatives in some regions (Ellstrand *et al.*, 1999). Based
on ecological experience and theory, the rate of response is often very slow,

so that most attempts at detection will be conducted over too short a period. Environmental monitoring of GMOs can – within 10 years – only target simple cause–effect and feedback responses at the population level.

Even when not anticipating any realistic hazard effect, detecting and quantifying introgression is a very labour-intensive procedure. We have only a poor understanding of 'baseline' population genetic structure and diversity. More than 80 years of ecological studies of genes under selection in natural populations have indicated that:

- populations frequently make a genotypical response to habitat (the power of ubiquity of natural selection);
- response varies between species depending on life history, population structure, genetic systems, etc; and
- the pattern and rate of response often closely correlate with scale and intensity of environmental variation.

At the same time, ecological studies of 'fitness' traits are in their infancy (Saeglitz and Bartsch, 2002). Almost 50 years of empirical studies of neutral genes (mostly allozymes) in natural populations have indicated that genetic structure and introgression of populations also depend on intrinsic biological properties such as breeding system, ploidy level, meiotic behaviour, pollination mode, seed dispersal, life form and history; interacting with extrinsic dynamic processes such as population size, fluctuations, bottle-necks, successional changes and historical events (from Pleistocene glaciation patterns to modern agriculture).

An interesting example of a modelling study, integrating several of these aspects, was developed by Marvier *et al.* (1999). They indicated that a monitoring plan should have a very intensive sampling strategy to enable early detection of increased invasiveness of a GMP and to detect transgene spread by hybridization and introgression. Assessment should also be made of whether the observed (invasiveness) effect is causally linked to the transgene event.

The EU Guidance Note for Monitoring (2002/811/EG)

In order to gain insight into the monitoring aspects related to consequences of gene flow from GM crops into wild relatives, a short survey of EU regulations for monitoring of marketing of GM crops is given. Appendix VII to the Directive 2001/18/EC on the deliberate release of GMOs into the environment gives outlines for the monitoring requirements, and subsequent to this document, a GN (2002/811/EG) for the design of a monitoring plan has been approved. In this note, a set of principles and practical instructions are given for the actual design of a monitoring plan after the GMP has obtained general consent for marketing. Therefore, all the following applies to post-market monitoring ('Part C' of the Directive) and not to aspects of monitoring that may be part of the pre-market RA procedure. The main aim of the monitoring programme is to identify and evaluate the potential negative effects of the GMP on the environment or human health, these effects being either direct or indirect, immediate or delayed. In this contribution, only environmental aspects will be mentioned.

The instructions for the design of the monitoring plan clearly distinguish two separate approaches. First, the risks that are identified during the assessment procedure are to be checked. This means that on the basis of the RA outcome, hypotheses can and will be generated that are to be tested in the monitoring activities. This phase of the monitoring will deal with already known direct and/or short-term effects and to a lesser extent with foreseen indirect and/or delayed effects, and it is called case-specific monitoring (CSM). The details of the programme design further depend on taxon, trait and region.

The second part of the overall monitoring plan is to observe any unforeseen and unexpected effects. By definition, these will not have been distinguished by the RA procedure, so no scientific hypotheses can be put forward. The instructions admit that it will be very difficult to account for such effects, but they see no reason not to screen for these. In contrast to the first part, this monitoring for the unexpected is prescribed to be very general; it will be performed under the name general surveillance (GSur). The foreseen period of GSur monitoring is medium to long term. According to the GN, it could be performed in cooperation with, or through existing networks for monitoring several other aspects of environmental and bio-diversity changes and developments, such as monitoring of plant health and protection, veterinary and medical products, and pollutants. Figure 27.3 shows the relationship, along the time axis, of the development and (step-wise) introduction of a GMP, the investment for RA and the diverse monitoring aspects. In Part C, the CSM component is expected to decrease in

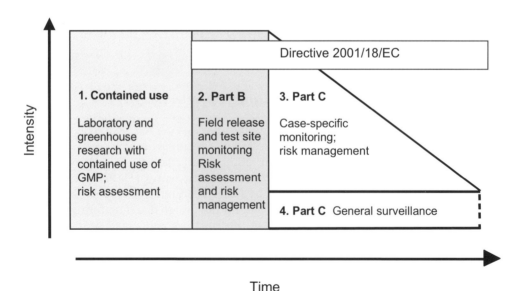

Fig. 27.3.   The different phases of risk assessment monitoring and release-related monitoring.

intensity with time, as the size of any effect becomes apparent. The GSur will last for a much longer period, due to the fact that the unexpected may only become apparent after a considerable period. According to the Directive and the GN, the responsibility for the establishment of the monitoring plan is with the consent holder.

One main problem of the GN is that there is no clear differentiation between elements that can be used in either in CSM or GSur. CSM should be concentrated on a limited number of parameters and a limited number of sites, where a more intensive data collection takes place. On the other hand, GSur is less experimental but more 'keeping an eye out' for any unexpected phenomenon. If unusual observations are reported, more in-depth studies can be added.

Prescribed elements of the monitoring plan

The note elaborates three main aspects that must be part of the monitoring plan: (A) strategy, (B) methods and tools and (C) analysis and report.

A. Monitoring strategy

With emphasis on direct, indirect, immediate and delayed effects. This should be based on the following.

A.1 *Risk assessment procedure*: this will produce testable hypotheses on effects and consecutive risks.

A.2 *Background information*: all available scientific and RA data should be used for the design, including those from experimental release in the developmental phase; detailed agronomic data on growing, field treatment, etc. are also requested.

A.3 *Monitoring approach*: this is dependent on the aspect of monitoring at hand. For the first approach, the design must of course be case specific, and capable of measuring the effects predicted in the RA. Depending on the situation, it further should be designed on a step by step basis, following the scaling-up from small trials and growing areas to full commercial growing over large areas (that may encompass several biogeographical areas, with consequent variation in response of local biota). The system should be designed for early warning so that the feedback is direct and short term, and management strategies can be changed adequately if necessary. For the second approach, GSur, it is particularly recommended that existing long-term observation networks be integrated in the recording system.

A.4 *Baseline data*: this is an area that will be difficult to comply with, since the fundamental idea is that the situation prior to the market introduction of the GMP serves as a reference for any eventual changes, so knowledge of these systems and the pertaining biota is critical. Explicit aspects are: natural environment and changes therein, and the analysis of the causes of such changes. To be able to evaluate any GMP-associated changes fully, the influence of non-GMP varieties on environment and ecosystems needs to be identified. This will include the effects of current agricultural practices and the effects of 'conventional' management. In certain cases, depending on RA expectations, it may be necessary to measure the genetic status of wild relatives, including the effect of past introgression from conventional crop varieties.

A.5 The monitoring plan should also present a clear time scale for the assessments; it is further explicitly stated that third parties and existing surveillance systems may contribute to the data.

B. Methods and tools

B.1 The methods to be used depend on what parameters are being measured. They must be developed, i.e. approved, using a case by case approach. Further, in-depth study is sometimes necessary to gain data on the dispersal of the transgene, as well as its capability for outcrossing with wild relatives, if the RA outcome predicts an adverse effect as a consequence of gene flow. Similarly, in-depth studies may be needed to determine changes in the ecology of associated taxa, and in the biodiversity of the region where the GM crop is to be grown, if the RA predicts adverse effects.

B.2 Apart from the area where the GM crop will be grown, a control/reference area needs to be identified. In addition, general information on the habitat vulnerability and invasibility of regions where the GM crop will be grown and the rate of disturbance is important, since these are crucial in case of increased invasiveness and weediness of the GM plant and introgressed species.

B.3 The frequency and timetable of surveying must also be described in the monitoring plan, to make clear that the scheme fits the expected effects; the same holds true for the sampling locations and the intended statistical analysis.

C. Analysis and report

C.1 A special chapter is required showing the methods for analysis and evaluation of collected data and for reporting results. The report should also describe the methods that will be used to review the RA in the light of new knowledge and for generating changes to the risk management, in line with the current market consent.

These elements for monitoring are fairly comprehensive, and it is beyond the scope of this contribution to review all of them. In this chapter, we will focus on those aspects that relate to introgression events from the GM plants into wild relatives and their possible consequences.

The Guidance Note and GMP Introgression

It is evident from the literature that where co-occurrence of crop and compatible wild relatives exists, gene flow is likely to occur. Frequencies may be high or low, but RAs conclude that gene flow will take place. Factors determining frequency are discussed later in this chapter. There will also be regional differences because wild relatives or progenitor(s) do not co-occur in all areas where the GM crop is grown. However, there are nearly no crops that do not co-occur growing with close relatives in some parts of their cultivation area (see, for example, Ellstrand *et al.*, 1999; Ellstrand, 2003). Crops that are widespread are likely to have greater opportunities of outcrossing with a wider range of wild relatives (see, for instance, Klinger, 2002). Thus, not only in RAs where the frequency of an event (= gene flow) is high, the key question is: what is the effect of this gene flow?

Although the formation of F_1 hybrids does not necessarily lead to introgression of crop genome elements, various chapters in this volume make clear that further hybridization and backcross generations are very likely to occur in many crop–wild relative complexes (see, for example, Hauser *et al.*, Chapter 4, this volume for *Daucus*; Van Dijk, Chapter 5, this volume for *Beta*; Westman *et al.*, Chapter 7, this volume for *Fragaria*; Papa and Gepts, Chapter 10, this volume for *Phaseolus*; Slyvchenko and Bartsch, Chapter 14, this volume for *Beta*; Pilson *et al.*, Chapter 17, this volume for *Helianthus*; Jørgensen *et al.*, Chapter 19, this volume for *Brassica napus/rapa*; Chèvre *et al.*, Chapter 18, this volume for *Brassica* and allies; see also Metz *et al.*, 1997; Ellstrand *et al.*, 1999; Snow *et al.*, 2002; Messeguer, 2003). So, the issue is rather how do we monitor for the effects of this gene flow?

Indeed it is important to consider whether monitoring gene flow is useful if we cannot translate it into population- and community-level consequences (i.e. effects) (A. Gray, Amsterdam, 2003, personal communication). Currently, only a few research projects have concentrated on studying the effects of gene flow, and it is important that future RA studies should deliver more data on impacts. From the list of potential effects presented in the section 'Risk Assessment and Monitoring', the following events of gene flow may apply and therefore be a target for monitoring: (i) movement of genes by pollen; and/or (ii) dispersal of seeds and other diaspores, including vegetative reproduction units, that after successful germination and establishment lead to feral transgenic plants. The latter event would be a manifestation that, after initial escape followed by gene flow, a sustainable population of transgenic plants has become established (eventually away from the original growing field) that might consequently have some ecological effect.

What is the sort of effect one might expect to occur after initial crop–wild relative hybridization, depending on the transgenic trait? The expected effects are given below.

1. Increased fitness of the recipient wild taxon, including possible increased weediness. The F_1 might be interpreted as showing 'hybrid vigour'; in further generations, this increased fitness may be maintained. Alternatively, restoration of fitness to higher levels than in the parental plants (e.g. different from a heterosis effects in general), after a F_1 drop in fitness, is also stipulated (van Raamsdonk and Schouten, 1997). Here, we

have to study not only individual plants, but also the demography of the populations and the individual plant contributions to it, in order to understand and predict their development. Such increased fitness and subsequent increase in populations may actually lead to changes in the competitive relationships within plant communities. It can also influence higher order trophic interactions. Thus, in this train of events, effects can also be expected on plant-dependent biota (pollinators, herbivores, parasites, detrivores, etc.).
2. Decreased fitness of the recipient taxon, due to disadvantageous combinations of genes and/or gene complexes ('outbreeding depression'). In general, this effect will only occur if the pollen pressure on the receiving taxon is severe, so that many of the disadvantageous combinations are formed. Thus, the relative size of the populations of source (crop) and sink (the wild relative) is important (Klinger, 2002). In current agricultural settings, cultivated fields generally contain enormous populations relative to the commonly small, literally marginal ones of the wild relatives. The first step in this process is loss of genetic identity (as in the example of *Medicago falcata* L. in Switzerland, after hybridization with *M. sativa* L.; Rufener al Mazyad and Ammann, 1999). Ultimately, extinction of the introgressed taxon may follow (*Oryza rufipogon* ssp. *formosana*, a rare subspecies in Taiwan; see Ellstrand *et al.*, 1999).

So, in order that monitoring programmes assess harmful effects, we will often need to add studies and data that give insight into such phenomena, and which will mostly have baseline research character. These will be discussed further below.

Monitoring strategy aspects

Logically, the main targets for a monitoring programme will be based on the outcome of the RA studies done prior to the market introduction phase. From these studies, it must be clear which sort of (undesired) effects may follow from the introduction, so testable hypotheses to check for these will be at the basis of the monitoring. However, apart from these hypotheses, a large amount of species- and case-specific baseline and background data is also necessary in order to evaluate the events and the subsequent risks. Following the GN instructions, below, a list of study fields to be considered is presented and discussed. All of these contribute to the baseline data in connection with the process of hybridization and introgression.

Two monitoring models for the establishment of baselines can be discussed: (i) Model 1 – comparison of effects occurring in GM crops with any effects previously observed in conventional crop plantings ('pre–post' GMP introduction model); and (ii) Model 2 – effects observed in 'parallel' growing of GM and non-GM crops using reference areas/populations where no GM crops have been grown (Bartsch *et al.*, 2003). Model 2 has the advantage of taking dynamic changes into account. Model 1 is superior if gene flow is so strong that unaffected areas/populations will not be found after introduction of a GMO, and natural variation has been measured as baseline before. In both cases, we need baselines in order to detect deviations from natural variation of certain end-point parameters.

Type of baseline data

Biogeography

Co-occurrence of crop and wild relatives may involve taxa which are weeds as well as taxa occurring in less disturbed (more natural) habitats surrounding agricultural fields or neighbouring (semi-)natural communities. Distribution patterns of the taxa concerned need to be studied or drawn from databases.

Examples on a regional scale are shown for The Netherlands (Figs 27.4 and 27.5), where the spontaneous distribution of species such as *Lactuca serriola* and *Daucus carota*, respectively, largely coincides with the regions where the crops derived from these wild species are grown.

Undisputed taxonomy is an important prerequisite for this sort of data to be reliable. In the European and North American situation, this will be (quite) feasible, although in regions rich in narrow endemics and rich in genetic variation for a lot of crop-related species, such as the Mediterranean area, data on taxonomic incidence and distribution are still meagre. This might be even more the case in developing regions, where species richness is generally much higher and less well known.

Reproductive biology

Wild relatives should be genetically closely related enough to show sexual compatibility. Phylogenetic data may help in assessing the chances of hybrid formation.

Further, the breeding system should allow cross-fertilization; selfing strategy and apomixis will restrict cross-fertilization. However, one must be cautious in interpreting these systems, because selfing is very rarely absolute, and in apomictic taxa sexual reproduction is occurring at variable rates. Further, regional and temporal changes in compatibility and/or asexuality should be accounted for.

Phenology

Reproductive ecology must show enough synchrony for cross-fertilization to be possible, and the distances between donor and receptor populations should be assessed relative to the distances the pollen vector is able to cross. The same holds true for seed dispersal agents. A special aspect is pollen viability, which may be limited, but detailed data are largely lacking.

Fig. 27.4. (A) Cultivation area of *Lactuca sativa* (lettuce), and (B) distribution area of wild *Lactuca serriola* (prickly lettuce) in The Netherlands. From de Vries (1996).

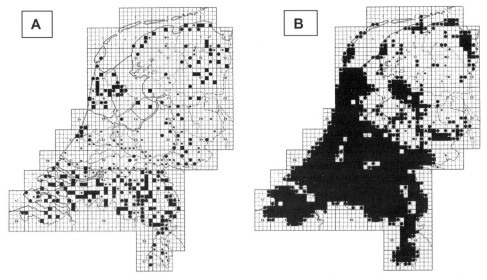

Fig. 27.5. Distribution maps of *Daucus carota* in The Netherlands. (A) Cultivation area of the crop (carrot). (B) Growing area of wild-type conspecific. From de Vries (1996).

Demography

Information on the degree of fitness of the recipient taxon is needed; can establishment capability be predicted sufficiently on the basis of the traits present? Data on the demographic characteristics of the wild taxon will help in modelling the population development (D. Claessen, Amsterdam, 2003, personal communication; C. Lamb, Amsterdam, 2003, personal communication), but weediness and invasiveness are notoriously difficult to predict (Pysek *et al.*, 1995; Kareiva *et al.*, 1996).

Hybridization and introgression events in the past

In order to predict any future event and possible consequences, it is important to know how the crop and its relatives have interacted in the past. Ellstrand gives impressive data from past hybridization analyses, which have become available only recently (Ellstrand *et al.*, 1999; Ellstrand, 2003).

To help provide basic data, so-called Botanical Files have been promoted as very useful (de Vries, 1996; Jacot and Ammann, 1999). In such files, much of the above information is integrated, leading to an assessment of the crops' capabilities for escape, hybridization and further (gene) spread. An important component of this assessment is the screening of herbaria for deposited hybrid specimens. Although this data mining from such large collections may be very indicative, it may also lead to underestimates of what has really happened in 'nature' in the past, particularly in crop–wild relative complexes that are phylogenetically very close. In the extreme case where crop and wild relative taxa are conspecific, morphology may be misleading when hybrids are difficult to detect and may easily be overlooked. In a study of the crop–wild relative complex *Lactuca sativa–L. serriola* (lettuce and prickly lettuce, respectively), it was found that F_1 hybrids between prickly lettuce and lettuce show marked similarity to the maternal prickly lettuce (see Fig. 27.6; D.A.P. Hooftman, Amsterdam, 2003, personal

Fig. 27.6. *Lactuca serriola* (A), *L. sativa* (C) and their F_1 hybrid (B), grown in the greenhouse, University of Amsterdam, December 2002.

communication). Consequently, one could very well imagine that botanists have not distinguished such specimens from wild types in the field, so that the herbaria might give a false-negative indication. In principle, Botanical Files may indeed be a very useful tool from which to draw baseline data. However, precisely in the most critical situations, where crop and wild relative are (almost) conspecific, they generally lack robust data. A plea must be made to improve this situation by collecting experimental (genetic) data.

Historical and current agronomical and environmental effects

It is also essential to determine the impact of current agricultural practices on biota in and around the fields, because they are important elements of the reference base for evaluating new developments.

   If introgression is expected to lead to changes in ecological interactions in the surrounding communities, we would need biodiversity data based on data banks built up over sufficient time periods. For a balanced evaluation of new effects, we would also need information about changes not (primarily) caused by agriculture, such as effects of climate change, or changes in overall land use in a given region (e.g. increased urbanization) (see, for instance, Kjellsson and Strandberg, 2001). Thus much of these data will be based on ecological studies conducted for other purposes which are beyond the scope of GMP notifiers. In the *Lactuca* studies, analysing a large set of vegetation records from about 1940 to the present, D.A.P. Hooftman (Amsterdam, 2003, personal communication) found that the ecological position of the wild-type *L. serriola* in The Netherlands has changed with time. The species nowadays co-occurs significantly more with species typical for more dense plant communities, which is interpreted to indicate a tolerance to higher soil moisture levels and lower temperatures than *Lactuca* showed before. Indeed, in analysing time trends, Kjellsson and Strandberg (2001) indicated that reference data should be based on sufficiently long recording periods since short-term studies may show completely different patterns from the real long-term development trends (see also Fig. 27.2).

Background data

In addition to the set of baseline data requirements, information on the background of the crop and the cropping system is also needed. These form part of the reference base for the evaluation of any effect of a GM introduction. Although the GN itself produces few instructions for this aspect, for instance the European Enforcement Project MON (Monitoring the environmental effects of genetically modified plants, an EU-funded programme by the European Joint Enforcement Group of Deliberate Release of GMOs) lists an exhaustive set of parameters that should be registered, independent of plant species and the GM traits (European Enforcement Project MON, 2003).

The data should comprise information on:

- Climatic conditions on the growing area/sites: wind, rainfall, temperature;
- Size and location of cultivation field(s);
- Density of the cultivation: number of plants per square metre;
- Occurrence of plant diseases: caused by bacteria, viruses and fungi;
- Time of sowing;
- Herbicides application(s): timetable as well as the chemicals applied;
- Treatment of the field: ploughing, tilling, etc.;
- Fertilizer application;
- Crop rotation scheme;
- Soil characteristics: a set of characteristics such as soil types, pH, organic matter, C/N, basal respiration.

There is debate as to whether or not the full set of parameters should be determined in all fields and all seasons, or only in a selection of fields instead. On farmland, most of these data are already collected by farmers since they know their fields best. A following question is who will be able to analyse all these raw data, e.g. for 50,000 farmers that grow sugarbeet in Germany. Not listed, but useful for pre-assessment of the potential for outcrossing of the local cultivation, are data on the position of the growing area relative to the biogeography of the wild relatives. Is it distant from occurrence of wild relatives, or is it in the centre of genetic variation of the wild taxa, or in the direct neighbourhood of rare and/or endemic species/populations? Here, the background data overlap with some data from the baseline records.

Monitoring methods and tools: desiderata and bottle-necks

In order to provide data on all the aspects listed, and to set up a proper monitoring programme, sufficient methodology and tools are needed (see Kjellsson, 1999). If we concentrate on the main elements of the interplay between species *inter se*, and the environment, we may need basic non-GMP-specific research data to incorporate the following factors in the analyses: (i) with respect to gene flow by either pollen (driven by insects, possibly other animal activity and wind) or seed (through wind, vertebrates and humans). All of these vectors may lead to (ii) stochasticity, causing effects in tails of leptokurtic distributions, the tail effects that form the biologically extremely important long-distance events (which, as a matter of fact, are inherently rare and difficult to analyse). Finally, (iii) the environmental conditions, biotic (relative size of populations with possible source–sink relationships), as well as abiotic (with spatial and temporal variation patterns with respect to region, substrate, climate, etc.), must be taken into account. There is quite a lot of experience with relevant methods, but these are generally expensive in terms of both time (ecology) and money (molecular markers).

The monitoring programme should further plan an intensive sampling design (either collection intensive in specific local areas within a CSM, or generally extensive on a large area covering Gsur), since it is the only way to detect increased invasiveness early enough to manage it successfully. Marvier *et al.* (1999) presented a simulation study for the timely and efficient discovery of GMP escapes from arable fields. They made clear that only intensive sampling of surrounding areas will suffice in order to fulfil the need of early warning. This is due to several characteristics of plant invasiveness, among others, to the lag time period that is very commonly met in the invasion process. Escaped specimens tend to stay rare for an unforeseeable period and then suddenly may gain invasive behaviour (see Kowarik, 1995). For monitoring effects of fitness changes, data on population development and growth models are necessary. Full knowledge of the life cycle and demographic data of target taxa (wild relatives and weeds) are needed to assess and predict population growth on the basis of matrix modelling (Bullock, 1999).

An additional point to consider is the sometimes surprising patterns in genetic variation and the connected adaptation patterns within species. Regional within-species differences in, for instance, disease resistance, substrate preference and/or tolerance are well known; a recent example from this field is wild-type *D. carota* from Iran showing resistance to the indigenous fungal pathogen *Alternaria*, whereas Dutch wild *Daucus* does not exhibit such resistance (Schouten *et al.*, 2002).

Identification of transgenes

Crucial for all monitoring to be effective is that GMPs and/or transgenes can be detected and followed after their release and eventual escape. Thus identification is an essential point. At the very start of production of a GM crop variety, the large majority of GMP constructs had, in addition to the target traits, marker genes that were specifically inserted for selection purposes. These were mostly one of the antibiotic genes. In Europe, it is foreseen, due to the relevant regulation, that antibiotic resistance genes that have a significant therapeutic use will not be allowed in market introductions. Therefore, some antibiotic resistance gene will no longer be an option for detection of GMPs or introgression events.

There are various options for identification based on morphological or physiological traits. There are many possibilities within the latter field, which mainly focus on proteins and (secondary) metabolites, but they all appear to be rather laborious and (thus) expensive. Recently, *in vivo* markers have been proposed, such as green fluorescent protein; specimens that express this gene fluoresce green under blue and UV light. Its use has been advocated for analysing gene flow (Stewart, 1996; Leffel *et al.*, 1997; Tian *et al.*, 1999; Hudson *et al.*, 2001; Halfhill *et al.*, Chapter 20, this volume). Polymerase chain reaction (PCR) approaches are increasingly feasible due to the availability of high throughput sequence equipment at low cost, for example for the detection of GM traits themselves by trait-specific probes or by specially built-in (short, non-coding) GM sequences (Simonsen, 1999; Kapuscinski, 2002). Given the importance of being able to track genes easily for RA and monitoring and for commercial property rights, it is a viable option. However, we are still far from the ideal situation, in which methods should:

- be easily applicable on large numbers of plants and populations;
- be very reliable;
- be inexpensive;
- be non- or only a little destructive; and
- result in unambiguous data (based on Simonsen, 1999).

Implementation of the monitoring Guidance Note; an example

From the above, it will be clear that, even in the limited scope of monitoring gene flow and its consequences, a monitoring plan that fits the GN instructions will be very complex and laborious. To give an illustration of this, an example is shown in Table 27.1 for *Brassica napus* (oilseed rape). This plan

Table 27.1. Checklist of parameters relating to exposure within environmental monitoring of GM oilseed rape. Shown are the gene flow items only.

What?	Where?	When?	How?
Transgene dispersal			
Hybridization with wild relatives	Field surroundings (1 km)	Annually, related species	Topographic mapping, molecular analysis of leaves
Seed dispersal	Roads, neighbouring fields	Transport period	Molecular analysis of leaves
Adventitious presence of crops conventional/ organic	Agencies, harbours, breeders	Annually before sowing	PCR random samples
Establishment of wild GMP population			
Plant establishment	Roads, neighbouring rural areas, railways, fields	Annually	Topographic mapping, molecular methods
GM and hybrid volunteers	Roads, neighbouring rural areas, railways, fields	Annually	Topographic mapping, molecular methods
Multiple GM volunteers and hybrids	Random in surrounding areas	Annually	Topographic mapping, molecular methods
Ecological effects			
Species composition, weed density	Field sites, surroundings	Growing period	Topographic mapping

From European Enforcement Project MON (2003).

has been developed by the European Enforcement Project (2003) as a CSM plan. The table shows the parameter checklist as far as transgene dispersal is concerned. In order to be successful, the spatial heterogeneity and the relative size of the source and sink populations of crop and wild relatives, respectively, should be taken into account. In addition, the results from the simulation study by Marvier *et al.* (1999), which points to the need for an extensive sampling network, further increase the time involved. The consequences of this will be discussed further below.

Conclusions and Suggestions

The GN prescribes in much detail the design of the monitoring plan. It is clear that a huge investment of time and money is needed to test the identified risks in order to provide feedback to the original RA. In addition, monitoring for unexpected developments will be necessary in the longer term. From the GN, it is as yet unclear how to design this second element of the process.

It would be very useful if tools were available that help to predict the effect of certain traits on plant fitness and that translate individual plant effects into impacts at the population level. The further development of the introgression effects could be predicted using population modelling algorithms. In order to achieve this, deliberately designed genetic and demographic studies of target species and populations are needed. However, these appear to be seen as outside the scope of most monitoring studies, being issues of broader public research, whereas the above data are designated basic ecological knowledge.

Another currently underdeveloped area is the spatial analysis of meta-population structures and dynamics. In order to predict the spread of any (unwanted) taxon across the landscape, it is necessary to utilize GIS expertise in monitoring and, consequently, also in risk management. It goes without saying that a fully equipped database which holds the biogeographical, phylogenetic and breeding system data of all the crops and related species of an area would be an important and efficient tool for such evaluations. More specifically, geographically and biosystematically developed 'Botanical Files of Europe', on the basis of the concept as worked out in The Netherlands (de Vries, 1996) and Switzerland (Jacot and Ammann, 1999), are badly needed.

Monitoring using the methods described above is likely to be very laborious and time consuming. Also, the likelihood that certain effects will only be recognized after long periods raises doubts about the effectiveness of monitoring. Therefore, it may well be a better solution to produce GMPs in which gene flow is completely prevented. This will certainly pose a problem in many species, but pleas are nevertheless being made for this approach (Kuvshinov *et al.*, 2001). Among the options to achieve this are male sterility, apomixis and terminator constructs in seeds. There has been broad discussion on these options already, and it is unlikely that there is a panacea among them. A recent further alternative has been advocated by Kuvshinov *et al.*

(2001), called recoverable block of function (RBF). This is a construct that completely prevents sexual reproduction and consequent pollen and seed flow unless a (chemical or other) trigger is applied to the GMO. One may expect that this solution will also lead to a lot of politically driven discussions, since the technique resembles the terminator concept.

Much of the above discussion primarily relates to the first section of the prescribed monitoring, the case-specific elements. The GSur part, i.e. to uncover and register unexpected effects from the GM plant release, is even less easy to programme and design, because it is inherently difficult, in fact principally impossible, to give a priori answers to questions such as: what, where and when will the unexpected be? Some of the aspects dealt with in this contribution may adequately be assessed, others need to be developed and are still under discussion. However, what is quite clear is the fact that data reporting from the monitoring of any GM crop as such is, according to the EU regulation, solely the responsibility of the consent holder, which is most probably a private company. There is the danger that overloaded monitoring plans cannot be managed by small and medium sized companies, and also big companies may hesitate to accept expensive monitoring plans. A potential solution exists particularly for the GSur part, if existing agricultural monitoring activities, such as for land use, pollution, pathogens, etc., are permitted to contribute to the generation of the data. However, the regulatory framework seems to put the responsibility for this GSur also with the consent holder. This means that the consent holder will need to access and coordinate these other sources of information, which may be difficult since some of the GSur systems normally are state controlled and access to the data is restricted due to confidentiality, trade and other requirements. Judging from this complex situation, it could be argued that member states should consider making appropriate information available for monitoring or taking some of the responsibility for general monitoring of agricultural systems.

It will be of great practical value if diagnostic methods for the detection, identification and quantification of GMOs and their products were developed. The Organization for Economic Cooperation and Development (OECD) has been working towards establishing unique identifier systems for transgenic plants, which will identify the producer and the insert (OECD, 2002). Monitoring is also meant to give feedback to a competent authority (CA), who accepted the notifier's RA. For this, it is important to follow the fate of GMOs in the environment. It will be impossible to track down any environmental effect to a particular farmer, but this is another story. Also, for co-existence between GMO- and non-GMO-using farmers, there will be no 'simple' solution. Here, the identification of an event may play no practical role, since the source of the adventitious presence is likely to be unknown if there is more than one farmer. Co-existence is not part of the monitoring plan. However, the introduction of notifier-specific molecular nucleotide sequences would make GMO screening much easier.

There is a need to define what impacts of gene flow are unacceptable in terms of: (i) nature conservation; (ii) protection of economic value; and

(iii) responsibility of stakeholders in cases where environmental or economic harm has occurred (Braun, 2002). In relation to (i), it is important that plant genetic resources (gene pool) are protected from loss of diversity (e.g. for beet, Frese *et al.*, 2001). In order to protect beet crops in relation to (ii), the development and recombination of certain transgenic traits that may enhance fitness in natural wild beet habitats or develop aggressive weed beet genotypes by enhancing invasiveness should be avoided. Herbicide tolerance genes seem to play a minor ecological role (Bartsch *et al.*, 2003; Dawson, 2003).

The latest EU draft on environmental liability with regard to the prevention and remedying of environmental damage (Proposal COMMON POSITION (EC) No. 58/2003) defines 'damage' as being a measurable adverse change in a natural resource and/or measurable impairment of a natural resource service that may occur directly or indirectly and that is caused by any of the activities covered by this Directive, in which the use of GMOs is included. Environmental damage means:

- Damage to protected species and natural habitats, which is any damage that has significant adverse effects on reaching or maintaining the favourable conservation status of such habitats or species. The significance of such effects is to be assessed with reference to the baseline condition.
- Water damage, which is any damage that significantly adversely affects the ecological, chemical and/or quantitative status and/or ecological potential, as defined in Directive 2000/60/EC, of the waters concerned, with the exception of adverse effects where Article 4(7) of that Directive applies.
- Land damage, which is any land contamination that creates a significant risk of human health being adversely affected as a result of the direct or indirect introduction in, on or under land, of substances, preparations, organisms or microorganisms.

So far, there are no plausible hypotheses that link gene flow from GMPs already placed on the market to any of these three natural resources listed above.

The ecological and genetic consequences of gene flow depend on the amount and direction of gene flow as well as on the fitness of hybrids. The assessment of potential risks of GMPs has to be performed taking into account that conventional crops also hybridize with wild plants. Monitoring must demonstrate, on a larger scale, the prognostic assumptions made by the biosafety research that underpins the risk assessment. However, case-specific monitoring in agricultural and peri-agricultural habitats to test identified risks from RA is certainly feasible. General surveillance for indirect, delayed or unforeseen effects of gene flow in semi-natural environments is extremely difficult, very costly and has to be done against a background of constant environmental change. Introgression of GM sequences into wild plant populations is only one issue among several other environmental exposure pathways. Possible introgression effects are to be characterized on the basis of a sound understanding of the biology and ecology of the wild species. Given the well-known problems with predictability and monitoring of other ecological phenomena such as plant invasiveness, it is also concluded that the effective monitoring programmes will necessarily be intensive and therefore expensive. Financial resources for this work will be limited, so that we will need to select monitoring and surveillance tools accordingly. Full use should be made of existing environmental surveillance networks. In addition, integrating spatial and temporal dimensions and the use of (population growth) modelling may also be part of the solution. Additional monitoring elements can be integrated in existing (state-owned) monitoring networks, which are outside the regulatory requirements and therefore do not need to be paid by the notifier.

The best possible chance of detecting an unanticipated adverse effect would be ensured by having an adequate number of people, with relevant experience, involved in the surveillance process. It follows, therefore, that those persons or organizations normally involved in agriculture, or whose activities are connected to agriculture, the environment, and human and livestock health, will be in the best position to participate in a general surveillance plan. In order to allow detection of the broadest possible scope of unanticipated adverse effects, we propose that general surveillance monitoring is incorporated into company stewardship programmes and is performed by existing networks, as well by specific measures in the stewardship programmes.

Acknowledgements

We wish to thank Dr Ullrich Ehlers and Jeremy Sweet for helpful comments on this manuscript.

References

Bartsch, D., Cuguen, J., Biancardi, E. and Sweet, J. (2003) Environmental implications of gene flow from sugar beet to wild beet – current status and future research needs. *Environmental Biosafety Research* 2, 105–115.

Braun, R. (2002) People's concerns about biotechnology: some problems and some solutions. *Journal of Biotechnology* 98, 3–8.

Bullock, J.M. (1999) Using population matrix models to target GMO risk assessment. In: Thomas, M.B. and Kedwards, T. (eds) *Challenges in Applied Population Biology*. Association of Applied Biologists, Wellesbourne, UK, pp. 205–212.

Dawson, H. (2003) *GM Crops, Modern Agriculture and the Environment*. Conference report. AgBiotechNet 5, ABN 109.

de Vries, F.T. (1996) Cultivated plants and the wild flora. Effect analysis by dispersal codes. PhD Thesis, Rijksherbarium, Hortus Botanicus, Leiden.

Ellstrand, N.C. (2003) *Dangerous Liaisons? When Cultivated Plants Mate with Their Wild Relatives*. Johns Hopkins University Press, Baltimore, Maryland.

Ellstrand, N.C., Prentice, H.C. and Hancock, J.F. (1999) Gene flow and introgression from domesticated plants into their wild relatives. *Annual Review of Ecology and Systematics* 30, 539–563.

European Enforcement Project MON (2003) *Monitoring the Environmental Effects of Genetically Modified Plants.* Action 19: monitoring oilseed rape. Contact: Dr Thomas Engelke, Ministry for the Environment, Nature Conservation and Agriculture, Schleswig-Holstein Land, Germany and Dr Birgit Corell, Bezirksregierung Braunschweig, Germany. Available at: http://eep-mon.iitb.fhg.de/

Frese, L., Desprez, B. and Ziegler, D. (2001) Potential of genetic resources and breeding strategies for base-broadening in *Beta*. In: Cooper, H.D., Spillane, C. and Hodgkin, T. (eds) *Broadening the Genetic Base of Crop Production*, IPGRI/FAO, Rome, pp. 295–309.

Hoffman, C. (1990) Ecological risks of genetic engineering of crop plants. *Bioscience* 40, 434–437.

Hudson, L.C., Chamberlain, D. and Stewart, C.N. Jr (2001) GFP-tagged pollen to monitor pollen flow of transgenic plants. *Molecular Ecology Notes* 1, 321–324.

Jacot, Y. and Ammann, K. (1999) Gene flow between selected Swiss crops and related weeds: risk assessment for the field releases of GMO's in Switzerland. In: Ammann, K., Jacot, Y., Simonsen, V. and Kjellsson, G. (eds) *Methods for Risk Assessment of Transgenic Plants. III. Ecological Risks and Prospects of Transgenic Plants*. Birkhäuser Verlag, Basel, pp. 99–108.

Kapuscinski, A.R. (2002) Controversies in designing useful ecological assessments of genetically engineered organisms. In: Letourneau, D.K. and Burrows, B.E. (eds) *Genetically Engineered Organisms: Assessing Environmental and Human Health Effects*. CRC Press, Boca Raton, Florida, pp. 386–415.

Kareiva, P., Parker, I.M. and Pascual, M. (1996) Can we use experiments and models in predicting the invasiveness of genetically engineered organisms? *Ecology* 77, 1670–1675.

Kjellsson, G. (1999) Methodological lacunas: the need for new research and methods in risk assessment. In: Ammann, K., Jacot, Y., Simonsen, V. and Kjellsson, G. (eds) *Methods for Risk Assessment of Transgenic Plants. III. Ecological Risks and Prospects of Transgenic Plants*. Birkhäuser Verlag, Basel, pp. 185–194.

Kjellsson, G. and Strandberg, M. (2001) *Monitoring and Surveillance of Genetically Modified Higher Plants. Guidelines for Procedures and Analysis of Environmental Effects*. Birkhäuser Verlag, Basel.

Klinger, T. (2002) Variability and uncertainty in crop-to-wild hybridization. In: Letourneau, D.K. and Burrows, B.E. (eds) *Genetically Engineered Organisms: Assessing Environmental and Human Health Effects*. CRC Press, Boca Raton, Florida, pp. 1–15.

Kowarik, I. (1995) Time lags in biological invasions with regard to the success and failure of alien species. In: Pysek, P., Prach, K., Rejmanek, M.J. and Wade, M. (eds) *Plant Invasions: General Aspects and Special Problems*. SPB Academic Press, Amsterdam, pp. 15–18.

Kuvshinov, V., Koivu, K., Kanerva, A. and Pehu, E. (2001) Molecular control of transgene escape from genetically modified plants. *Plant Science* 160, 517–522.

Leffel, S., Mabon, S.A. and Stewart, C.N. Jr (1997) Application of green fluorescent protein in plants. *Biotechniques* 23, 912–918.

Marvier, M.A., Meir, E. and Kareiva, P.M. (1999) How do the design of monitoring and control strategies affect the chance of detecting and containing transgenic weeds? In: Ammann, K., Jacot, Y., Simonsen, V. and Kjellsson, G. (eds) *Methods for Risk Assessment of Transgenic Plants. III. Ecological Risks and Prospects of Transgenic Plants.* Birkhäuser Verlag, Basel, pp. 109–122.

Messeguer, J. (2003) Gene flow assessment in transgenic plants. *Plant Cell, Tissue and Organ Culture* 73, 201–212.

Metz, P.L.J., Jacobsen, E. and Stiekema, W.J. (1997) Aspects of biosafety of transgenic oilseed rape (*Brassica napus* L.). *Acta Botanica Neerlandica* 46, 51–67.

National Research Council (2002) *Environmental Effects of Transgene Plants: the Scope and Adequacy of Regulation.* National Academy Press, Washington, DC.

OECD (Organization for Economic Cooperation and Development) (2002) *OECD Guidance for the Designation of a Unique Identifier for Transgene Plants. Series on Harmonization of Regulatory Oversight in Biotechnology,* no. 23. OECD, Paris ENV/JM/MONO. Available at: www.olis.oecd.org/olis/2002doc.nsf/LinkT0/env-jm-mono(2002)7

O'Neil, R.V., De Angelis, D.L., Waide, J.B. and Allen, T.F.H. (1986) *A Hierarchical Concept of Ecosystems.* Princeton University Press, Princeton, New Jersey.

Pool, R.P. and Esnayra, J. (2001) *Ecological Monitoring of Genetically Modified Crops. A Workshop Summary.* National Academy Press, Washington, DC.

Pysek, P., Prach, K., Rejmanek, M. and Wade, M. (eds) (1995) *Plant Invasions: General Aspects and Special Problems.* SPB Academic Publishing, Amsterdam.

Rissler, J. and Mellon, M. (1996) *The Ecological Risks of Engineered Crops.* MIT Press, Cambridge, Massachusetts.

Rufener al Mazyad, P. and Ammann, K. (1999) The *Medicago falcata/sativa* complex, crop–wild relative introgression in Switzerland. In: van Raamsdonk, L.W.D. and den Nijs, J.C.M. (eds) *Plant Evolution in Man-made Habitats.* Hugo de Vries Laboratory, University of Amsterdam, Amsterdam, pp. 271–286.

Saeglitz, C. and Bartsch, D. (2002) Plant gene flow consequences. *AgBiotechNet* 4, ABN 084.

Schouten, H.J., van Tongeren, C.A.M. and van den Bulk, R.W. (2002) Fitness effects of *Alternaria dauci* on wild carrot in The Netherlands. *Environmental Biosafety Research* 1, 39–47.

Simonsen, V. (1999) Molecular markers for monitoring transgenic plants. In: Ammann, K., Jacot, Y., Simonsen, V. and Kjellsson, G. (eds) *Methods for Risk Assessment of Transgenic Plants. III. Ecological Risks and Prospects of Transgenic Plants.* Birkhäuser Verlag, Basel, pp. 87–93.

Snow, A.A., Uthus, K.L. and Culley, T.M. (2002) Fitness of hybrids between cultivated radish and weedy *Raphanus rapistrum*: implications for rapid evolution of weeds. *Ecological Applications* 11, 934–943.

Stewart, N.C. Jr (1996) Monitoring transgenic plants using *in vivo* markers. *Nature Biotechnology* 14, 682.

Tian, L., Lavée, V., Mentag, R., Charest, P.J. and Séguin, A. (1999) Green fluorescent protein as a tool for monitoring transgene expression in forest tree species. *Tree Physiology* 19, 541–546.

van Raamsdonk, L.W.D. and Schouten, H.J. (1997) Gene flow and establishment of transgenes in natural populations. *Acta Botanica Neerlandica* 46, 69–84.

Index